Environmental Science

Series editors: R. Allan · U. Förstner · W. Salomons

Springer
Berlin
Heidelberg
New York
Barcelona
Hong Kong
London
Milan
Paris
Tokyo

Luiz Drude de Lacerda (Ed.)

Mangrove Ecosystems

Function and Management

With 56 Figures and 55 Tables

Springer

Editor

Prof. Dr. Luiz Drude de Lacerda
Department of Geochemistry,
Universidade Federal Fluminense,
Niteroi, 24020-007, RJ, Brazil

ISSN 1431-6250
ISBN 3-540-42208-0 Springer-Verlag Berlin Heidelberg New York

Die Deutsche Bibliothek – CIP-Einheitsaufnahme

Mangrove ecosystems : function and management ; with 55 tables / Luiz Drude de Lacerda (ed.). - Berlin ; Heidelberg ; New York ; Barcelona ; Hong Kong ; London ; Milan ; Paris ; Tokyo : Springer, 2001
ISBN 3-540-42208-0

Springer-Verlag Berlin Heidelberg New York
a member of BertelsmannSpringer Science + Business Media GmbH

Printed in Germany

Cover Design: Struve & Partner, Heidelberg
Dataconversion: Büro Stasch, Bayreuth (*www.stasch.com*)

– 3020 – 5 4 3 2 1 – Printed on acid-free paper

This book is dedicated to Rozane and Ana Luiza,
who kept me vivid even when this book seemed so far from reality.

Preface

Protection of the environment has nowadays become a major challenge and a condition for survival of future human generations and life on Earth in general. Yet it is still far too much of a dream or hope rather than a reality in the policy of our societies. Presently we are experiencing an unprecedented exponential growth of demography combined with a race for profit, resulting in excessive consumption particularly of energy, and a serious impact on the world ecosystems. Various types of pollutants and emerging new diseases not only disrupt the normal course of life, but also above this some of the atmospheric pollutants are most likely involved in the changing climate. We fear and literally shiver at the thought that the "changing climate" would ultimately disrupt the fragile thermodynamic equilibrium between the atmosphere and the oceans. Are we insensitive to these facts to the point of pushing our descendants, some generations ahead, into a new glacial period after a first period of warming up, at least, in northern Europe, like the one that took place 13 to 14 millennia ago? Surely the planet's nature is not prepared to be dominated by man and will go its way, whether humanity will be alive or dead.

More concretely, in the new worldwide movement for environmental concern, which emerged in the second half of the 20th century, coastal environments had been left aside for a long time. The coastal areas were considered restricted zones for national security, or waste and unhealthy areas or just sandy beaches recognized only for leisure, swimming and as tourist resorts under the sun. Even the scientific community neglected the coastal zone and gave priority either to the terrestrial environments or to the open ocean, leaving the coastal areas as a kind of no-man land in between the two. It is a matter of fact that the encounter of seawater with the land and with fresh waters produces special ecosystems having their own physico-chemical and biological peculiarities. This requires combining understanding and competence of both terrestrial and marine environments for adequate development and management. In addition to indirect benefits, the mangroves' material products are multifold, including food, materials for constructions and medicines. Among the coastal ecosystems, the mangroves were about the last to receive attention, compared for example, to the coral reefs and seagrass beds. It is a matter of fact that it is much more pleasant to enter and work in the crystal-clear and clean waters of a coral reef than to suffocate in the inextricable and muddy mangrove ecosystem full of mosquitoes. Therefore one has to pay tribute to those scientists and managers who have understood the importance of the mangroves and who are devoting their career to them. Hence the authors who have contributed to the present volume are specially remembered.

The relations of the human society to mangroves are basically twofold. On one side, there is the traditional mangrove dwellers who live in the mangroves or who are in their daily dependence, and on the other side there is the industrial society which sees the mangroves from a distance, either as a nuisance or eventually as a source of profit. Unfortunately, the latter are gradually taking over from the first, while dramatically omitting to learn from the accumulated traditional wisdom of the elders, including the knowledge of plants and animals and the use of single resources without destruction of the ecosystem. It is not surprising to read this concern in the present volume particularly under the pen of the African authors. Unfortunately, the political weight of traditional villages is usually negligible. Even the very numerous villagers along the tropical mangrove coasts, how can they carry weight against big international trusts? Though it is rather easy to produce plans, laws and rules for the protection of mangroves, it would appear to be much more difficult to enforce the law because of conflicting interests, compromises and too much fragmentation of the responsible administrations. Nevertheless, while villagers cannot always hope to influence the decisions of the society, the mangrove dwellers and specialists should, untiringly, and unremittingly aim at convincing authorities at all levels, of the importance of the mangrove ecosystems for the coastal and coastal related marine and terrestrial ecosystems. These ecosystems are crucial for their economies, and we still have much to learn from the knowledge and wisdom of traditional societies of the coastal zone. Mangrove ecosystems offer a multitude of direct and indirect benefits, under the form of material produce, such as timber, fire wood, fisheries and minor forest products such as medicines, in addition to offering efficient protection against destructive and at places frequent episodic, catastrophic events, such as cyclones, tidal bores or floods. Protection to the mangrove ecosystem, regular monitoring and rational management, including the relationships with terrestrial and marine ecosystems, must be part of national, regional and local planning. This is the principal message of this volume.

Present in the estuaries and shores, mangroves were probably used by humanity early in its development. From the literature, the Greek historians contemporaries of Alexander the Great as well as Eratosthenes, already recorded the presence of mangroves, which they recognized as something special, they marveled at their growth along arid coasts and noticed that their presence coincided with rich coastal fisheries. For instance, they called "ichthyophagi" the fish eating dwellers of the Makran coast (present day Pakistan) while it is also known that Greeks and Romans traded with the "Gangarides", or people of the Ganga River Delta, the Sundarbans, already as early as few centuries B.C. As forests and scrub formations all along the tropical coasts, mangroves are composed of about 80–90 species, of which the Rhizophoraceae and Avicenniaceae are pan tropical and exclusive to the intertidal zone. They encompass all continents and therefore are a matter of common concern for most of the tropical countries and, in particular, for the scientific community and planners. Cooperation is therefore a must for planning in this discipline as in most coastal sciences. This volume is a brilliant example of cooperation and policy directives during the last twenty years. From a few specialists at the beginning, the community of mangrove scientists and managers is now including all tropical countries. There is a growing network of regular contacts among the specialists form different countries, which operates directly or through national or multinational organizations. However the level of mangrove training varies from country to country and the emerging cooperation is still fragile and needs support and reinforcements. While the book gives credit to a number of on-going mangrove rehabilitation initiatives, the land coverage of mangroves in the world is still on an alarming descending curve, clearly due to human mismanagement.

In the present volume, scientists from Africa, Asia and the Pacific, and Latin America are covering their region in three respective chapters. In itself, this is already an achievement and a demonstration that in each region permanent nuclei of responsible managers and confirmed scientists are now at work. The differences in the chapter contents and presentation reflect the variety of approaches and form richness in itself. The interregional cooperation makes it easier to compare similarities and differences among mangrove ecosystems in time and space, of different places and between regions. Although the ultimate goal is the protection and management of mangroves, the uniqueness of the ecosystem makes scientific research itself particularly interesting, such are: bio-ecology, bio- and geo-chemistry, water and sediments dynamics, taxonomy and phylogeny, paleogeography, paleontology and genetics. The place and time of emergence of mangrove species on Earth is still a matter of debate. It is commonly believed that they first appeared in the equatorial waters of south-east Asia around 60 to 80 million years ago and then further spread towards Africa and the Americas, probably following the climatic variations associated with the glacial/interglacial periods and corresponding regressions and progressions of the coast lines. However, *Nypa* is already recorded in the late Tertiary on both sides of the South Atlantic and also in Europe. Plate tectonics, particularly regarding the evolution of the Tethys Sea, may have played a major role in the development of the mangrove formations. The book summarizes various theories, all of them inconclusive. During the last few years researchers have not only progressed in this area, but also in other advanced fields such as physiology, geo-chemistry, phenology and genetics, species variability, hybridism, speciation, palynology, microbiology and nutrients turnover rate in soils and waters, and soil sciences, for a better understanding of the ecosystem as a whole.

Most of the efforts are nevertheless concentrated on the mangrove forest, which is the one that creates the ecosystem, though it is not the most valuable component in terms of money return area-wise. Much effort goes into the analysis of coverage and evolution of the forest in time, on conversion of mangrove ecosystems to agriculture or aquaculture, on impacts on the environment, on the inter-relations of the mangrove ecosystem to off-shore water systems, particularly regarding fisheries, and the control and rehabilitation of mangrove formations. We learn that the recent estimates of world mangrove cover vary and range from 14 to 30 million ha, with an average of about 17 million ha, of which about half are in the Asia-Pacific region and the rest about equally distributed in Africa (25%) and the Americas (25%). We also learn, unhappily, that the estimated total surface of the world mangroves is probably already over-estimated due to continuing misuses. Once destroyed, mangrove rehabilitation is a costly, lengthy and labor-intensive process, while we do not yet know fully the structure and function of the component parts of the mangrove adaptation to the tidal environment. In addition, when Nature is injured, in order to recover, it imposes a time factor, which is often out of range with the perception of time by man. Therefore mangroves cannot be managed properly if the role played by its functional components is not understood and advance research must continue to go hand in hand with management. Each site is unique resulting from the encounter of given species with the specific environmental factors and research is needed on the adaptations to local conditions. This is another message of the book. Nature in the mangroves does not make the work of environmental managers any easier. In an extreme environment like that of mangrove ecosystems, more than anywhere else, man has to conform his behavior to the peculiarities imposed by nature.

The three regional chapters are somewhat different in their contents and presentation. In a way it reflects some fundamental difference between the three regions in terms of their various aspects, distribution and relative importance of the mangroves along the tropics, of the species composition combined with specific climatic, oceanic and coastal environmental factors. The relative importance of certain factors like droughts, tidal range, coastal currents and fresh water supply are different from one region to another. It also reflects to a certain extent, a different perception of the mangrove environment and resources in terms of social and economic interests, although there is a clear common belief that the mangrove ecosystem's protection and rational management is mandatory. Differences are also related to the interests and concerns of the coastal dwelling populations, often in conflict with external economic interests.

In the Indo-West Pacific chapter, readers are referred to the literature for detailed description of the country-wise species composition and distribution, while the attention is focused on different types of coastal environments and mangroves, given the importance of these factors for the rational management of the mangrove ecosystems themselves. Thus are included arid coastal mangrove systems, estuarine and deltaic systems, coastal lagoons and special attention is given to small islands, which are less covered by the scientific literature. Extensive description is given of the adaptation and role played by the mangroves in these various environments. Hence, a land desert with a coastal mangrove belt associated to ocean upwelling can result in rich coastal fisheries. The mangrove associations and their evolution over time are discussed at length for the main estuaries and deltas like those of the Indus, the Ganga-Brahmaputra, the Irrawaddy and the Mekong. Already since the time of Nearchus (4th century B.C.) the large amount of silt carried by the Euphrates-Tigris Rivers proved to be an ecological hazard for the establishment of seeds and propagules which become silted under, on the extensive mud-banks, though silt is normally a favorable element due to the amount of nutrients it carries. As for the Indus River system, the main ecological hazard is the man-made large number of barrages built across the river, which have drastically reduced the discharge of river water in the delta and has imperiled the existence of the mangroves downstream. The immense low-laying plain created by the delta of the Ganga-Brahmaputra system which was built over geologic times by silt carried down from Himalayan slopes, has been host to human populations since, at least, historical times as confirmed by the trade of the Greeks and Romans with the "Gangarides". An extensive discussion is made of the monsoon regime and of the consequences of its gradual shifting eastwards over South Asia. Recent years have shown how particularly destructive is the modern clear felling of mangroves for conversion of land to agriculture and aquaculture. Therefore, particularly fascinating is the story of the wisdom of the traditional societies which, for centuries, have smoothly rearranged the land while at the same time maintained the stability of nature by using special silvi-agri-aquaculture techniques under names such as tambaks and tumpang-sari in Indonesia and pokkali and bheri in India.

An extensive discussion on the origin, paleogeography, diversification and distribution of the mangroves in the Gulf-Caribbean Region and along the American Continent is given in the Latin-American chapter. Detailed description is given of the mangrove flora and associated flora and fauna. Particularly instructive are the paragraphs on hydrology and mangrove soils, as well as the discussions concerning the effects of hurricanes, sea level rise, global warming and oil pollution. The relation of man with mangroves during pre-Colombian and colonial times concludes this chapter.

The picture of the African mangroves shows important differences between the western and eastern coasts, including Madagascar. Indeed the floristic composition of East-African mangroves is much higher compared to that of the West, most probably due to the fact that the Atlantic is a relative young Ocean. The two coasts are also characterized by quite different hydro-climatic conditions. While a large part of the eastern coastline is dominated by a monsoon regime, the western coast is bordered by strong coastal upwelling, like the Benguela Current, enhancing coastal and offshore fisheries. West Africa has gone through a period of intense land desertification during the past 25–30 years, with phenomena such as "inverted-estuaries", as for instance in the Saloum River where marine high salinity waters had penetrated up to several hundreds km inland. In Africa the traditional way of life is still the rule along the coastal zone and conflicts of interests between the people and external forces applying pressure to submit the mangroves to destructive activities such as intensive agri- or aquaculture, clearance for salt pans, for urban and industrial development and for diversion of freshwater by building barrages, has reached calamitous levels.

To conclude, the mangrove ecosystems are not wastelands. They represent a real asset for tropical coastal economies, either for the traditional societies living therein or of it, even for modern planners anxious to perpetuate the productivity of the coastal areas, while at the same time providing the hinterland with a protective belt against storms of all kinds. They are part of the heritage of the tropical world and it is up to the decision makers and the scientific community of the interested tropical countries to preserve this unique element of their heritage. It gives great pleasure to note that the present book strongly supports this viewpoint. The text also emphasizes the need and reasons for cooperative ventures among the people, scientists and managers from all over the world. Let us hope that this book will be followed by others equally minded.

Marc Steyaert

Contents

Contributors

Alarcón, C.
PDVSA Intevep
Apdo 76343, Caracas 1070-A, Venezuela
alacon@pdvsa.com

Alvarez-León, R.
Calle 142AN.52-36, E1-/-6a-201
Santa Fé de Bogotá, Colombia
Alvarez_leon@hotmail.com

Aizpuru, M.
Université Paul Sabatier, laboratoire d'Écologie Terrestre (UMR 5552)
13 Avenue du Coronel Roche, BP 4403, 31405 cedex 4, Toulouse, France
François.Blasco@cict.fr

Ayivor, J. S.
Volta Basin Research Project, University of Ghana
P.O. Box 209, Legon, Accra, Ghana
chrisgordon@ighmail.com

Blasco, F.
Université Paul Sabatier, laboratoire d'Écologie Terrestre (UMR 5552)
13 Avenue du Coronel Roche, BP 4403, 31405 cedex 4, Toulouse, France
François.Blasco@cict.fr

Conde, J. E.
Centro de Ecologia, Intituto Venezolano de Investigaciones Cientificas (IVIC)
Ministerio de Ciencia y Tecnologia, P.O. Box 21827, Caracas 1020A, Venezuela
jconde@oikos.ivic.ve
http://oikos.ivic.ve/jconde.cveng.html

Diallo, A.
Department of Biology, University of Dakar
Senegal, Dakar, Fann, Senegal
alguisse@syfed.refer.sn.com

Diop, E. S.
SPO – WATER, UNEP
P.O. Box 47074, Nairobi, Kenya
Salif.Diop@unep.org
esdiop@africaonline.co.ke

Diouf, M.
Department of Biology, University of Dakar
Senegal, Dakar, Fann, Senegal
alguisse@syfed.refer.sn.com

Duke, N. C.
The University of Queensland, Botany Department
Brisbane, Queensland 4072, Australia
n.duke@mailbox.uq.edu.au

Gordon, C.
Volta Basin Research Project, University of Ghana
P.O. Box 209, Legon, Accra, Ghana
chrisgordon@ighmail.com

Guissé, A.
Department of Biology, University of Dakar
Senegal, Dakar, Fann, Senegal
alguisse@syfed.refer.sn.com

Kjerfve, B.
Marine Science Program, EWS 508, University of South Carolina
Columbia, SC 29208, USA
bjorn@geol.sc.edu

Lacerda, L. D.
Department of Geochemistry, Universidade Federal Fluminense
Niteroi, 24020-007, RJ, Brazil
ldrude@fortalnet.com.br

Lal, P.N.
School of Graduate Studies in Environment Management and Development
Australian National University, GPO Box 1527, Camberra, ACT 2601, Australia
plal@aciar.gov.au

Polanía, J.
Universidad Nacional de Colombia, Sede San Andrés
Carretera Circunvalar de San Luis, Sector Free Town # 52–44
San Andrés Isla, Colombia
jhpolanv@dnic.unal.edu.co

Shafiu, M.
Senior Field Officer, Ministry of Fisheries, Agriculture and Marine Resource
Ameeru Ahmed Magu, Ghazee Bldg. Malé, 20-25
Republic of Maldives

Semesi, A.
Agricultural University of Norway
Noragric, P.O. Box 5001, N-1432 ÅS, Norway
adelaida.semesi(noragric.nlh.no

Siddiqqi, N. A.
Bangladesh Forest Research Institute - BFRI
P.O. Box 273, Chittagong, 4000, Bangladesh
bfri@aspnetctg.com

Soumaré, A.
Department of Geography, University of Dakar
Senegal, P.O. Box 22087, Dakar, Ponty, Senegal
a_soumare@hotmail.com

Stayaert, M.
10, rue du Docteur ROUX, Paris 75015
marc.steyaert@wanadoo.fr

Vannucci, M.
International Society for Mangrove Ecosystems
a/c College of Agriculture, University of the Ryukyus
Nishihara, Okinawa, 903-0129 Japan
drvannucci@hotmail.com

Acknowledgements

Mangroves are one of the most exquisite forest ecosystems from the tropics and have attracted attention from naturalists since the antiquity. Books on these biological rich ecosystems have been, however, rarely written. Available textbooks on mangroves are with few exceptions written by scientists who studied mangroves but do not live in the regions where these forests are the dominant feature of the sea-continent interface. Therefore, although most are excellent scientific literature, almost all fail to cover the actual social and economic importance of these ecosystems. The present volume do not aim to solely advance on the pure scientific endeavor of mangroves, rather it put together the present state of the art ecological knowledge on mangroves with their management and utilization by local populations.

The idea behind this book was born during the course of the ITTO/ISME projects on "Conservation and Sustainable Utilization of Mangrove Forests", in Africa, Asia and Latin America. After evaluating the reports from those projects, it became clear that the huge amount of new and exciting knowledge produced during the 5 years of the projects, should be summarized into a more widespread way. The key came from the profound understanding of scientific divulgation of Springer Verlag, particular from their editors and friends Wim Salomons and Wolfgang Engel. Upon acceptance of the task, the two-years long process was carefully followed by Miss Luisa Tonarelli at Heidelberg, without whom this task would have been nearly impossible.

Such a book can not, and is not, an individual enterprise, scientific colleagues throughout the world were gathered to contribute, sometimes just with a picture or a comment, sometimes actually writing significant parts of chapters, or carefully reviewing the text. My friends Dr. Salif Diop, UNEP, Nairobi and Dr. Marta Vannucci, ISME, Okinawa, exponents of mangrove research worldwide, acted as truly co-editors, by dealing directly with authors in Africa and Asia respectively. Among the contributors, some also helped in many ways to the whole book. I am specially grateful to Dr. Chris Gordon, University of Ghana, Dr. Jesus Eloy Conde, IVIC, Venezuela and Dr. Bjorn Kjerfve, University of South Carolina, USA, for their permanent help.

Other contributions to this volume also deserve special thanks particularly Dr. S. Baba, ISME, Okinawa, Dr. L. C. Pires, Universidade do Amazonas, Brazil and Dr. F. R. Mochel, Universidade Federal do Maranhão, Brazil, who kindly provided original pictures.

Many institutions provided direct and indirect support during this two-years period; I am particularly indebt to the International Society for Mangrove Ecosystems and the International Tropical Timber Organization, by organizing many international meetings of experts worldwide allowing ample and personal discussion of this book. The National Research Council of Brazil, for its permanent personal support and the Universidades Federais Fluminense and of Ceará, for releasing me for the necessary time to conclude this project.

Luiz Drude de Lacerda
Fortaleza, November 2001

Chapter 1

American Mangroves

L. D. Lacerda · J. E. Conde · B. Kjerfve · R. Alvarez-León · C. Alarcón · J. Polanía

1.1 Origin, Palaeogeography, Diversification and Distribution of the New World Mangroves

1.1.1 Origin and Palaeogeography

It is quite difficult to conclusively pinpoint the dates when mangroves first appeared and diversified in the Neotropics, thus a number of hypotheses have transpired, and the topic is still a matter of debate. These explanations can be classified into two categories: those that propose a unique centre of origin followed by radiate dispersal (through three routes), and those based on the fragmentation of a common widespread ancestral population or biota. As an alternative to those all-inclusive schemes and given the wide taxonomic variety of mangroves, *ad hoc* explanations should be implemented for each species, as essayed by Duke (1995) with the genus *Avicennia*. This author has also cautioned that the present-day co-occurrence of mangrove species should not be interpreted as a measure of common origin.

The oldest mangrove record for this region is *Nypa*, a mangrove palm that dates back to the Late Cretaceous, but that is presently restricted to southeast Asia (Rull 1998), with the exception of Panama, where it has been introduced (Duke 1991; Spalding et al. 1997). The more diversified mangrove flora of the Indo-Pacific acquired its mangrove habitat about 80 million years (Ma) ago (Field 1995b), although a figure of 100 Ma has been proposed (Duke 1995). The New World mangrove flora is believed to have evolved much later. The genera *Rhizophora* and *Avicennia* are thought to have evolved earlier and therefore may have spread through the Tethys Sea into what is presently the Mediterranean Sea and thence to the East coast of the Americas and the West coast of Africa. They may have reached the Caribbean by the early Eocene (some 55–50 Ma), when the distance from the Tethys sea was considerably smaller and the warmest interval of the Cenozoic occurred (Haq 1981). The palaeo-circulation of the oceans during the Palaeocene, when the Tethys Current was a dominant feature of surface circulation, lends credence to this scheme (Haq 1984). Prior to the late Eocene, however, one of the most abundant pollen types in coastal deposits was *Brevitricolpites variabilis*, which had unknown affinities, but dominated the same habitat now occupied by *Rhizophora*, which is clearly equivalent to that genus. It eventually disappeared from the palynofloras in the middle Eocene (Graham 1995).

Mangrove pollen of *Rhizophora*, *Avicennia* and *Pelliciera* has been reported from the Gatuncillo Formation, Panama, from the middle to late Eocene (ca. 40 Ma) (Graham 1989). Pelliceroid pollen has been found in river delta sediments of northern South America dating back to the Oligo-Miocene (Fuchs 1970). Pelliceroid pollen was also present in different locations in northern South America and the Caribbean during the Eocene: e.g., tertiary of the Guiana Basin (van der Hammen and Wijmstra 1964) and along the Pacific Ocean between Colombia and Costa Rica (Johnston 1949; Wijmstra 1968; Jaramillo 1998) and northeastern Colombia (Dueñas 1983), middle Eocene of Jamaica, late Eocene of Panama; Oligocene of Puerto Rico; Oligo-Miocene of Chiapas, Mexico (Graham 1977, 1995). Contemporaneous to *Pelliciera* extensive *Rhizophora* and *Avicennia* forests covered the northern South American coast. Later on, pollen registers along the Americas show abundant mangroves (Graham 1989, 1992) that confirm the early Eocene as the probable date of mangrove arrival to the New World.

By the time of the arrival of the first mangrove genera to the American continent, the Panama isthmus was still open, making the colonization of the western coast of the Americas possible. The closure of the Tethys Sea route and the decrease of the temperature of the South African coast by the Benguela Current would have prevented the migration of more recently evolved mangrove genera. This hypothesis has been endorsed by Chapman (1976) and others, and has been widely accepted.

Hou (1960), van Steenis (1962) and Muller and Caratini (1977) provide a contrasting view on the origins of New World mangroves. These authors proposed that at least for the Rhizophoraceae, the family originated in the Eocene in Southeast Asia together with inland relatives (e.g., *Carallia* and *Anisophylla*). This proposal agrees with the most accepted views of mangrove origin. From southeast Asia, the Rhizophoraceae would have extended their distribution eastward, and not westward, through trans-Pacific land bridges to tropical South America, reaching the Caribbean before the rise of the Panama Isthmus and therefore being able to cross to the Atlantic and colonise North and South America and West Africa. However, the absence of mangroves in several Pacific islands, mainly those encompassed by meridians 90 and 170° W (Woodroffe 1987; Tomlinson 1994), creates an unexpected situation if the migratory route to the New World had crossed this ocean, and casts some doubts on this theory. In the particular case of Rhizophoraceae, Woodroffe and Grindrod (1991) suggested that idiosyncrasies of distribution, nonetheless, could be understood in light of sea-level and other environmental fluctuations during the Quaternary, causing the disruption of a widespread ancestral population that was present throughout the Pacific.

Duke (1995), who renders this route plausible, has tentatively postulated a third path around Antarctica or southern Africa. Several authors have accepted evidence of the presence of *Avicennia* in Gondwana (Specht 1981; Mepham 1983). If *Avicennia* was present there, it could have migrated from Australia to South America around the then subtropical-warm coast of the Antarctica prior to the Eocene. Unfortunately, there are no fossil records to support this idea.

Different views on the origin of mangroves and some of the proposed routes to the New World, showing major fossil sites and possible origin are depicted in Fig. 1.1. The configuration of the continents is, as far as possible, believed to have occurred at the end of the Cretaceous (100–70 Ma) after Mepham (1983), Haq (1984) and Degens (1990). It seems quite clear that the view of Chapman (1976), on a migration route through the Tethys Sea and then to the New World, is supported by the fossil record and is more readily acceptable than a trans-Pacific migration. However, different centres of dispersion and speciation along the Tethys Sea proposed by Mepham (1983), remain a possibility worth considering.

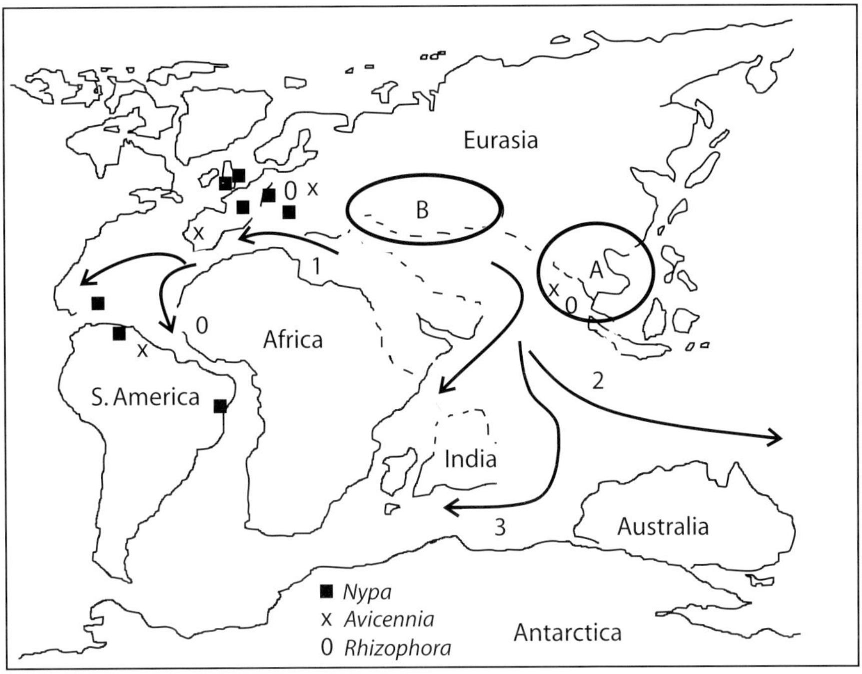

Fig. 1.1. Palinological record and proposed origin centres and migration routes of New World mangroves. *A* Southeast Asia origin centre; *B* Tethys Sea origin centre. *1:* Migration through the Tethys Sea to the Atlantic and trans-Pacific migration; *2:* migration from the Tethys Sea to SE Asia and the Atlantic and through the Panama Isthmus; *3:* Antarctica/South Africa route

Mepham (1983) argued against the hypothesis of the origin of mangrove flora in Southeast Asia and their distribution by radiate dispersal from a single hypothetical source in this area. He proposed a broader Tethyan region as the place where angiosperms first acquired the mangrove habit and where most contemporary mangrove genera originated. However, he also agrees that migration through Tethyan distribution routes by the early-middle Tertiary was probably the major route of colonization of North and South America, the Caribbean and the West coast of Africa for the early mangrove genera *Rhizophora* and *Avicennia*. This route is believed to be confirmed by the widespread pollen of *Nypa* in late Eocene deposits (ca. 40 Ma) along the present Mediterranean coast, England, North America, West Africa and northeastern Brazil (Muller 1961; Macnae 1968; Dolianiti 1955).

It is interesting to note that the distribution of fossils, which is widely used as an argument in favour of a single centre of dispersion, may also be interpreted as an indication that mangroves arrived in this region from a Tethyan origin as proposed by Mepham (1983), rather than originating there. But the fossil evidence for Pleistocene mangroves is so scarce, that the reconstruction of mangrove distributions remains largely speculative (Woodroffe and Grindrod 1991).

Duke (1995) has dealt specifically with the origin and evolution of *Avicennia*. According to this author, the earliest *Avicennia* ancestor appeared during the early Cretaceous, along the northwestern coast of Gondwanaland, in a region where a slow expanding huge estuary would eventually appear. This estuarine setting, held during a protracted period, is considered to be essential for the selection and ensuing development of plants with mangrove-like features. The events, which allowed the speciation of *Avicennia* probably unfolded at different times. The first phase occurred 80–90 Ma, during the late Cretaceous. The formation of an isthmus between North and South America might have isolated the future *A. bicolor* populations. Further ancestral populations of *Avicennia* were isolated, as the Gondwanaland breakdown proceeded into the late Cretaceous. South American populations remained as they are currently, but others sustained major changes, leading to further speciation. During the late Cretaceous and Palaeocene (60–80 Ma), populations in northwestern Africa would eventually spread to North America through southern Europe, given warmer climatic conditions and the connections at that time. By the late Eocene, some 50 Ma, *A. germinans*, which had appeared as such 50–80 Ma, was split by the formation of the North Atlantic Ocean. The isolation of Atlantic and Pacific populations on the coasts of America probably resulted from the formation of a modern isthmus in the Late Pliocene, 2–3 Ma.

Based on fossil data, restrictions to long distance dispersal capabilities and present-day associated biota distribution, McCoy and Heck (1976) also contended the "centre of origin" explanation for mangroves and other shallow-water tropical marine communities. They argue that the biogeographic patterns of mangroves (as well as hermatypic corals and seagrasses) are better explained by the existence of an ancient, widely distributed biota, later modified by speciation, extinction and tectonic events. They further suggest that the association of mangroves, corals and seagrasses developed during the Cretaceous, allowing for the diversification of scleractinian corals and the evolution of seagrasses and mangrove lineage. Those groups co-occurred along the Tethys Sea. Woodroffe and Grindrod (1991) have refined this scheme by applying Daly's "glacial control" theory, developed to explain former and present distributions of coral reefs. This theory emphasises broad quaternary environmental fluctuations as the driving force behind reef growth and distribution. Particularly, lowered temperatures and sea levels, as well as increased water turbidity during glacial stints might have led to local disappearances of coral reefs. Woodroffe and Grindrod (1991) suggest that those environmental fluctuations, especially sea level lowering, probably played an influential role in the palaeogeography of mangroves, in a similar way to the role played in coral reefs biogeography.

In general, the distribution of mangroves along the American continent prior to the full development of glaciations in the Pleistocene seems to reflect the climatic conditions of that time (Wijmstra 1969). During the Eocene period, the mangrove flora extended to higher latitudes. *Nypa*, for example, was very common along the Caribbean basin from northeastern Brazil to.

Southeastern USA (Muller 1961; Dolianiti 1955; Thanikaimoni 1987), as well as *Pelliciera rhizophorae*. The distribution of both species was drastically reduced later, probably as a result of more pronounced seasonal rainfall, longer dry conditions, higher salinity (Germeraad et al. 1968; Muller 1980; Jiménez 1984), and sea-level and coastline geomorphological changes (Woodroffe and Grindrod 1991). Eventual competition with the newly-arriving genera *Rhizophora* and *Avicennia* may have also played some role in this distribution reduction (Graham 1977). *Nypa* disappeared from the Americas and the West coast of Africa, while *P. rhizophorae* presents a scattered and very restricted distribution from the Pleistocene to present. Along the coast of what is today Suriname, mangrove forests spread over large areas of the continental platform following regressions of the sea and colonised large plains landward during transgressive periods (Wijmstra 1969). During these landward incursions between 15 and 23 Ma, mangrove representatives from the Caribbean, in particular *Rhizophora*, reached the higher Solimões formation near the Colombian-Peruvian-Brazilian border in the Amazon basin (Hoorn 1993; Lovejoy et al. 1998).

1.1.2 Mangrove Diversification in the Gulf-Caribbean Region

Whatever the provenance and time of arrival of mangroves to the Americas might be, a trend is clear in the Gulf-Caribbean region: a gradual increase in diversity of true mangrove and associate species (Graham 1995). In this region, coastal areas were occupied during the early Eocene by four principal genera, *Acrostichum*, *Nypa*, *Pelliciera*, and the now extinct *Brevitricolpites variabilis*. By the middle Pliocene, six true mangrove and three associated species were present: *Acrostichum*, *Avicennia*, *Crenea*, *Laguncularia*, *Pelliciera*, *Rhizophora*, and *Acacia*, *Hampea*, *Hibiscus* and *Pachira*, respectively. *Conocarpus* appeared during the Quaternary, as well as eight possible non-mangrove associates. Twelve other genera were added during the Cenozoic that constitute the present-day community of 27 genera (Graham 1995).

However, this was not an orderly linear process, but was interspersed with extinction, introductions and adaptations. The Eocene mangroves suddenly disappear at the end of this epoch, being substituted by post-Eocene species, which are quite different taxonomically (Rull 1998). The two types of guilds never coexisted, with the only exception of *Pelliciera*, which was present in post-Eocene mangles where *Rhizophora* prevailed, and the associate *Acrostichum*. These significant shifts in the vegetation composition and distribution of mangroves in the Americas happened during the so-called Eocene-Oligocene Crisis (Rull 1998), when a meaningful decrease in the number of families of marine organisms took place (Benton 1995). These variations seem to have been a consequence of the same factors that produced die offs and extinctions at the end of the Eocene on a global scale (Rull 1998); that is, intense short-term fluctuations. *Nypa* and *B. variabilis* disappeared almost simultaneously, while *Pelliciera* imploded gradually, although it did not disappear entirely and currently is still present in isolated areas of the Caribbean (Rull 1998) and along the Pacific coast from Colombia to Costa Rica (Jimenéz 1994). From the Oligocene to the present, no extinction of Neotropical mangroves occurred. However, the rate of taxonomic additions was not constant. During the Oligocene and Miocene, the rate was low, but it picked up the most over the past five million years, during the Pliocene and Quaternary, when 2/3 of the contemporary species were incorporated into the guild (Rull 1998).

Other factors and events that could have determined cardinal shifts in the distribution and vegetational composition of mangroves in the Americas, including impacts of meteorites, are examined point by point by Rull (1998).

1.1.3 Modern Distribution of New World Mangrove Forests

During the Late Quaternary sea level rise starting about 18 000 B.P., landward migrations of the various facies zones paralleling the American coasts occurred. The most impressive of these migrations has been the retreat, followed by a landward invasion, of mangroves (Seibold and Berger 1993). The present distribution and composition of mangrove forests along the American continent is a result of these recent sea level changes. During Pleistocene dry periods, mangrove species seemed to occur only in climatically wetter areas and were more equatorial restricted than they are today. The occurrence of coastal refugees promoted accelerated speciation, in particular along the Pacific coast of South and Central America, where endemic species (e.g., *Avicennia tonduzi* Moldenke and *A. bicolor* Standl) occur (Gentry 1981; Jiménez 1984). Also, the fluctuating climate of the Pleistocene may have led to the natural selection of adaptive gradients in mangroves for the diverse habitats available, in particular at the latitudinal extremes of their distribution (Sherrod and McMillan 1981).

Today, mangrove forests are unevenly distributed along the American continent's coastline, with the Atlantic and Caribbean coasts harboring nearly 70% of the total mangrove area. Along the Atlantic coast they form a nearly continuous belt from southern United States to southern Brazil. Along the Pacific coast they have a more restricted distribution due to climatic constrains generated by the oceanographic conditions along the Peruvian and California coasts. In these regions, the upwelling of cold waters of the Humboldt Current waters suppresses connective activity and results in extremely arid climates, high soil salinity and nearly no freshwater input, restricting the extension of mangrove forests (Lacerda and Schaeffer-Novelli 1992).

Along the Pacific coast of South America, mangroves extend only to latitude 5°32' S, at the Piura River estuary in northern Peru, because of cold water upwelling resulting from La Niña events. South of this point, lack of suitable landforms, extreme aridity, and extraordinary fluctuations in river flow particularly during *El Niño* events hamper their establishment (Cintrón-Molero and Schaeffer-Novelli 1992). Their northern Pacific limit, however, reaches latitude 30°15', in Puerto de Lobos, Mexico. On the Atlantic coast, mangroves extend northward to Bermuda (latitude 32°20' N) and St. Augustine, Florida (28°50' N); and southward to Laguna, Santa Catarina, in South Brazil (latitude 28°30' S). With the exception of the South Pacific coast, the frequency, duration, and intensity of cold winter temperatures, rainfall and/or frost determine the latitudinal limits of mangroves. The western limit of mangroves on the American continent is the Galapagos Islands, off the coast of Ecuador (longitude 91° W, latitude 0°), where *A. germinans*, *L. racemosa*, *C. erecta* and *R. mangle* have been reported (West 1977; Chapman 1976). The eastern limit is Fernando de Noronha Islands, off the coast of Pernambuco, Brazil (longitude 32°24' W; latitude 3°50' S), where a small forest of *L. racemosa* covers approximately 0.15 ha at the estuary of Maceio River, Sueste Bay, which drains the largest watershed of these islands (Herz 1991).

1.1.4
Extension of Mangroves on the American Continent

On the American continent, mangroves occur in all maritime countries except the three southernmost nations Chile, Argentina and Uruguay, and northernmost Canada. Only recently, however, have efforts been undertaken to survey mangrove cover in these countries. Table 1.1 shows the most recent, and probably most reliable estimates available of mangrove area on the American continent.

The first comprehensive estimate of mangrove area in this region was published by Saenger et al. (1983), based on available information given by forest and environmental authorities and from scientific reports from the countries. This estimate shows that the total mangrove area in the Americas is approximately 6.7 million ha, with 70% (4.8 million ha) occurring along the Atlantic Ocean and the Caribbean and only 30% (1.9 million ha) distributed along the Pacific coast. The largest forests are located in Brazil, Mexico, Cuba, and Colombia.

Although this estimate has been widely accepted as representative, recent surveys suggest that it is an overestimate. A second comprehensive estimate of mangrove cover in the Americas was published by Lacerda (1993), under an international effort coordinated by the International Society for Mangrove Ecosystems (ISME). Following this effort, Spalding et al. (1997) published the first World Mangrove Atlas under the support of the same society (ISME).

Yet for several countries the estimates by Saenger et al. (1993) are in accordance with the most recent ones and in some cases even lower. The data presented in Table 1.1, include figures from Saenger et al. (1983) for countries where no other information is available and updated estimates by Lacerda et al. (1993) and Spalding et al. (1997). According to recent estimates, mangrove cover in Latin America and the Caribbean is approximately 4.26 million ha, which is only about 60% of the total given in Saenger et al. (1983). For example, in Brazil during the late 1980s, a complete survey by satellite imagery and aerial photography of all coastal states showed that mangrove cover is approximately 1.01 million ha. Nearly 85% of the total forests are on the north coast of the country (Herz 1991), including the largest single mangrove stretch of about 700 000 ha along the cost of the Brazilian states of Maranhão and Pará (Mochel 1999) (Fig. 1.2, 1.3). Considering that this is possibly underestimated by 20% to 30% (Kjerfve and Lacerda 1993), this is still less than half of the 2.5 million ha proposed by Saenger et al. (1983).

In Venezuela, mangroves occur along nearly 35% of the coastline. Saenger et al. (1983) estimated that mangrove cover reaches 673 600 ha, while the most recent figure is only 250 000 ha (MARNR 1986), which is only 37% of the previous estimate (Conde and Alarcón 1993). Similarly in Colombia, a detailed study using remote sensing techniques (radar INTERA and LANDSAT satellite) covering the evolution of mangrove areas during the past three decades also showed mangrove extension differently from previous reports. On the Pacific coast mangrove forests cover 292 725 ha, and 86 311 ha on the Caribbean coast (Sánchez-Páez et al. 1997a,b) For several countries, however, the estimates by Saenger et al. (1983) are in accordance with the most recent ones and in some cases even lower.

In Ecuador, a detailed study by remote sensing techniques covering the evolution of mangrove areas during the last two decades (CLIRSEN 1984; MAG 1991), showed mangrove cover to be 161 770 ha, with nearly 70% of total mangrove area (110 000 ha) located in Guayas Province. This number is approximately the same as that proposed by Saenger et al. (1983) of 160 100 ha. For other countries such as Cuba, where mangrove forests cover approximately 5% of the total country's surface, a recent estimate from the Ministry of Agriculture gives a total mangrove area of 529 700, which is nearly 25% larger than the estimate proposed by Saenger et al. (1983).

Taking into consideration the more recent data presented in Table 1.1, total mangrove cover in the Americas reaches about 4.5 million ha, about 70% of the earlier estimate by Saenger et al. (1983).

Table 1.1. Recent estimates of mangrove cover and the respective percentage of total countries' area and length of the coastline in the American continent

	Area (ha)	% of surface	Mangrove area/coast	Reference
Continental countries				
USA	190 000	0.02	10	Odum et al. (1982)
Mexico	524 600	0.27	56	Yáñez-Arancibia et al. (1993)
Belize	71 900	3.03	181	Spalding et al. (1997)
Guatemala	16 040	0.15	40	Jiménez (1992)
Nicaragua	60 000	0.50	66	Saenger et al. (1983)
Honduras	231 600	2.08	308	Spalding et al. (1997)
Costa Rica	41 330	0.08	32	Jiménez (1992)
El Salvador	35 235	1.65	45	Jiménez (1992)
Panama	171 000	2.22	69	D'Croz (1993)
Colombia	389 036	0.31	148	Sanchez-Paez et al. (1997a,b)
Ecuador	161 770	0.60	72	MAG (1991)
Peru	4791	0.01	2	Echevarria and Sarabia (1993)
Venezuela	250 000	0.27	76	MARNR (1986)
French Guyana	5 500	0.06	15	Saenger et al. (1983)
Guyana	80 000	0.37	179	Spalding et al. (1997)
Suriname	115 000	0.70	298	Saenger et al. (1983)
Brazil	1 376 255	0.13	134	Kjerfve and Lacerda (1993)
Insular countries				
Trinidad and Tobago	7 150	1.40	20	Bacon (1993)
Jamaica	10 624	1.02	7	Bacon (1993)
Cuba	529 700	4.80	142	Padron et al. (1993)
Haiti	13 400	0.60	9	Spalding et al. (1997)
Dominican Republic	69 600	1.51	48	Spalding et al. (1997)
Puerto Rico	6 500	0.71	–	Saenger et al. (1983)
Bahamas	141 957	10.18	40	Bacon (1993)
Bermuda	20	<0.01	<1	Ellison (1993a)
Guadeloupe	8 000	4.49	20	Saenger et al. (1983)
Martinique	1 900	1.73	7	Saenger et al. (1983)
Cayman Islands	7 268	27.60	45	Bacon (1993)
Antilles[a]	24 571	–	–	Bacon (1993)

[a] Includes only the islands where reliable mangrove surveys have been reported (Anguilla, Antigua, Aruba, Barbados, Barbuda, Bonaire, Curaçao, Dominica, Grenada and Grenadines, Montserrat, Nevis, St. Kitts, St. Lucia, St. Vincent, Turks and Caicos).

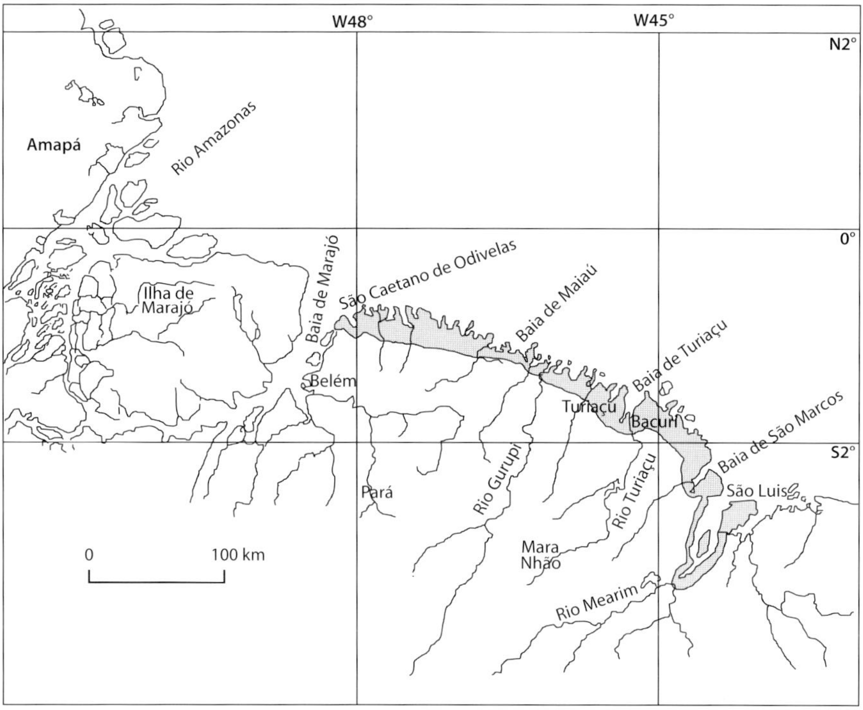

Fig. 1.2. Distribution of the world's largest mangrove system, covering the coast of Pará and Maranhão states in northern Brazil, and measuring at least 700 000 ha (Kjerfve et al. 2001)

The relative importance of mangroves to each country in tropical America is illustrated by comparing the ratio between total mangrove area and total surface area of each country and with the ratio between mangrove surface and coastline length (Table 1.1). Among the continental countries, the percentage of the total surface covered by mangroves ranges from 0.01% in Peru to 3.03% in Belize. Mangrove area to coastline length ratios range from a minimum of 2 in Peru, to a maximum of 308 in Honduras. For insular countries, the percentage of total mangrove area can be as high as 27.6% as in the Cayman Islands and 10.2% in the Bahamas, to less than 0.01% in Bermuda. On the other hand, mangrove area to coastline length ratio ranges from a minimum of less than 1 for Bermuda to a maximum of 142 in Cuba. This rough representation shows that mangroves are the most important forest formation in certain countries, particularly in the insular Caribbean, and they should take priority in these countries' management and conservation policies.

Notwithstanding the differences between the recent estimates of mangrove cover in each country, presented in Table 1.1 and those from, Saenger et al. (1983), both estimates agree upon the relative distribution of mangroves between the Pacific and the Atlantic coasts of Latin America and the Caribbean region. The Pacific coast, with approximately 1.21 million ha, has 26.6% of the total mangrove cover in tropical America, while the Atlantic coast has approximately 2.52 million ha (55.6% of the total) and the Caribbean insular countries approximately 0.81 million ha or 17.8% of the total (Table 1.2).

Fig. 1.3. Typical stands of mangrove forest along the estuarine shore near Bragança, Pará State, northern Brazil, with both *Rhizophora* and *Avicennia* trees on the average exceeding 30 m in height (*photo:* B. Kjerfve)

Table 1.2. Mangrove forests area distribution in the Atlantic and Pacific coasts of the Americas and the Caribbean Islands

Region	Mangrove area ($\times 10^6$ ha)	(% of the total)
Atlantic coast	2.52	55.6
Pacific coast	1.21	26.6
Caribbean Islands	0.81	17.2
Americas (total)	4.54	100

Table 1.3. Mangrove forests cover in the Americas compared to World mangrove forests areas

Region	Mangrove area ($\times 10^6$ ha)	% of the World total	Reference
Americas	4.54	26.6	This work
Asia	7.51	44.0	Spalding et al. (1997)
Africa	3.26	19.1	Diop (1993)
Arabia	0.03	0.2	Saenger (1995)
Australasia	1.72	10.1	Field (1995b)
World total	17.06	100	This work

Nearly all countries of the Americas are mapping their mangroves to estimate area and density distribution, due to a growing concern of the importance of these forests and their health and survival in the coastal region. In several countries, mangroves are being cut and replaced by other biological or engineering activities, such as large scale shrimp farming in Ecuador, tourism in southeastern Brazil and harbour and industrial complexes. On the other hand, replanting and recuperation programmes are being carried out in degraded mangrove areas, such as those presently occurring in Cuba and Brazil. Therefore, although the data given in Table 1.1 are, to our knowledge, the most recent, they are not final, and will certainly be changing in the near future, as well as the relative contribution of the continent's mangroves to the world total.

Recent estimates of world mangrove cover vary and range from 14 to 30 million ha, with an average of about 17 million ha (Field 1995b; Lacerda 1993; Twilley et al. 1992; Lugo et al. 1990; Saenger et al. 1983; Spalding et al. 1997). American mangroves represent approximately 27% of the world total mangrove cover based on the figures presented in Table 1.3. African mangroves cover about the same area, whereas mangrove cover in the Indo-Pacific region represents nearly half of the world's mangroves.

1.2 Flora Composition and Distribution

Mangrove flora belongs to at least 17 families and has developed different morphological and physiological adaptations to survive the harsh land-sea interface. Mangrove flora is operational classified in two different groups according to the occurrence of a given species in other ecosystems. Obligatory or "true" mangroves are those species that occur exclusively in this type of ecosystem, such as the genus *Rhizophora*, and *Avicennia*. About 70 taxa of higher plants belong to this group. The second group, called mangrove associates, includes those species whose occurrence is mostly in mangrove ecosystems, but are also found in other coastal vegetation types. Among the higher plants, certain species of *Hibiscus* and various Palmae are typical. The group also includes many species of benthic algae, epiphytes and lichens (Hutchings and Saenger 1987; Vannucci 1999).

1.2.1
True Mangrove Trees

Contrary to southeast Asia, which has about 60 true mangrove plant species (Rao 1987; Mepham and Mepham 1985), New World mangroves have a very low number of tree species. Further systematic research is needed, due to high population variability among species of each genus and to the length of the American coasts harbouring mangroves. Basic biological surveys are yet to be done for areas of northern South America and some Caribbean Islands. Also, systematic work on the genetics of these species is at the onset, and with time more species may appear as varieties, and/or existing recognised species may prove only to be varieties.

New World mangroves include only 11 tree species and one variety. The *Rhizophora* (Rhizophoraceae) and *Avicennia* (Avicenniaceae) are dominant with 4 species each. Other genera are *Laguncularia* and *Conocarpus* (Combretaceae) and *Pelliciera* (Pelliceriaceae), all with only one species each. *Conocarpus* seems to have an endemic variety (Bacon 1993) on northern Caribbean Islands (Table 1.4).

The genus *Rhizophora* (Fig. 1.4) is represented by *R. mangle* L.; *R. harrisonii* Leechman, *R. racemosa* G. F. W. Meyer and *R. samoensis* (Hochr.) Salvosa. *Rhizophora mangle* is the most widespread; it occurs along all of the tropical and sub-tropical Atlantic coast from Bermuda to Santa Catarina, South Brazil, and although restricted by extreme climatic conditions, it is present from Baja California, Mexico to north Peru, at the Tumbes River estuary (Cintrón-Molero and Schaeffer-Novelli 1992; Breteler 1969; Savage 1972). *Rhizophora racemosa* and *R. harrisonii* have more restricted distribution. Along the Atlantic coast, both species occur from Central America to the Guyanas, and from Trinidad to northern Brazil (Bacon 1970). The southern limit of these two species was established by Prance et al. (1975) at the equator, and was later extended to Maranhão (2°40' S) by Santos (1986). In this area, large (up to 20 m tall) fringes of both species occur generally landward of *R. mangle* belts. Along the Pacific coast, these two species occur from southern Mexico to Ecuador (Flores-Verdugo et al. 1992; Jiménez 1992). The fourth species of *Rhizophora* is *R. samoensis*, which was first reported for the Pacific coast of Colombia, ca. 5° N (West 1956; Hueck 1972). Later, Horna et al. (1980) and Twilley et al. (1990) recorded *R. samoensis* for the coast of Ecuador (latitude 1° N to 3°20' S). A recent survey of the mangroves of Ecuador (MAG 1991), however, does not include this species. *Rhizophora samoensis* may be a variety of *R. mangle* (Hou 1960). However, Mepham (1983) does consider it to be a true species. The occurrence of *R. samoensis* along the Pacific coast of South America is far from established and still requires more systematic work. As for nearly all other New World mangroves, genetic work is necessary for solving such taxonomic problems.

Table 1.4. Updated list of species of "true mangrove" trees in American continent and their distribution limits (*sources:* Araújo and Maciel 1979; Flores-Verdugo et al. 1992; Jiménez 1984, 1992; Chapman 1974; Cintrón-Molero and Schaeffer-Novelli 1983, 1992; Calderon 1983; Breteler 1969; Santos 1986; Prance et al. 1975; Bacon 1970, 1993; Stoffers 1956; West 1956, 1977; Roth 1992; Savage 1972; Horna et al. 1980; Hueck 1972; Roth and Grijalva 1991; Winograd 1983; Rincón and Mendoza 1984)

Species	Atlantic Ocean		Pacific Ocean	
	North	South	North	South
Family Rhizophoracea				
Rhizophora mangle L.	Bermuda, 32°20'	Praia do Sonho, Brazil, 27°53'	Puerto Lobos, Mexico, 30°15'	Tumbes River, Peru, 3°34'
R. harrisonii Leechman	Estero Real, Nicaragua, 13°00'	Preguiças River, Brazil, 2°40'	Chantuto, Mexico, 15°15'	Guayas River, Ecuador, 2°20'
R. racemosa G.F.W. Meyer	Estero Real, Nicaragua, 13°00'	Preguiças River, Brazil, 2°40'	Chiquirí River, Panama, 10°00'	Guayas River, Ecuador, 2°20'
R. samoensis (Hochr.) Salvosa	No occurrence	No occurrence	Chiquirí River, Panama, 10°00'	Guayas River, Ecuador, 2°20'
Family Avicenniaceae				
Avicennia germinans L.	Bermuda, 32°20'	Atafona, Brazil, 21°37'	Puerto Lobos, Mexico, 30°15'	Piúra River, Peru, 5°32'
A. schaueriana Stapf. and Leech	St. Kitts Island, Caribbean, 17°30'	Laguna, Brazil, 28°30'	No occurrence	No occurrence
A. bicolor Standl.	No occurrence	No occurrence	Restricted to the Pacific coast of Central America	Restricted to the Pacific coast of Central America
A. tonduzii Moldenke	No occurrence	No occurrence	Restricted to the Pacific coast of Central America	Restricted to the Pacific coast of Central America
Family Pelliceriaceae				
Pelliciera rhizophorae Pl. and Tr.	Prinzapolca River, Nicarag., 13°00'	Cartagena Bay, Colombia, 9°00'	Gulf of Nicoia, Costa Rica, 9°30'	Esmeraldas, Ecuador, 1°00'
Family Combretaceae				
Laguncularia racemosa Gaertn.	Florida, USA, 28°50'	Laguna, Brazil, 28°30'	Estero Sargento, Mexico, 29°17'	Piúra River, Peru, 5°32'
Conocarpus erecta L.	Florida, USA, 28°50'	Araruama Lagoon, Brazil, 22°55'	Estero Sargento, Mexico, 29°17'	Tumbes River, Peru, 3°34'
Conocarpus e. sericeus	Endemic to the Northern Caribbean Islands	Endemic to the Northern Caribbean Islands	No occurrence	No occurrence

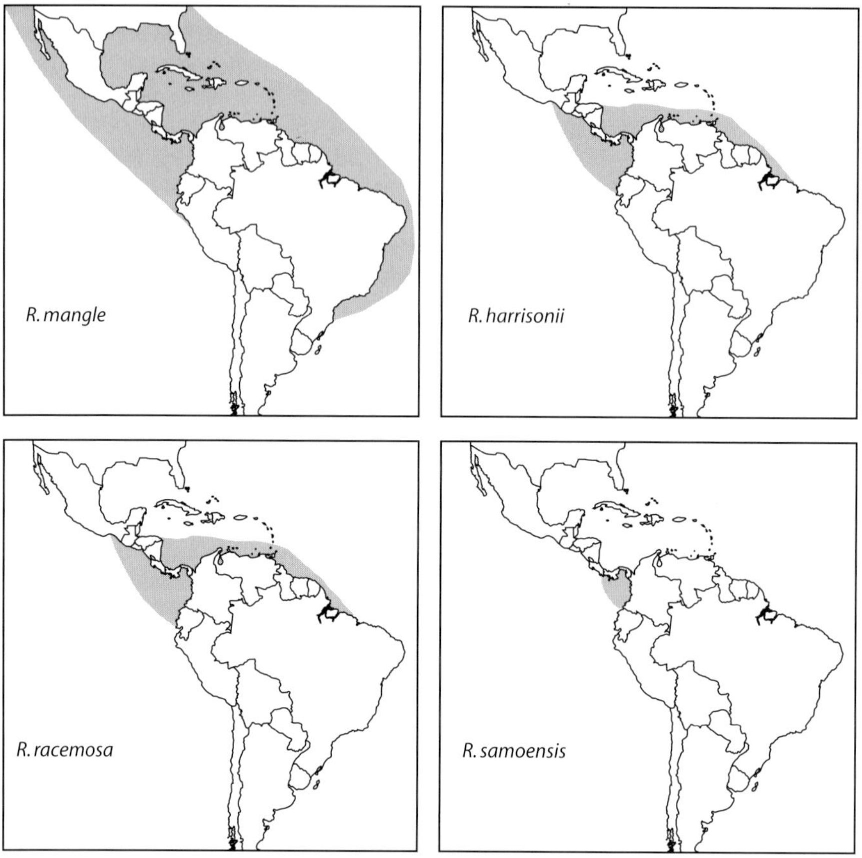

Fig. 1.4. New World distribution of the genus *Rhizophora*

The geographical distribution of the genus *Avicennia* (Fig. 1.5) is similar to that of *Rhizophora*. One species, *A. germinans* L., has a pan-continental distribution. It occurs from Bermuda, latitude 32°20' N to Atafona latitude 21°37' S, in southeastern Brazil on the Atlantic and from Puerto de Lobos, Mexico, latitude 30°15' N to Piúra River estuary, Peru, latitude 5°32' S on the Pacific (Ellison 1993a; West 1977; Cintrón-Molero and Schaeffer-Novelli 1992; Echevarria and Sarabia 1993; Lacerda and Rezende 1993). At the latitudinal distribution extreme, *A. germinans* is represented by populations selected for survival of colder winter temperatures, compared to more tropical populations of the same species. *A. germinans* from the coast of Texas, USA for example, can withstand exposure to chilling temperatures (2–4 °C), which produce considerable damage to populations from more tropical latitudes along the Gulf of Mexico (McMillan 1975).

Avicennia schaueriana Stapf. and Leech. is also widely distributed, endemic to the Atlantic coast, and spreads from the Caribbean Island of St. Kitts, latitude 17°30' N to

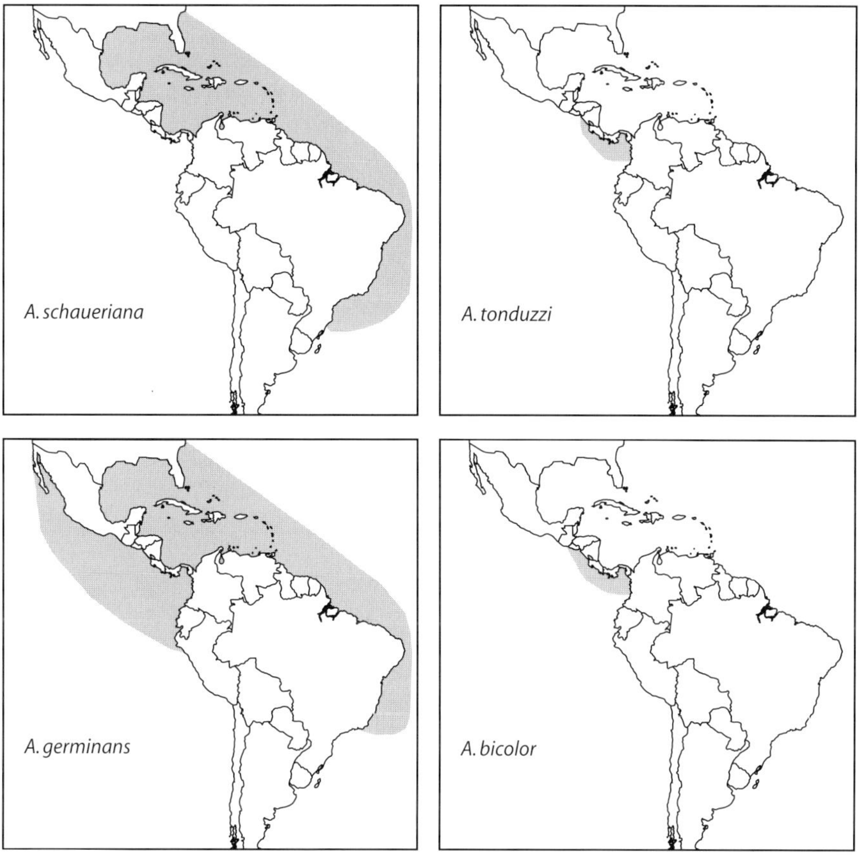

Fig. 1.5. New World distribution of the genus *Avicennia*

Laguna, Brazil, latitude 28°30' S (Cintrón-Molero and Schaeffer-Novelli 1992; Bacon 1993). The other two species *A. bicolor* Standl. and *A. tonduzii* Moldenke have been recorded only on the Pacific coast of Central America from Costa Rica to Panama, and probably resulted from speciation in climatic refugees during the Pleistocene (Jiménez 1992; Cintrón-Molero and Schaeffer-Novelli 1983; D'Croz 1993).

The family Combretaceae (Fig. 1.6) has two widely distributed species: *Laguncularia racemosa* Gaertn. and *Conocarpus erecta* L. The first has a distribution similar to that of *A. germinans*; although it does not extend as far north as Bermuda, its northern limit along on the Atlantic coast is Florida, USA, latitude 30° N (Savage 1972). *Conocarpus erecta* has a smaller range of distribution on the Atlantic coast where it extends from Florida, latitude 28°5' N, to the coast of Rio de Janeiro, Brazil, latitude 22°55' (Araújo and Maciel 1979). A variety of this species *C. e. sericeus* has been recorded in the northern Caribbean Islands (Bacon 1993), although other authors consider this as synonymous to *C. erecta* (Barth 1982). Along the Pacific coast its distribution is similar to *L. racemosa*.

Fig. 1.6. New World distribution of the genera *Laguncularia* and *Conocarpus*

Fig. 1.7.
New World distribution of the genus *Pelliciera*

The family Pelliceriaceae (Fig. 1.7) is represented by *Pelliciera rhizophorae* Pl. and Tr., a species that probably evolved along the East coast of Africa by the Oligo-Miocene. Pelliceroid pollen has been found in sediments from this period at the Niger River Delta (Fuchs 1970). From Africa it has attained a wide distribution in the Caribbean, Central America and northern South America during the Miocene (Graham 1977). Today it is restricted to sites of the coasts of Central America and the Pacific coast of South America (Kobuski 1951; Roth and Grijalva 1991; Roth 1992; Winograd 1983). Significant stands of the species are found only on the very wet Pacific coast of Costa Rica, where annual rainfall may be as high as 6 000 mm (Jiménez 1984). These forests are probably relicts of a wide distribution of *P. rhizophorae*, and could be due to low tolerance of high water salinity caused by dry climatic conditions (Jiménez 1984). This species, however, has recently been discovered in many sites along the Caribbean coast of Colombia (Calderón-Sáenz 1983, 1984; Winograd 1983; Ramírez et al. 1996).

1.2.2 Associated Flora

1.2.2.1 *Terrestrial Plants*

Many plant species occur that are associated with mangrove forests in Latin America and the Caribbean. The diversity could be due to climatic conditions and proximity of other pristine ecosystems. A complete list of such flora, however, is a mere curiosity, since it varies from region to region and even from forest to forest in a given region. Some of these species appear to be associated with mangrove forests throughout their distribution range in Latin America and the Caribbean. Among them the fern *Acrostichum aureum* L. and the Malvaceae *Hibiscus tiliaceus* L. are the most widespread; they occur in most mangroves of the world. These species frequently form dense belts along the landward edge of mangroves, on more elevated sites and around dry and saline areas inside mangroves. *Acrostichum aureum* biology has been studied by Lamberti (1969) in Brazil and by Medina et al. (1990) in Puerto Rico. Notwithstanding this, the biology and ecology of these two species are poorly known.

In the very humid Pacific coast of Central America, many tropical forest species "invade" mangrove forests, such as the Leguminosae vine *Dalbergia brownei* (Jacq) Urban, and the Apocynaceae liana *Rhabdadenia biflora* (Jacq) Hull (Jiménez 1992). In Venezuela, mangroves are frequently invaded by the Lorantaceae *Phthirusa marina*. This liana has also been reported to be a typical component of mangrove forests in French Guyana (Lescure 1977). From the Amazon estuarine region to the coast of Maranhão State, north Brazil, various typically freshwater macrophytes occur among true mangrove species, such as the Araceae *Montrichardia arborescens* Schott and the Leguminosae *Mora oleifera* (Triana) Duke (Pantoja 1993; Mochel 1993). Several palm species, e.g., *Euterpe oleracea* Mart., are also common in these mangrove forests, where they are seldom exposed to high or even moderate salinity. *Mora oleifera* along the Pacific coast of Costa Rica to Ecuador shows many characteristics of a true mangrove species such as tabloid roots with lenticels, viviparity and halophylie, although more detailed investigation is needed (Pinto-Nola 1996).

The glicophytes that occur in association with mangroves of Venezuela are evergreen communities of *Symphonia globulifera* L. (Clusiaceae); *Virola surinamensis* (Myristicaceae); *Pterocarpus officinalis* Jacq. (Papilionaceae); *Mora excelsa* (Caesalpiniaceae) and *Pachira aquatica* (SW) Aubl. (Bombacaceae) and the endemic *Tabebuia aquatilis*. Among the palms, *Mauritia flexuosa* L., *E. oleracea* and *Phenankospermum guyanense* are frequent (Huber and Alarcon 1988).

From Rio de Janeiro to Santa Catarina State, south Brazil, mangroves occur adjacent to mountain rain forests, and are host to a great diversity of epiphytes of the families Bromeliaceae and Orchidaceae in particular *Tillandsia usneoides* L., *T. stricta* Solander and *Vrizia* spp. (Lacerda 1984). A High diversity of epiphytes is also common in *R. harrisonii* forests of southern Ecuador (MAG 1991) and Colombia. Typical species are *Vrizia grandiflora* (Wend) and *Guzmania musaica* (Londoño and Martín 1988). On the Pacific coast of Mexico and on the coast of Venezuela, *Pachira aquatica* typically occurs as a member of the mangrove community (Flores-Verdugo et al. 1992).

1.2.2.2
Salt Marshes and Sea Grasses

Typical salt marsh species also occur in New World mangroves, though they are in general restricted to pioneer formations at the seaward fringe of the mangrove forest, and natural gaps under the canopy, and occasionally as fringes of tidal creeks and channels. They occupy large areas of landward margins of mangroves in drier Caribbean Islands such as British Virgin Islands and Turks and Caicos (Bacon 1993). This marginal distribution is due to light attenuation by the mangrove canopy (Santos 1989; Costa and Davy 1992; West 1977; Reitz 1961; Bacon 1993).

The most common grass along the mangroves of the Atlantic coast of Latin America and the mainland Caribbean coast is *Spartina alterniflora* Loisel, a world-wide distributed salt marsh species. It has been recorded from almost all mangroves of the continent typically at the seaward fringe along the forests. This species may play an important role in the dynamics and environmental conditions of mangroves (Costa and Davy 1992; Lacerda and Abrão 1984), as it promotes sediment fixation and trapping of chemical elements in the mangrove environment. In Insular Caribbean, *Sporobolus virginicus* substitutes *S. alterniflora* in these habitats (Bacon 1993). In the Pacific, salt marsh-mangrove association has not been reported (D'Croz 1993).

Other salt marsh species specialise in colonising areas disturbed by hurricanes or anthropogenic activities. *Sesuvium portulacastrum* L. (Aizoaceae) and *Blutaparon vermiculare* L. Mears (Amaranthaceae), have been reported as typical of such areas forming very dense communities that may permanently occupy disturbed mangrove areas (Lacerda and Hay 1982). In many saline areas of the Venezuelan coast, these species are generally associated with *C. erecta* and other halophytes such as *Atriplex pentandra*, *Salicornia fruticosa* and *Batis maritima* (Huber and Alarcon 1988).

Common salt marsh species associated with mangroves include the Batidaceae, *Batis maritima* L., widely distributed throughout the Americas and the Caribbean; the Chenopodiaceae *Salicornia ambigua* Michx., along with *B. vermiculare*, *Portulaca pilosa* L., *Cakile lanceolata*, *Ipomea pes-caprae* and *S. portulacastrum* occupies high salinity areas inside mangroves; and the Gramineae *Sporobolus virginicus* L. Kunth and *Paspalum vaginatum* Swartz frequently occur in more sandy areas (Huber and Alarcon 1988; Pannier and Pannier 1985).

Other typical mangrove associates are submersed macrophytes, seagrasses and macroalgae and fungi. Among the seagrasses, *Holodule wrightii* Aschers, *Halophila baillonis*, Potamogetonaceae, *Thalassia testudinum* Konig, Hydrocharitaceae, *Syringodium filiforme* and *Ruppia maritima* L., Zannicheliaceae are recorded throughout the Atlantic and Caribbean mangroves (Phillips 1992; Acosta-Fabelo 1974). Seagrass beds are not commonly distributed on the Pacific coast (D'Croz 1993).

1.2.2.3
Seaweeds

Mangrove scientists throughout the Americas have been intensively studying the mangrove macroalgae. Cordeiro-Marino et al. (1992) listed over 150 taxa from the New World mangroves. The highest diversity occurs among the Rhodophyta with 78 species, and the lowest among the Phaeophyta, less than 15 species. In the Caribbean due to high water transparency, stable salinity, and association with coral reefs, algae di-

Table 1.5. Composition and distribution of benthic sea weeds in New World mangroves, based on Cordeiro-Marino et al. (1992) (percentage numbers are rounded)

Region	Chlorophyceae species (%)	Rhodophyceae species (%)	Phaeophyceae species (%)	Total species
North Pacific	4 (40)	06 (60)	0	10
South Pacific	11 (36)	18 (58)	2 (6)	31
Total Pacific	15 (37)	24 (59)	2 (4)	41
North Atlantic	13 (62)	8 (38)	0	21
South Atlantic	21 (34)	37 (60)	4 (6)	62
Total Atlantic	34 (41)	45 (54)	4 (5)	83
Caribbean Sea	54 (50)	48 (44)	7 (6)	109

versity is maximum, with 109 species and a high degree of endemism, nearly 70% among the Rhodophyta (Table 1.5).

Low water transparency, lack of suitable substrate and salinity variation due to estuarine conditions typical of most continental mangroves is probably the cause of the small number of species compared to the Caribbean (Oliveira Filho 1984; Por and Dor 1984). The North Pacific region, including Central America, presents the lowest algal diversity with only 10 species. However, this region is the poorest studied of all coasts on the American continent. Therefore, this low diversity may just represent a lack of systematic work in the region (Table 1.5).

Mangrove macroalgae are mostly found on the roots of mangrove trees (over 50% of the species). Other hard substrates like rocks, stones and large shell fragments account for nearly 30% of the species, while soft mud and sandy substrates account for the other 20% of the species (Cordeiro-Marino et al. 1992).

In most mangroves, the macroalgal community of trunks and aerial roots is dominated by the Bostrichietum association, which includes the genera *Bostrichia*, *Caloglossa* and *Catenella*. These include 12 typical species year round reproduction (Braga et al. 1990). Another typical association is in the sediment, the so called Rhizoclonietum association, formed mostly by over 10 species of green algae of the genera *Rhizoclonium*, *Enteromorpha* and *Cladophora* (Pedrini 1980). Typical tropical species of *Caulerpa*, *Acetabularia*, *Halimeda*, *Sargassum* and *Penicillus* are frequent in the clear-water mangroves of the Caribbean (Pannier and Pannier 1989).

At the sediment interface, cyanophyceae of the genera *Lyngbya* and *Anacystis* form, together with sulfate reducing bacteria and benthic diatom, a key microbial community. This "biofilm" is responsible for most of the mangrove sediment metabolism (Pires 1997).

1.2.2.4
Fungi

Marine fungi of mangroves belong to all groups of higher fungi of warm waters. They infest submersed roots, stems and twigs, as well as sessile animals and algae. Over 30 species of marine fungi occur in New World mangroves. However, only a few seem to be host-specific (Kohlmeyer and Kohlmeyer 1979).

The distribution and species composition of the mycoflora is determined by the duration of submergence of the parts of the tree infested by the fungus and its physiology. Submersed parts have a different mycoflora from the non-submersed parts. However, at high tide level, marine and terrestrial fungi overlap. Host-specific species generally have a restricted distribution, while more omnivorous species tend to have a wider distribution (Kohlmeyer 1969).

Among the New World mangrove marine fungi, typical host-specific species reported are the Ascomycetes; *Didymosphaeria rhizophorae* J. and E. Kohlm.; *Keissleriella blepharospora* J. and E. Kohlm.; *Leptosphaeria australensis* (Cribb et Cribb); *Phoma* sp.; and the Deuteromycetes, *Cytospora* sp., on *R. mangle*. On *A. germinans*, host-specific species of marine fungi are *Rhabdosphora avicenniae* J. and E. Kohlm., *Phoma* sp. and *Lulworthia* sp. and *Mycosphaerella pneumatophorae* Kohlm. Host specific species on *C. erecta* are *Halosphaeria quadricornuta* Cribb et Cribb and *Lulworthia* (Kohlmeyer 1968, 1969).

Most terrestrial fungi from New World mangroves that were described occur as parasites of living leaves, while only a few were described as wood inhabiting. Most frequent host-specific terrestrial fungi on *R. mangle* are the Acomycetes *Anthostomella rhizophorae* Visioli, *A. rhizomorphae* Berl ex Voge; *Physalospora rhizophorae* Bat. and Maia and *P. rhizophoricola* Bat. and Maia (Batista et al. 1955; Stevens 1920; Vizioli 1923). Among the Deuteromycetes, the genus *Pestalotia* (Guba 1961) and *Cercospora* (Craeger 1962) show the highest species diversity. Host-specific terrestrial fungi were also reported for *L. racemosa* and included the Ascomycetes *Irene laguncularie* (Earle) Toro, *Micropeltis laguncularie* Wint and *Physalospora laguncularie* Rehm (Kohlmeyer 1969). Among the Deuteromycetes, *Helminthosporium glabroides* F. L. Stevens has been reported (Seymour 1929).

Marine fungi play a key role in litter decomposition processes and nutrient cycling. Important species such as *Pestalotia*, *Nigrospora* and *Gliocidium* show marked succession during the different phases of the decomposition process (Fell and Master 1973). Studies in the Laguna de Tacarigua (Venezuela) showed that Deuteromycetes and Phycomycetes, in particular *Aspergillus* and *Penicillum* were the dominant fungi involved in the litter decomposition process of mangrove organic matter (Barreto 1988).

1.3 Mangrove Fauna

1.3.1 Introduction

Given the tidal location of many mangrove woodlands where terrestrial and marine ecosystems interpenetrate, a transitional quality to those habitats is rendered, and the mangrove fauna thus comprises elements from both marine and terrestrial habitats. Indeed, few species can be considered exclusive inhabitants of mangroves. For instance, of the 358 macroinvertebrates and vertebrates recorded from a 4 000 ha mangal in Trinidad, only the barnacle *Chthamalus rhizophorae* was strictly confined to this habitat, although many of the other species were most commonly found associated with mangroves, and it is only in this sense that they can be regarded as mangrove fauna (Lacerda et al. 1993). Most of the animals to be found in mangroves also appear elsewhere in other coastal ecosystems, even in areas hundreds of kilometres apart from the coastal strip, as in the case of the scarlet ibis, *Eudocimus ruber* (Conde and Alarcón 1993), but some endemics have been reported, such as the Trochilidae birds *Lepidopyga lilliae* and *Amazilia tzacatl*, the latter being responsible for the pollination of *Pelliciera* (Prahl et al. 1990; Lemaitre-Velez and Alvarez-León 1993). The presence of a given species depends on a number of factors, such as rainfall regime, tides and life cycle stage, among others. Hence the difficulty of characterizing a "true" mangrove fauna. Besides a few permanent residents, animals can be classified as "visitors" (occasionally seen) or "associated", that is, those that utilise mangroves but are not restricted to them (Saenger et al. 1977; Tomlinson 1994). However, this classification is somewhat misleading, since some species of "visitors" can depend strictly on mangroves to accomplish their life cycles.

Whether transient or permanent, the mangrove fauna is large and diversified. In the Americas, over 140 species of birds and 220 species of fishes and hundreds of species of terrestrial and marine invertebrates create high diversity assemblages along mudflats and other intertidal settings. Alike numbers are presented by Saenger et al. (1977) for Australian mangroves and tidal salt marshes; their list includes 95 species of mollusks, 65 crustaceans, 97 worms and 242 species of birds.

In barren areas that otherwise are deprived of any vegetation, such as coastal strips in dry regions, mangrove woodlands offer exceptional opportunities for faunal elements to rest, feed, find shelter, and nurse. Due to the accelerated destruction of inland forests in some Latin American and Caribbean countries, mangrove stands have become important sanctuaries and stepping stones in the migratory routes of various species, which otherwise would be confined to small vegetational patches and thus threatened to extinction. For example, the monkey *Chipodes satanas*, the manatee *Trichechus manatus* and the scarlet ibis *Eudocimus ruber* found refuge in the extensive mangrove forests of Maranhão, northern Brazil. In Venezuela and many Caribbean Islands, as well as in Peru, small populations of the endangered American crocodile, *Crocodylus acutus*, inhabit mangrove swamps, which have become their remaining primary shelters. In Ecuador, at the Churute Mangrove Ecological Reserve, the only population west of the Andes of the horned screamer, *Anhima cornuta*, finds refuge.

1.3.2 Crabs

Crustaceans are among the most noteworthy taxa associated with mangrove swamps. Among them, crabs are one of the most important animal groups, so much for their high diversity in some regions, as for their importance in a series of ecological processes that unfold in these forests and bushes. Crabs are very abundant and may represent 80% of the macrofaunal biomass (Golley et al. 1962). The families Grapsidae and Ocypodidae dominate crab lists (Jones 1984). Over a hundred brachyuran species are known from mangroves in Peninsular and East Malaysia (Tan and Ng 1994). However, the much lower diversity of crabs in American mangroves contrasts with those high numbers. For instance, of a total of 23 species of *Sesarma* and *Armases* reviewed by Abele (1992) only five species are associated with mangroves. In Rookery Bay, Florida, USA, only five species of crabs were caught during a protracted period of nocturnal samplings by means of pitfall traps (McIvor and Smith 1995).

The fauna associated with mangroves in the Pacific coast of Latin America comprises species living on the branches and leaves of mangrove trees, the ground and the brackish waters of the tidal channels. The most common arboreal crab is *Aratus pisonii*, although other species as *Goniopsis pulchra* can be observed occasionally on mangrove trees. The ground of mangals is the habitat for other crabs, such as *Cardisoma crassum*, *Ucides occidentalis*, and several species of *Uca*. The tidal channels support large populations of hermit crabs of the genus *Clibanarius*, the portunids *Callinectes arcuatus* and *C. toxotes*, and xanthidae as *Panopeus purpureus*, among others. Most of those genera are also found in the Caribbean and the Atlantic coasts of Latin America. *Goniopsis cruentata* can be observed occasionally on trunks and branches of mangrove trees, although most of the time they can be sighted on the ground. In the same habitat and in sand flats close to mangrove forests, many species of *Uca* are present; among them, *U. mordax* y *U. rapax* are very common (Conde and Díaz 1985). *Ucides cordatus* is also very abundant in some mangrove forests, where it is exploited by crabbers, such as it happens in many localities of Brazil. Several species of the blue crab *Callinectes* are harvested by fishermen in mangrove lagoons. Another species usually associated with the most terrestrial zone of mangrove forests of the Caribbean is *Cardisoma guanhumi*, a commercial species which has been severely depleted in some localities and is a very important staple in the economies of some villages next to mangrove forests.

Mangrove crabs frequently show patterns of zonation related to sediment type, depth of tidal flooding, and mangrove forest type. In Jamaica, the crabs *Panopeus herbstii*, *Uca thayeri*, *Pachygrapsus gracilis*, and *Goniopsis cruentata* showed lower abundances along transects from the sea to landward in mangrove zones (Warner 1969). However, in Florida, increased tidal elevation leads to enhanced crab abundance (McIvor and Smith 1995). Thus, crab abundance and guild composition might result from the interplay between dissecation, nektonic predation and other factors. Some of these decapods ingest detrital material from mangrove's litterfall, as reported for *Uca*, *Sesarma*, *Cardisoma* and *Ucides* (Abele 1992). Other species are filter feeders (*Petrolisthes*), predators and scavengers (*Callinectes*) and some might include mangrove leaves in their diet.

During recent years, the ecology of mangrove crustaceans and particularly crabs has received a great deal of attention. Progress has been made in the knowledge of their feeding behaviour, nutrient cycling role, impact on substrate topography and mangrove community structure, population dynamics, and life histories (Lacerda 1981; Giddins et al. 1986; Robertson 1986; Warren and Underwood 1986; Smith 1987a,b; Camilleri 1989; Conde and Díaz 1989a,b; Conde et al. 1989, 1995, 2000; Díaz and Conde 1988, 1989; Neilson and Richards 1989; Robertson and Daniel 1989; Smith et al. 1989; Wilson 1989; Warren 1990; Lacerda et al. 1991; Micheli et al. 1991; Emmerson and McGwynne 1992; Micheli 1993a,b; Davie 1994; Frusher et al. 1994; McIvor and Smith 1995; Kwok and Lee 1995; McGuinness 1997; Arruda Leme and Negreiros-Fransozo 1998; Lee 1998; Stieglitz et al. 2000). As a consequence of these studies, crabs are currently regarded a key ecological element in mangrove forests (Smith et al. 1991; Lee 1998). Some of the main ecological processes influenced by crabs are nutrient cycling, litter degradation and vegetational structure. Most of the knowledge on these features, though, springs from the Indo-West-Pacific, hence these functions are still to be confirmed, pondered and fully understood in American mangroves. McIvor and Smith

(1995) have found great contrasts between the role played by crabs in Australia and Florida regarding crab abundance, leaf removal rate and trophic guild composition. In Florida, where the assemblage was dominated by carnivorous xanthids and deposit-feeding ocypodids, the abundance of crabs was higher than in northesat Australia, and no tethered leaves were taken down to crab burrows (McIvor and Smith 1995).

In spite of the enlarging body of information on mangrove crustaceans, it is considered that there is little quantitative information on community structure and population dynamics (Twilley et al. 1996), and thusly the factors that regulate populations of mangrove crustaceans are still poorly known. One of the few exceptions is the grapsid *Aratus pisonii*, which is deemed as the only true Neotropical marine arboreal crab and one of the most common crustacea in American mangroves. This brachyuran, a highly plastic species that is found in mangroves that grow in marine, estuarine, hypersaline and freshwater habitats shows substantial ecological differences on local and regional scales, even in populations separated by a few hundred metres. Several of its life history and population dynamics features have been shown to be correlated to productivity, rainfall patterns and stress induced by exposure to low tides (Conde 1989; Conde and Díaz 1989a,b, 1992a,b; Conde et al. 1989; Díaz and Conde 1989). Its body size and life history traits have been shown to be closely related to mangrove structural characteristics (Conde and Díaz 1989b 1992a; Conde et al. 1989, 1993). In low productivity mangroves that grow in hypersaline lagoons in the western coast of Venezuela, populations are characterised by undersized individuals of *A. pisonii*, including unusually small ovigerous females (Conde et al. 1989; Conde and Díaz 1992b), while the largest individuals and highest reproductive efforts appear in mangroves that grow in river mouths (Conde 1989; Conde and Díaz 1992b). Based on these relations, Conde and Díaz (1989b) have sketched a conceptual model that relates a series of population features of *A. pisonii*, such as maximum body size, reproductive efforts, and sex ratios to mangrove productivity, as inferred from vegetational structural characteristics (Conde and Díaz 1989b; Conde et al. 2000). In addition, Conde et al. (1995) have also reported correlations between concentrations of nitrogen and tannins in mangrove trees and body size in this species.

1.3.3 Mollusks

Mangrove mollusks are dominated by gastropods and bivalves. They are commonly found attached to the roots of mangrove trees; examples of these are some scavenger snails as *Littorina*, *Nerita*, and filter feeders such as the oysters *Ostrea cortezensis* and *O. columbiensis*. The muddy bottoms of the mangrove swamps are inhabited by such bivalves as *Chione subrugosa*, *Tellina ecuatoriana*, *Tagelus* spp., *Anadara* spp., and *Mytella guyanensis* (Lacerda et al. 1993). In the Caribbean coast, the mangrove oyster *Crassostrea rhizophorae* is ubiquitous on prop roots and so is the flat tree-oyster *Isognomon alatus*. Few ecological studies, however, have been done on mangrove mollusks.

The impact of mollusks on mangroves is a promising research line; the mangrove periwinkle (*Littorina angulifera*) and the coffee bean snail (*Melampus coffeus*) are known to eat *Avicennia germinans* propagules. The latter has been found to be an herbivore in controlled experiments with tethered leaves (McIvor and Smith 1995).

1.3.4 Fish

Mangrove lagoons provide nursery grounds and refuge for many species of fishes. In these lagoons, large predators are not common, and hence the predation pressure over juveniles, including those belonging to species whose adults live in deeper waters, is reduced. Although many species can be found in these environments, a small number of species – six or seven – concentrate most of the total catch. In the western Atlantic mangroves, three groups of fishes have been identified by Stoner (1986): resident small flatfish; several species of transient juveniles; and small planktivores. In some localities, a fourth group – marine catfish – appears (Phillips 1981; Yáñez-Arancibia et al. 1980). A high percentage of juveniles, up to 55%, belongs to species that are not resident in the lagoons and that live offshore as adults (Stoner 1986) and 46% (Yáñez-Arancibia et al. 1980).

The complex habitat that results from the imbrication of numerous red mangrove prop roots provides refuge and nursery grounds for many species of fishes. The number of species, density and biomass of fishes found in that environment can be up to several orders of magnitude larger than nearby ecosystems, such as beds of *Thalassia* and other species of seagrasses (Thayer et al. 1987). In a fringe mangrove in south Florida, 36 species were collected exclusively in the mangrove prop root system, while 24 species were sampled solely in the adjacent ecosystems (Thayer et al. 1987). Furthermore, juveniles of commercially important fishes such as the gray snapper *Lutjanus griseus* feed exclusively in the prop root habitat.

However, surveys of mangrove utilization by fish are highly variable. Sheridan (1992) found opposite results; overall, fish densities were higher in non-vegetated open waters than in flooded *Rhizophora mangle* prop roots, although this habitat was sporadically utilised by resident and transient fishes at densities similar to those in seagrasses and open waters. With a share of nearly 75%, the spotfish mojarra (*Eucinostomus argenteus*) was the dominant species among prop roots, where 13 species of fishes were collected.

Among the most common fishes in mangrove swamps of the Pacific coast are: the white or silver or white mullet *Mugil curema*, the mojarra *Eucinostomus californiensis*, and several snooks: *Centropomus armatus*, *C. robalito*, *C. nigrescens* and *C. unionensis*. All of these fishes are found as juveniles in the mangrove channels. In coastal lagoons of the Caribbean coast of Venezuela, the most common species are the catfish *Arius herzbergii* and *Cathorops spixii*; the snooks *Centropomus undecimalis*, *C. ensiferus*, *C. pectinatus*, and *C. parallelus*; the mullets *Mugil liza*, *M. brasiliensis*, and *M. curema*; and the mojarras *Eugerres plumieri*, *Diapterus rhombeus* and *Gerres cinereus* (Ginés et al. 1972; Pagavino 1983; Cervigón and Gómez 1986). Less abundant, but present, are the Atlantic tarpon (*Megalops atlanticus*) and juveniles of the crevalle jack *Caranx hippos*, the ladyfish *Elops saurus*, and the hogchoker *Trinectes maculatus brownii* (Pagavino 1983). In the Caribbean, a great diversity of typical coral reef fishes are frequently found in mangroves (Alvarez-León 1993). Additional species lists and analyses of community structure are provided by Austin (1971), Phillips (1981), Stoner (1986), Yáñez-Arancibia et al. (1980), and Sheridan (1992).

1.3.5 Insects

Even though insects are very abundant and could have a great impact on mangroves through their herbivorous activity, thorough inventories of this group are still pending. The most evident groups are ants, mosquitos, midges and bees. In mangroves located in dry areas where mangals can be the only forests available, africanised bees are common (Conde 1995). In Cuba, during the flowering season of mangroves, some 40 000 beehives of the domestic honey bee, *Apis mellifera*, are taken to mangrove forests fringing the southern coast, where they remain for four months every year, so that they can take advantage of the flowering mangrove trees, and especially of *Avicennia germinans* (Padrón et al. 1993).

Mangroves provide a wide gamut of opportunities for insects. Many lay their eggs in the fruits of mangrove species. Tidal ponds are also a favourite place to lay eggs, and several species of mosquitos (*Aedes* spp.) use holes in the trunks of mangrove trees to oviposit. Insects are particularly interesting, since they could provide alternative pathways to detritus feeders, which are held as the main via for biomass into food chains (Tomlinson 1994; Hogarth 1999).

1.3.6 Birds

In addition to the groups already described, birds, reptiles and mammals also inhabit mangrove forests. Birds have been considered the terrestrial group with the most adaptations to mangroves (Hutchings and Saenger 1987). Most of these adaptations are related to feeding activities. As compared to closely related terrestrial species, Australian mangrove birds have longer bills, a trait that has been considered to be an important adaptation in avoiding the clogging of bristles around the mouth and soiling of the forehead while foraging on mudflats (Hutchings and Saenger 1987).

Because of their high mobility, the degree of association between birds and mangroves is also highly variable and shows a gamut of modalities. Many species of birds are residents, or in one or other stage of their life cycles, can invade mangrove forests, as a part of any of their daily activities or during their migrations, such as many species of Nearctic long-distance migrants or seasonal visitors from adjacent habitats.

The high diversity in species and life histories can be illustrated through the bird fauna that inhabits Venezuelan mangrove forests. More than 150 species of birds have been recorded from these mangroves. Among the birds that have been reported as typical of Venezuelan mangroves are the yellow warbler (*Dendroica petechia*), the bicolored conebill (*Conirostrum bicolor*), the clapper rail (*Rallus longirostris*), the great-tailed grackle (*Cassidix mexicanus*), the spotted tody-flycatcher (*Todirostrum maculatum*), and the rufous crab-hawk (*Buteogallus aequinoctialis*) (MARNR 1986).

Shared by all the seven localities on the Venezuelan coast where detailed inventories have been undertaken are the common egret (*Casmerodius albus*), the black vulture (*Coragyps atratus*), the brown pelican (*Pelecanus occidentalis*) and the scarlet ibis (*Eudocimus ruber*) (MARNR 1986). The latter, an endangered species in many places of the world, is a common sight in Venezuelan mangroves. This is regarded as the only country where there are still considerable reproductive colonies of this bird (Gremone et al. s.d.). Other common species are the brown pelican (*Pelecanus occidentalis*), the magnificent frigatebird (*Fregata magnificens*), the great-tailed grackle (*Cassidix mexicanus*), the tricoloured or Louisiana heron (*Hidranassa tricolor*), the great or common egret (*Casmerodius albus*), the white ibis (*Eudocimus albus*), the heron *Ardea cocoi*, the cattle egret (*Bubulcus ibis*), the egrets *Egretta tricolor, E. caerelea, Egretta alba*, the anhinga (*Anhinga anhinga*), the american wood ibis (*Mycteria americana*), the glossy ibis (*Plegadis falcinellus*), the limpkin (*Aramus guarauna*), the wattled jacana (*Jacana jacana*), the collared plover (*Charadrius collaris*) and the common stilt (*Himantopus himantopus*) (Figueroa and Seijas 1986). Nonreproductive visitors comprise the egret (*Egretta rufescens*), the great blue heron (*Ardea herodias*), the flamingo (*Phoenicopterus ruber*), the blue winged tea (*Anas discors*), the laughing gull (*Larus atricilla*), the gull billed tern (*Gelochelidon nilotica*), the Caspian tern (*Hydroprogne caspia*) and numerous species of Nearctic Limicolae (Figueroa and Seijas 1986).

Many species of winter migrants capitalise on the resources available in mangrove woodlands. Among the species that hibernate in Venezuelan mangroves are *Calidris mauri, Micropalama himantopus* and the osprey (*Pandion haliaetus*) (Figueroa and Seijas 1986). The Neartic migratory species comprise *Egretta rufescens, Butorides virescens virescens, Anas discors, Pluvialis squatarola, Numenius phaeopus, Tringa melanoleuca, T. flavipes, Actitis macularia, Catoptrophorus semipalmatus, Arenaria interpres, Limnodromus griseus, Calidris canutus, C. alba, C. mauri, C. minutilla, Larus atricilla*, and several species of *Gelochelidon, Hydroprogne* and *Sterna*, among them *Hydroprogne caspia* and *Sterna maxima* (Figueroa and Seijas 1986). Nonaquatic common species are the orange-winged parrot (*Amazona amazonica*), which forms great groups, the yellow-headed parrot (*Amazona ochrocephala*), the pale-vented pigeon (*Columba cayennensis*), and dense populations of the macaws *Ara chloroptera* and *A. severa* (Figueroa and Seijas 1986).

Venezuelan mangroves and mudflats are very important as feeding grounds for thousands of flamingos (*Phoenicopterus ruber*) that reproduce in the nearby Lesser Antilles. Furthermore, flamingos nest in mangals in western Venezuela, which is only one the four locations in the Caribbean where this species breeds.

On the Pacific coast, many of these genera, and in some cases the same species, are present; instances are the pelicans *Pelecanus erythrorhinchus* and *P. occidentalis*, the spoonbill *Ajaia ajaja*, the kingfisher *Chloroceryle americana*, and the egret *Bubulcus ibis*.

As shown for Panamanian mangrove forests, number of bird species, guild composition and abundance can vary substantially among mangrove stands mostly due to differences in the availability of supplementary invertebrate prey and adjacent habitat profiles (Lefebvre and Poulin 1997). Only 34% of the bird species were shared by the two mangrove stands, located barely 65 km apart. The assemblages were dominated by gleaning insectivores (33 species), followed by flycatching insectivores (15) and bark-foraging insectivores (12). Other functional groups represented were hovering insectivores, frugivores, granivores, as well as piscivores, carnivores and 3 species that foraged directly on the mudflats. Feeding guilds also varied considerably between sites. On the Caribbean coast (Galeta), bark-foraging, flycatching and hovering insectivores, as well as frugivores were the dominant groups; whereas at the Pacific mangrove (Juan Díaz), gleaning insectivores and granivores prevailed. These differences are considered to be consistent with adjoining habitat divergence, which in turn brings about different distributions of food resources. The Pacific mangrove is bordered by a grass field, while the Galeta mangal is contiguous to a forested habitat where diverse fruited plants were present.

1.3.7 Reptiles

Reptiles can be very common in mangroves. They comprise snakes, lizards and crocodiles that usually are not mangrove specialists but enter these habitats from adjacent areas in procurement of food. Reptiles that frequent mangroves show some adaptations to face osmoregulatory problems (Dunson 1970). Those include keratinised or impermeable skin, valvular nostrils, salt-secreting glands and selective drinking. Young individuals of the American crocodile *Crocodylus acutus*, a species that has functional lingual salt glands, are capable of selectively drinking water (Mazzotti and Dunson 1984).

Among the reptiles commonly found in American mangrove forests are: the spectacled caiman *Caiman crocodylus fuscus*, the American crocodile *Crocodylus acutus*, the arboreal snake *Corallus hortulanus* and the iguana, *Iguana iguana* (Figueroa and Seijas 1986). Marine turtles, such as the green sea turtle *Chelonia mydas*, are very common in the Caribbean, mainly in those places where *Thalassia* beds are associated with mangroves. Other species frequently sighted in the waterways of the huge riverine mangroves of eastern Venezuela is the yellow-spotted river turtle *Podocnemis unifilis* (MARNR 1986).

1.3.8 Mammals

Most, if not all, of the terrestrial mammals observed in mangroves are visitors from nearby habitats or species that have disappeared elsewhere or whose habitats have been fragmented or devastated. Mammals may forage mangrove seeds or foliage at low tide (e.g., the agouti, *Dasyprocta*, and the key deer, *Odocoileus*) or in search of crabs or fish, such as the racoon, *Procyon* spp., does.

Representative mammals in the Americas are the opossum *Didelphis marsupialis*, the crab-eating racoons *Procyon lotor* in the Pacific coast and *P. cancrivorus* in the Atlantic; the otters *Lutra annectens* in the Pacific and *L. longicaudis* in the Atlantic, the weasel *Mustela frenata*, the reeds *Odocoileus virginicus* and *Mazama mazama*, and many monkey species (Aveline 1980; Figueroa and Seijas 1986; MARNR 1986; Álvarez-León 1993). Other mammals reported for Venezuelan mangrove forests are the crab-eating fox (*Cerdocyon thous*), the eastern cottontail rabbit (*Sylvilagus floridanus*); the jaguar (*Panthera onca*) and the South American tapir (*Tapirus terrestris*), the ocelot (*Felis pardalis*), the giant anteater (*Myrmecophaga tridactyla*), the red howler monkey (*Alouatta seniculus*), the capuchin monkey (*Cebus* sp.), the paca agouti (*Agouti paca*), the kinkajou (*Potos flavus*), the agouti (*Dasyprocta guamara*) and several species of bats; many of those species are intermittent or regular visitors (Salvatierra 1983; Figueroa and Seijas 1986; MARNR 1986; Bisbal 1989). Among the aquatic mammals that have been observed in the pristine mangroves of the Orinoco Delta are several endangered species including the West Indian manatee (*Trichechus manatus*), the river dolphin (*Sotalia guianensis*) and the Amazon dolphin (*Ina geoffrensis*).

1.3.9 Mangrove-Root Epibiontic Community

In some localities, where waters are extremely transparent, the red mangrove (*Rhizophora*) roots provide colonization space for the attachment of many invertebrates. This variegate and highly diversified community, where many species and taxa are represented, although sponges, bivalves, and algae predominate, can reach a huge biomass on some roots. However, in some places of Venezuela such as Boca de Caño Lagoon and Tacarigua Lagoon, where high loads of sediments are in suspension, the adherent community of the mangrove roots is depauperate, being represented by a few specimens of *Crassostrea rhizophorae* and other bivalves. Similarly, algae coverage is scant, and sponges are isolated and poorly developed.

Prop root communities are typically zoned in relation to tidal levels and tolerance to desiccation during tidal exposure. Sessile community distribution in Trinidad showed an upper *Chthamalus* zone, a mid-tidal *Balanus* zone and a lower tunicate/sponge zone (Bacon 1970). The adherent community of the mangrove roots has been considered very stable (Sutherland 1980); however, it can be removed almost entirely by sediment resuspension and salinity changes produced by the pass of hurricanes or unusual torrential rains (Orihuela et al. 1991). Recovery of biodiversity in such a community can be slow; these authors observed that after twenty months had elapsed from dieoffs, only 45% of the species originally adhered to the red mangrove roots had recolonised the habitat, far from the original structure. Variability of water temperature, wave exposure, root type, and proximity of roots to the barrier reef affect epibiont distribution and richness in Belize mangroves (Ellison and Farnsworth 1992). Biological factors also play a role in the structuring of this complex community; the spatial heterogeneity probably results from differential larval recruitment (Sutherland 1980; Bingham 1992). The biomass of epibionts can be so large that it can affect root growth and production (Perry 1988; Ellison and Farnsworth 1990, 1992). The precipitation regime has also been observed to be related to biodiversity of colonizing epibionts (Reyes and Campos 1992). Proximity to other marine habitats is also a foremost factor in colonization patterns (Reyes and Campos 1992). Several species of root-fouling sponges have been shown to be facultatively mutualistic with red mangroves, whose growth is enhanced by those species (Ellison et al. 1996).

Among the more common species of this community is the mangrove oyster *Crassostrea rhizophorae*, which can be a very important staple in the economies of the mangrove forest villagers. In Jamaica, 17 species of sponges and 17 species of tunicates are commonly found in this habitat and can support dense masses of the commercially important mangrove oyster *C. rhizophorae*, *Isognomon bicolor* and the mussels *Brachidontes exustus* and *B. citrinus*. In Venezuela, 33 species of sponges and 12 species of tunicates have been cited for mangrove roots (Sutherland 1980; Díaz et al. 1985; Orihuela et al. 1991). Among the sponges of the adherent community the more common are: *Tedania ignis*, *Haliclona viridis*, *Spongia zimocca*, *Halichondria magniconulosa*, *Desmacellia jania*, and *Lissodendoryx isodictyalis*. The tunicates are: *Botrylloides nigrum*, *Botryllus niger*, *Didemnum* sp., *Diplosoma listerianus*, *Ecteinascidia conklini*, *E. turbinata*, *Microcosmus exasperatus*, *Phallusia nigra*, *Polyclinum constellatum*, *Pyura momus*, *Styela canopus*, and *Symplegma viridae*. Other organisms present are the octocoral *Telesto riisei*; the bivalves *Brachydontes* sp., *Crassostrea rhizophorae* and *Isognomon alatus*; the cirriped *Balanus eburneus*; the anelid *Spirorbis* sp.; the briozaries *Schizoporella* sp., and *Bugula* sp.; the polychaete *Sabellastarte magnifica*; and the actinid *Aiptasia pallida* (Sutherland 1980; Díaz et al. 1985; Orihuela et al. 1991). Species lists are also provided for Quintana Roo, Mexico (Inclón-Rivadeneyra 1989); Puerto Rico (Rodríguez and Stoner 1990); Florida, USA (Bingham 1992); Belize (Ellison and Farnsworth 1992); and Bahía de Chengue, Colombia (Reyes and Campos 1992). A comparison of methods estimating the number of species and coverage has been presented by Díaz et al. (1992). In some localities (for example, Cocinetas Lagoon, Venezuela), the mangrove oyster reported is *Crassostrea virginica*, a larger species than *C. rhizophorae* and apparently more tolerant to extreme salinities (MARNR 1991).

1.3.10 The Sediment Fauna

The complex habitat that results from the entanglement of numerous red mangrove prop roots provides refuge and nursery ground for many species of fishes. The number of species, density and biomass of fishes found in that environment can be up to several orders of magnitude larger than nearby ecosystems, such as beds of *Thalassia* and other species of seagrasses (Thayer et al. 1987). In a fringe mangrove in south Florida, 36 species were collected exclusively in the mangrove prop root system, while 24 species were sampled solely in the adjacent ecosystems (Thayer et al. 1987). Furthermore, juveniles of commercially important fishes such as the gray snapper *Lutjanus griseus* feed exclusively in the prop root habitat (Thayer et al. 1987).

Benthic epifauna and infauna in mangroves frequently show patterns of zonation related to sediment type, depth of tidal flooding, and mangrove forest type. In Port Royal, Jamaica, the crabs *Panopeus herbstii*, *Uca thayeri*, *Pachygrapsus gracilis*, and *Goniopsis cruentata* showed a reduction in numbers along transects from the sea landward in relation to the mangrove vegetation zones (Warner 1969).

The fauna associated with mangroves along the Pacific coast of Latin America is composed of numerous species living in the mangrove trees, the ground and the brackish waters of the tidal channels. Crustaceans are among the most noteworthy taxa related to the mangrove swamps; this group has been very well described. The most common arboreal crab is *Aratus pisonii*, although other species such as *Goniopsis pulchra* can be observed occasionally in mangrove trees. The soil of the mangroves is the habitat for other crabs such as *Cardisoma crassum*, *Ucides occidentalis*, and several species of *Uca*. The tidal channels hold large populations of hermit crabs of the genus *Clibanarius*, the portunids *Callinectes arcuatus* and *C. toxotes*, and xanthidae such as *Panopeus purpureus*, among others. Some of these decapods feed on detrital material from the mangrove's litterfall, as reported for *Uca*, *Sesarma*, *Cardisoma* and *Ucides*. Other species are filter feeders (*Petrolisthes*), predators and scavengers (*Callinectes*), and some include mangrove leaves in their diet.

Many of the aforementioned genera are also found in the Caribbean and on the Atlantic coast of Latin America. Likewise, *Aratus pisonii* is one of the most common crustaceans in the New World mangroves, and perhaps the only true marine arboreal crab in that region. This species is found in mangroves that grow in marine, estuarine, hypersaline, and freshwater habitats (Conde and Díaz 1989a,b, 1992a,b; Conde et al. 1989; Díaz and Conde 1989), where it shows size and life history traits gradation closely related to mangrove productivity (Conde and Díaz 1989b, 1992a; Conde et al. 1989, 1993). The population dynamics of *Aratus pisonii* is related to rainfall and tide regime (Conde 1990; Conde and Díaz 1989a; Díaz and Conde 1989). *Goniopsis cruentata*, apparently a predator of *A. pisonii* (Warner 1967), can be observed occasionally on mangrove trees; although most of the time it can be sighted on the mangrove ground. In the same habitat and in the sand flats close to mangrove forests, many species of *Uca* are present; among them, *U. mordax* y *U. rapax* are very common (Conde and Díaz 1985). *Ucides cordatus* is very abundant in some mangrove forests. Several species of the blue crab *Callinectes* are caught by fishermen in mangrove lagoons. Another species usually associated with the most terrestrial zone of mangrove forests of the Caribbean is *Cardisoma guanhumi*, a commercial species that has been severely depleted in some localities and a very important staple in the economies of some mangrove forest villagers.

1.3.11 Mangroves and Fisheries

The importance of mangroves to fish, shrimp or prawn fisheries has been repeatedly highlighted, and it is regarded as one of the most cherished paradigms in mangrove conservation and management (Edwards 1978; Chong et al. 1990; Hogarth 1999), mostly based on the fact that many species of commercially-important aquatic animals spend part of their life cycles in mangrove swamps. Chong (1995) has profusely illustrated the tight association between penaeids and mangroves in the Indo-West Pacific region. Also, for the Indo-West Pacific region and Australia, empirical logical models that relate shrimp catch and mangrove area, mangrove-fringed coastline or intertidal vegetation (including salt marshes) and latitude have been established (Turner 1977; Chong 1995). Similarly, based on computer simulations, Wolanski and Sarsenski (1997) have established using Australian prawn fisheries data that mangroves sustain prawn fisheries by providing a hydrodynamic trap. The level and quality of water in mangroves is regarded as key elements to these fisheries.

These paradigms are still to be confirmed for Neotropical mangroves; however, the data at hand hint at similar relationships. Several species of penaeid shrimps are related to the Pacific coast mangroves, most importantly *Penaeus occidentalis*, *P. vannamei*, *P. stylirostris* and *P. californiensis*, although as much as nine species have been reported to use the mangrove swamps as nursery areas (D'Croz and Kwiecinski 1980). Some of these shrimps enter the mangrove swamps, where they live for four to five months as small post-larvae seeking the protection and food available in this brackish habitat; later, they move offshore as juveniles. These penaeids are the major component of the coastal shrimp fishery in the Pacific coast of Latin America. A similar pattern is found in the Caribbean. Mangrove lagoons are instrumental in the life cycle of the four species of *Penaeus*: *P. brasiliensis*, *P. notialis* (= *P. duorarum notialis*), *P. schmitti* and *P. subtilis* that are commonly harvested throughout most of the Caribbean basin. Most (if not all) of the species of penaeids that support great-scale fisheries in Venezuela are associated with mangrove-fringed coastal lagoons during one or another stage of their life cycles (Alió 2000). For instance, La Restinga, a coastal lagoon located on the Isla de Margarita, is regarded as the largest nursery ground in Venezuela of *Penaeus brasiliensis*. In Laguna Joyuda, Puerto Rico, as much as 955 kg wet weight of shrimp can be produced yearly by hectare (Stoner 1985). Recruitment of juvenile shrimps to Laguna Joyuda, where they find a highly productive and almost predator-free environment, is discontinuous throughout the year, but highly correlated to the rainfall pattern (Stoner 1988).

1.4 Mangrove Forest Structure and Development

1.4.1 Physiognomy of Mangrove Forests

Mangrove forests are best developed in a tropical climate, where the coldest winter temperature is above 20 °C and temperatures are fairly constant throughout the year (<5 °C of variation). Highly developed forests are also associated with low wave energy and protected shorelines with abundant freshwater supply, allowing the deposition and accumulation of fine organic muds, and keeping the salinity range between 5 and 30 ppt. Under such conditions, a large tidal amplitude will also allow the progradation of mangrove forests further inland, forming large forest belts that can extend over 60 km landward from the sea (Walsh 1974; Lugo and Snedaker 1974).

Under these optimal environmental conditions in these dynamic and humid regions, mangrove forests attain their maximum growth. Red mangrove (*Rhizophora*) forests 40–50 m in height and more than 1.0 m in diameter have been reported in Ecuador and Colombia (West 1956; Hueck 1972; Lacerda and Schaeffer-Novelli 1992). At the Southern coast of Costa Rica and several areas of the Panamanian coast, where seasons are less pronounced and annual rainfall ranges from 2100 to 6400 mm, mangrove trees exceed 35 m in height and a biomass of 280 t ha^{-1} (Jiménez 1992). Well-developed black mangrove forests, with trees up to 30 m in height and 0.7 m in diameter, occur on the coasts of Suriname, French Guyana and northern Brazil, frequently with a biomass of over 200 t ha^{-1} (Lacerda and Schaeffer-Novelli 1992) (Fig. 1.8 and 1.9).

Lugo and Snedaker (1974) have classified the structure of mangrove forests along the American continent. These authors recognised 6 different forest types: Fringe, Riverine, Basin, Overwash, Dwarf and Hammock forests. The last three types are supposed to be specific cases of the first three types (Cintrón-Molero et al. 1985). Briefly the major characteristics of these forest types are:

Basin forests grow inland in drainage depressions channeling terrestrial runoff toward the coast. Water flow velocities are slow, and extensive areas of low topography are flooded. They are particularly sensitive to inundation, and export of mangrove litterfall to coastal areas is minimal. They export carbon mostly in dissolved form (Twilley 1985).

Fringe forests occur along the borders of protected shorelines and islands and are periodically flooded by tides. Due to their greater exposure to waves and tides and their well-developed root system, they are particularly sensitive to erosion and marine contamination. Nutrient cycling and litterfall dynamics are very dependent on episodic climatic events rather than the ecophysiology of the forest itself. Variable and important amounts of nutrients and carbon of marine origin participate in the nutrient cycling processes of these forests (Lacerda et al. 1988a; Ovalle et al. 1990; Silva et al. 1991; Rezende et al. 1990).

Fig. 1.8.
Avicennia trees along the coast of Maranhão in northern Brazil occasionally grow to massive sizes with trunk diameter exceeding 1.0 m (*photo:* courtesy of F. C. Rebelo-Mochel)

Riverine forests occur along rivers and creeks and are flooded daily by tides. They generally consist of tall straight-trunked trees, and the low surface water flow velocity precludes redistribution of ground litter. Fluvial nutrients and constant freshwater frequently support high productivity rates in these forests. However, when freshwater flow dominates over the tidal prism, such as in the Amazon and Orinoco rivers estuaries, riverine mangrove forests may diminish or even disappear in the strong competition with freshwater macrophytes.

The classification described above, however, in not valid for certain mangrove areas and may change very rapidly following changes in coastal geomorphology. Also, the distribution of micro-variations in topography within a mangrove wetland often

Fig. 1.9.
The tallest mangrove trees along the Pará and Maranhão coasts in northern Brazil are typically *Avicennia*. This tree measures 39 m in height (*photo:* courtesy of F. C. Rebelo-Mochel)

controls the distribution of mangrove trees due to very small-scale spatial hydrological variations (Santos et al. 1997; Kjerfve et al. 1999). For example, in deltaic areas in Tabasco, Mexico, where distributionary diversion is quite common, the sequence of mangrove developments may be drastically altered following a shift in the centre of active sedimentation and freshwater discharge (Thom 1967). Along the Pacific coast of Colombia, high sediment loads brought in by rivers create depositional environments that are rapidly colonised by mangroves and completely change the pattern of forest distribution of already established mangroves (Alvarez-León 1993).

Several authors have described zonation and succession patterns in American mangroves. However, their omnipresent suitability has been questioned (Rodriguez 1987), and it is our view that these two parameters, although important in certain areas, are extremely site specific and no expected pattern can be safely forecasted.

Another important aspect of mangrove forest structure is the variability of major structural parameters according to a latitudinal gradient. Table 1.6 shows the variability of major structural parameters along the spectra of latitudes throughout the American continent where mangrove forests occur.

Apart from latitude, the major factors controlling mangrove forest structure are wave action, rainfall and freshwater runoff, which control important environmental variables such as erosion/sedimentation rates, aridity, salinity, nutrient inputs, and soil quality (Kjerfve et al. 1999).

Although mangroves preferentially occupy tidal fringes along marine coasts, inland mangroves are typical in some arid areas of the Caribbean. These forests occur as far inland as 15 km in Barbuda (Stoddart et al. 1973) and 50 km in Inagua Island in the Bahamas (Lugo 1981). Similar formations were also reported that were associated with freshwater plants in Lake Izabal in Guatemala (Brinson et al. 1974) and the South

Table 1.6. Structural parameters of New World mangrove forests. Forest types are *F:* fringe; *B:* basin; *R:* riverine

Location	Forest type	Latitude	Rainfall (mm)	Height (m)	DBH (cm)	Basal area (m^2 ha^{-1})	Density (trees ha^{-1})	Author
Estero Pargo, Mexico	F	18°30'N	1 680	6.0	5.6	23.3	7 510	Day et al. (1988)
La Lechuguila, Mexico	F	25°30'N	459	4.5	11.4	–	4 341	Flores-Verdugo (1986)
Marismas, Mexico	F	21°45'N	1 200	5.2	16.9	–	1 461	Flores-Verdugo (1986)
Agua Brava, Mexico	F	22°45'N	1 200	7.5	14.0	–	3 203	Flores-Verdugo (1986)
Isla Venado, Nicaragua	F	11°55'N	2 500	14.0	14.9	25.0	440	Roth (1992)
Ilha Comprida, Brazil	F	25°00'S	1 700	8.6	9.3	21.2	–	Adaime (1987)
Sepetiba Bay, Brazil	F	23°00'S	1 500	6.1	7.8	21.6	4 510	Silva et al. (1991)
Majana, Cuba	F	21°30'N	1 200	10.0	–	20.6	3 527	Padron et al. (1993)
Sipacate, El Salvador	F	13°00'N	–	9.0	–	9.2	3 400	Oxloj (1987)
Darién, Panama	F	8°00'N	2 200	22.0	10	35.0	320	Mayo (1965)
Tacarigua, Venezuela	F	10°50'N	990	9.5	2.7	–		MARNR (1986)
Morrocoy, Venezuela	F	10°40'N	1 065	11.0	–	10.2	440	MARNR (1986)
Cocinetas, Venezuela	F	11°50'N	277	7.0	–	30.0	4 000	MARNR (1986)
Barra Navidad, Mexico	R	19°11'N	750	4.9	–	14.0	2 090	Zamorano (1990)
Mona Island, Puerto Rico	R	18°00'N	810	13.0	–	27.3	–	Cintrón-Molero et al. (1978)
Boca Chica, Mexico	B	18°30'N	1 680	20.0	8.6	34.2	3 360	Day et al. (1988)
El Verde, Mexico	B	25°30'N	627	7.0	–	11.9	1 430	Flores-Verdugo et al. (1987)
Ilha do Cardoso, Brazil	B	25°00'S	2 269	8.7	9.5	25.9	3 735	Peria et al. (1990)
Tacarigua, Venezuela	B	10°50'N	990	15.3	10.4	30.1	790	Rodríguez (1984)
Morrocoy, Venezuela	B	10°40'N	1 065	15.0	–	–	1 320	MARNR (1986)
Cocinetas, Venezuela	B	11°50'N	277	7.5	20.0	16.8	1 280	MARNR (1986)
Orinoco Delta, Venezuela	R	9°00'N	2 290	28.0	–	65.1	1 000	MARNR (1986)
San Juan, Venezuela	R	10°10'N	2 055	28.0	–	27.5	350	MARNR (1986)
Paria Gulf, Venezuela	R	10°25'N	960	28.0	–	13.8	380	MARNR (1986)

coast of Cuba (Lacerda et al. 1993). In Paraguaná Peninsula, Venezuela, stands of *C. erecta* occur 10 km from the coast.

These formations attracted greater attention, since they seem to have no connection with the ocean. However, in general, they have high water and soil salinity, apparently due to some connection with the sea. In Lake Izabal, a long seawater wedge connects with the lake during the height of the dry season (Brinson et al. 1974). In Inagua and Barbuda, kartz formations possibly allow for the migration of tidal water very far inland. In Cuba, it seems that a salt water lens underneath the freshwater marshes goes far inland. There is also a possibility that these mangroves are relict forests from the Pleistocene, when the connection with the sea could have been permanent (Stoddart et al. 1973).

The colonization of these inland mangroves, at least where they are definitively relict forests, is still a puzzle. Since mangrove seedlings cannot float underground, the colonization of inland areas depends on episodic flooding events of abnormal sea level rises during hurricanes and strong storms when waves and swells usually reach far inland. This seems to be the case in Cuba and has been hypothesised by Lugo (1981) for other Caribbean mangroves. However, this colonization process has not been described in detail.

1.4.2
Primary Production, Biomass Distribution and Allocation

Net aboveground primary productivity of mangroves is the sum of wood growth and total litterfall. While many authors throughout the American continent have studied litterfall, wood growth has seldom been monitored. Therefore, estimates of net primary productivity of mangroves are scarce.

Wood growth seems to be influenced by the availability of freshwater and nutrients. Fringe and riverine mangrove stands in Laguna de Terminos, Mexico, under humid conditions (1 680 mm of rainfall) showed very different values (Day et al. 1988). Fringe forest wood growth was significantly smaller (772 g $m^{-2} yr^{-1}$) than riverine forests (1 206 g $m^{-2} yr^{-1}$). Under the more arid conditions found in Puerto Rico (810 mm rainfall), Golley et al. (1962), estimated a smaller wood growth rate of 307 g $m^{-2} yr^{-1}$.

Based on these values and of litterfall rates, net primary productivity estimates for these forests were 2 457, 1 606 and 781 g $m^{-2} yr^{-1}$, for the Mexican riverine, fringe and Puerto Rican forests respectively. Other estimates for mangrove primary productivity were provided through gas exchange experiments in various mature south Florida mangroves (Miller 1972; Lugo et al. 1975; Carter et al. 1973). These studies found much higher values ranging from 2 044 to 5 475 g $m^{-2} yr^{-1}$ probably due to the method used. No study to our knowledge has focused on the underground production and even underground biomass data are scarce.

Table 1.7 lists various mangrove litterfall data from different sites on the American continent. Leaf litter is the major component of total litterfall of mangrove forests, regardless of forest type, latitude or climate. In general, it sums up to over 70% of the total litterfall. At least from the available data, no clear relationship is found between rainfall and litterfall. The data suggest that a relationship between latitude and annual litterfall is not clear and that local, site specific differences seem to be more important in determining litterfall rates, and sites very close together show different litterfall values, such as in Puerto Rico Island (Musa 1986; Lugo and Snedaker 1975) (Table 1.7).

Table 1.7. Litterfall (g m^{-2} yr^{-1}), rainfall and location of New World mangrove forests. Forest types are: *F:* fringe; *R:* riverine; *B:* basin; *D:* dwarf

Location	Forest type	Latitude	Rainfall (mm)	Leaf litter (% of total)	Total litter	Author
Agua Brava, Mexico	F	22°45'N	1200	821 (81%)	1015	Flores-Verdugo (1986)
Estero Pargo, Mexico	F	18°30'N	1680	594 (71%)	834	Day et al. (1988)
Itacuruçá, Brazil	F	23°00'S	1500	697 (73%)	960	Silva (1988)
Bertioga, Brazil	F	23°53'S	2200	376 (90%)	416	Ponte et al. (1984)
Agua Brava, Mexico	R	22°45'N	1200	749 (59%)	1263	Flores-Verdugo (1986)
El Verde, Mexico	R	23°25'N	627	980 (89%)	1100	Flores-Verdugo et al. (1987)
Laguna Mancha, Mexico	R	21°30'N	1250	749 (59%)	1236	Rico-Gray and Lot (1983)
Boca Chica, Mexico	R	18°30'N	1680	881 (70%)	1252	Day et al. (1988)
Vacia Talga, Puerto Rico	R	18°00'N	890	931 (70%)	1322	Lugo and Snedaker (1974)
Chame, Panama	R	8°40'N	1500	–	900	D'Croz, 1993
Chiriqui River , Panama	R	8°30'N	2500	–	2000	D'Croz, 1993
Barra de Santiago, El Salvador	R	13°40'N	–	–	993	Ramirez and Nuñez (1988)
Barra Navidad, Mexico	B	19°11'N	750	1029 (80%)	1287	Zamorano (1990)
Laguna Joyuda, Puerto Rico	B	18°00'N	900	714 (77%)	919	Musa (1986)
Laguna Joyuda, Puerto Rico	B	18°00'N	900	415 (81%)	511	Musa (1986)
Majana, Cuba	B	21°30'N	1200	–	1060	Cuba (1993)
Florida, USA	B	25°–27°N	–	404 (75%)	538	Day et al. (1988)
Florida, USA	D	26°00'N	–	88 (72%)	122	Day et al. (1988)
Tacarigua, Venezuela	B	10°50'N	990	1000 (71%)	1400	Barreto (1988)

Table 1.8. Biomass allocation, forest type and rainfall in New World mangrove forests ($t\,ha^{-1}$). Forest types are: *F:* fringe; *R:* riverine; *B:* basin; *D:* dwarf

Location	Latitude	Type	Rainfall	Above-ground	Below-ground	Total	Author
Itacuruçá (BR)	23°00'S	F	1500	65	16 (20)	81	Silva et al. (1991)
Darien (PA)	8°00'N	R	2200	279	190 (41)	469	Golley et al. (1975)
Florida (USA)	25°00'N	D	–	8	8 (50)	16	Lugo and Snedaker (1974)
Florida (USA)	25°30'N	Island	–	8	14 (64)	22	Lugo and Snedaker (1974)
Puerto Rico (PR)	18°00'N	F	890	63	50 (44)	113	Golley et al. (1962)

Biomass distribution in mangrove forests was one of the most studied aspects of New World mangroves. However, most authors had studied aerial biomass and few data exists on underground biomass. Also, most studies are on *R. mangle*; *A. germinans* and *A. schaueriana*, and *L. racemosa*; nearly no study has dealt with less widely distributed mangrove tree species such as *R. harrisonii*, *R. samoensis* or *R. racemosa*, or *A. bicolor*.

Table 1.8 lists biomass distribution data for various mangrove forests throughout the American continent. The highest aboveground biomass is found roughly between 10° North and 10° South. No assumption can be made regarding underground biomass; however, it is very important in relation to aboveground parts ranging from 20% to 64% of the total forest biomass. The few data suggest that stressed mangrove communities such as those in arid climates tend to show higher percentages of underground biomass, but definitive conclusions are hampered by the existence of so little information.

1.5 The Physical Environment

1.5.1 Mangrove Soils

Dominant soils in continental mangroves are mostly entisols, but sometimes istosols. Riverine mangroves generally grow on immature clay soils (clay hydraquents). On tidal flats, mangroves grow on mineral (sulfaquents) and organic (sulfahemists) soils. Seaward fringe mangroves grow on acid sulfate soils, whereas in many Caribbean Islands, mangroves grow on bioclastic, coralline sands (Alvarez-León 1993).

A detailed study of mangrove soils was performed on the Guiana coast (Brinkman and Pons 1968; Augustinus 1978), where extensive mangroves have grown on the young coastal plains. Marine clays of the Comowine phase (later than 1000 B.P.) are found in a band along most of the Guiana coast with their surface at about high tide level. The sediments are saline with brown or reddish-brown mottles when leached at the surface. Original pyrite content is low to medium, but thin layers of pyrite clay occur in places. These clays on which pioneer *Avicennia* forest develops alternate with ridges of coarse or shell sand. Soils of river and estuary levees in areas dominated by *Rhizophora* show less variation in soil depth and degree of soil formation than marine clays.

In Venezuela, another detailed classification of mangrove forest soils was carried out by MARNR (1986). Mangroves are found mainly in istosols and entisols. In arid zones, however, mangroves grow in aridisols. In the alluvial plains of the Atlantic coast, soils saturated during long periods (hydraquents) prevail, associated with superficial tropohemists in the plain basins. In the rest of the landscape, tropofibrists and sulfaquents occur. In the Caribbean coast of Venezuela, mangroves are associated with coastal lagoons, growing in acid soils of sandy texture (tropofibrists) with high concentrations of organic matter (sulfaquents). In the arid zones, soils are predominantly loamy-sand texture to loamy-clay with torripsaments to sulphic hydraquents.

Rapid accretion of mud and sand characterises the coast of the Guianas, the dominant mud type being a reduced olive gray pelite. In Guadeloupe, mangrove soils were found similarly under reducing conditions, with redox potential of –300 mV, indicating strong anaerobiosis due to oxygen depletion by bacterial oxidation of organic mater (Febvay and Kermarrec 1978). Under such strong reducing conditions, soil metabolism is dominated by sulfate reduction, resulting in sulfide production, which frequently precipitates as framboidal pyrites (Fig. 1.10). The precipitation of pyrites may contribute to the accumulation of large amounts of sulfide-forming metals, in particular Fe, Cd, Zn, Cu, Pb and Hg (Lacerda 1998).

Studies by Thibodeau and Nickerson (1986), Nickerson and Thibodeau (1985) and Lacerda et al. (1993) showed that, at least close to the rhizosphere of mangrove plants, mangrove soils are highly influenced by the physiological activity of the roots, which are different depending on plant species dominating a given area. For example, con-

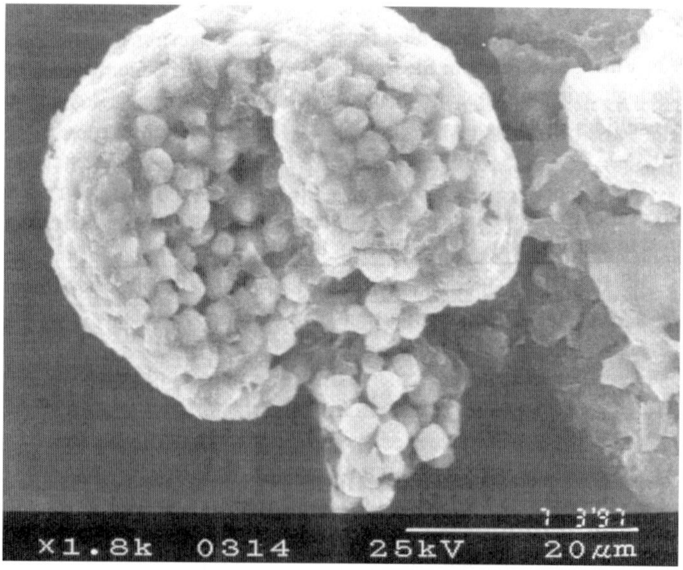

Fig. 1.10. Framboidal pyrite found in reduced mangrove soils. Itacu-russá Experimental Forest, SE Brazil (*photo:* courtesy of Dr. L. Pires)

sistently lower redox potential has been reported for *Rhizophora* soils when compared to *Avicennia* soils. Also, the organic content and composition of mangrove soils are directly related to the above vegetation. Carbon isotopic studies showed that over 95% of the carbon present in mangrove soils originate in mangrove litter (Lacerda et al. 1986a). Also, due to varying organic matter composition, the nutrient level of mangrove soils under different plant cover also varies (Lacerda et al. 1995).

1.5.2 Hydrology

The hydro-period of the fringe-overwash mangroves lasts only hours and occurs daily as a result of high tide wash, which is the only water input to this type of wetland. Riverine mangroves present hydro-periods, which may last from hours to days. The frequency of the hydro-periods may be daily or seasonal with deep waters. Water-level fluctuations within riverine mangroves are large and follow stream discharge patterns. Stream flow and tides represent the major water fluxes to this type of mangrove forest. Basin mangroves have hydro-periods of perennial duration, continuous frequency, and shallow waters. Fresh ground-water discharge, overland runoff, and surface and underground saltwater intrusion are the water inputs to this type of mangrove (Zack and Roman-Mas 1988).

In general, major transport of water occurs during short periods (1–2 h) of the tidal cycle immediately after the changing of flow direction (Kjerfve et al. 1999). Major hydrochemical changes also occur during these short periods (Ovalle et al. 1990; Rezende et al. 1990; Lacerda et al. 1988b).

Mangroves grow most prolifically on deltaic plains that are subject to regular flooding during high tides and have ample supply of freshwater via regular or episodic river flooding or rainfall. The hydro-period of inundation of fringe-overwash mangroves usually only lasts a few hours but occurs daily during high tides. This is the only water input to this type of wetland. Riverine mangroves, on the other hand, usually exhibit longer hydro-periods, which may last from hours to days and depend not only on tidal flooding but also on the river stage. Water-level fluctuations within riverine mangroves vary greatly and increase with increasing river and stream discharges. Basin mangroves often have hydro-periods of perennial duration, and are often covered in shallow water on a continuous basis. Groundwater seepage and overland flow, as well as some saltwater intrusion, are the main water sources for this type of mangrove system (Zack and Roman-Mas 1988).

In general, major rates of water transports in mangrove systems occur during short periods (1–2 h) of the tidal cycle, immediately before and after high water standstill, when the flow direction also changes. Although stream velocities can be substantial, the flow within the mangrove vegetation is slow and sluggish, and seldom exceeds 5 cm s^{-1}. Major hydrochemical water and soil changes can occur during these short periods (Lacerda et al. 1988b; Ovalle et al. 1990; Rezende et al. 1990; Kjerfve et al. 1999).

A critical factor for the well being of mangrove ecosystems is the availability of freshwater, usually indicated by the ratio of rainfall to evapotranspiration, R / E. Although mangroves are found in both humid ($R/E > 1$) and arid climates ($R/E < 1$) in Latin America and the Caribbean, mangrove structural development and growth rates are by far greater in humid equatorial areas with plentiful rainfall, which preferably is relatively evenly distributed during the year (Blasco 1984; Snedaker 1984; Kjerfve 1990). Most of the Latin American mangrove wetlands, as well as the mangrove wetlands on the larger islands in the Caribbean, are distributed along coasts where $R/E > 1$. Notable exceptions are the northern coast of Peru, portions of the Caribbean coasts of Colombia and Venezuela, the coast of Ceará in Brazil, and most of the smaller islands in the Caribbean, where $R/E < 1$. The most extensive and best developed mangrove systems in Latin America and the Caribbean exist in regions with ample freshwater supply and $R/E \gg 1$, including the Pacific coast of Colombia, the Caribbean coasts of Panama and Nicaragua, and the northern Brazil coasts of Para and Maranhão. Whereas mangroves flourish in the Orinocco River delta, this is not the case in the Amazon River where the discharge is so tremendous (average 175 000 $m^3 s^{-1}$) that the constantly fresh conditions in the mouth region of this world's largest river cause invasion of freshwater glycophytes, which successfully out-compete mangroves.

Rainfall by itself does not apparently limit the growth of mangrove wetlands, as these exist in arid as well as wet climates (Galloway 1982). However, rainfall does serve an important role as a primary control in leaching residual salts from mangrove soils, and thus acts to reduce soil salinity. Salts are deposited by tidal flooding in mangrove wetlands. In arid areas or regions with a strongly seasonal rainfall pattern, a barren salt flat often develops as a landward rim of the mangroves, where soil salinity often exceeds 70 ppt and restricts mangrove development (Kjerfve 1990). Such salt flats (in Brazil called *apicuns*), are the result of poor leaching of salt from the soil (Santos et al. 1997) and serve to stunt tree growth or eliminate trees altogether.

Other factors being equal, coasts with a great tidal range can be expected to have more extensive mangrove wetlands because of a greater potential for tidal flooding. Such conditions are encountered along the humid Pacific coast of Colombia with spring tides reaching 3.9 m at Buenaventura, and also along the humid coasts of Para and Maranhão in northern Brazil where in places semidiurnal spring tides exceed 7 m. On the other hand, in the inner part of Baja California, México, where the climate is arid, mangroves are poorly developed in spite of daily tides with a range greater than 7 m. In contrast, all of the Gulf of México and the Caribbean Sea is microtidal, sometimes diurnal and sometimes mixed, with a range less than 0.5 m (Seim et al. 1987; Kjerfve 1981). Here, the tide is of little consequence in terms of mangroves affecting mangrove distribution.

The difference in distribution of mangroves along the east and west coasts of South America is easily explained by the distribution of ocean currents. Temperature is the primary control of mangrove distribution (Kjerfve 1990) and the surface water temperatures, even in relatively near shore coastal waters; along the west coast of South America can be as low as 12 °C. The reason for this is the northward flowing cold-water Humboldt (or Peru) current and the presence of one of the most intensive wind-driven cold-water upwelling systems on earth along the coasts of Peru and Ecuador. As a result, mangroves only begin to appear at latitude 5°32' S in the estuary of Piúra in northern Peru and further north. Along the east coast of South America, the situation is completely different, and water temperatures are usually well above 20 °C except for the coast along the Cabo Frio upwelling system in the state of Rio de Janeiro. The warm-water Brazil current flows southward along the Brazilian coast, and mangroves are flourishing as far south as 32°20' S at Praia do Sonho, Santa Catarina.

1.6 Environmental Impacts on Mangrove Ecosystems

1.6.1 Natural Impacts

1.6.1.1 *Herbivory*

The role of mangrove leaves in marine coastal food chains has been emphasised repeatedly (Odum and Heald 1972, 1975; Lugo and Snedaker 1974). In New Guinea, up to 20% of the leaves are consumed by herbivores (Johnstone 1981). Insect herbivores remove up to 35% leaf area of Australian mangrove plants, but in general the leaf area losses are smaller than those reported for several terrestrial communities (Robertson and Duke 1989), although occasionally massive defoliations may occur (Whitten and Damanik 1986; Lee 1991). Similar percentages have been indicated for American mangroves (Lacerda et al. 1986b; Farnsworth and Ellison 1991). However, several authors have reported smaller proportions of damage to mangrove leaves, typically less than 8% of the total leaf area (Beever et al. 1979; Heald 1971; Lacerda et al. 1986b; Onuf et al. 1977; Saur et al. 1999), but higher defoliation rates (Farnsworth and Ellison 1991), including mass defoliation, do not seem to be uncommon. In a Belizean mangrove forest, Farnsworth and Ellison (1991) found that herbivores damaged 4.3 to 25.3% of *Rhizophora mangle* leaf area and 7.7 to 36.1% of *Avicennia germinans* leaf area. *R. mangle* was more frequently damaged than *A. germinans* (Farnsworth and Ellison 1991), even though the second species has a higher content of nitrogen (Ernesto Medina, personal communication). In mangroves of the Sepetiba Bay (Brazil) *A. schaueriana* had less leaf area eaten than *R. mangle* and *Laguncularia racemosa* (Lacerda et al. 1986b). Similar interspecific patterns and orders of magnitude were found by Saur et al. (1999) in Guadeloupe.

Herbivory rates vary widely from site to site and are related to several factors. These include species, leaf age, branch height, orientation of branches, seedling height and distance to nearest neighbour, presence of a canopy, and chemical composition of leaves, including ash, crude fiber, water contents, soluble carbohydrates and phenols (Lacerda et al. 1986b; Farnswoth and Ellison 1991). Herbivores in Belize include the common mangrove tree crab, *Aratus pisonii*; the gasteropod *Littorina angulifera*; Lepidoptera larvae, including *Megalopyge opercularis*, *Automeris* sp., *Phocides pygmalion*, and other unidentified species; Coleoptera larvae (mainly Chrysomelidae and Lampyridae); Homoptera (Aphididae, Blattidae, Cicadellidae, Cercopidae and Diaspididae); unidentified leaf miners; and larvae of unknown identity.

Damaged leaf area has shown a great variability (Lacerda et al. 1986b). In the Caribbean, the rate of direct grazing by crabs and insects on live mangrove leaves is small (Heald 1971; Beever et al. 1979; Onuf et al. 1977).

The isopod *Sphaeroma terebrans* has been claimed as the agent that has damaged many of the mangroves of southwestern Florida (Rehm and Humm 1973). This species bores into the tips of aerial roots of the red mangrove tree *R. mangle*, destroying root apices and, possibly, initiating lateral budding and the proliferation of lateral roots

(Gill and Tomlison 1977; Simberloff et al. 1978). *Sphaeroma terebrans* can affect up to 83% of the tips in some localities and frequently prevents the root from reaching the substrate (Rehm 1976). This process would stimulate a beneficial action of branching (Simberloff et al. 1978), although Ribi (1981) has not found any evidence to support this effect.

On mangrove cays off the coast of Belize, the isopod *Phycolimnoria clarkae* attacks the submerged roots of the red mangrove *R. mangle*, reducing root relative growth rate by 55% (Ellison and Farnsworth 1990). These isopods do not stimulate lateral branching, as was postulated by Simberloff et al. (1978) for *S. terebrans* in Florida. The epibionts that grow on submerged roots inhibit isopod colonization and thus facilitate root growth (Ellison and Farnsworth 1990).

Another species, *S. peruvianum*, can cause a 50% decrease in the growth rate of the prop roots of the red mangrove (*R. mangle*) in the Pacific coast of Costa Rica and a decrease of 52 to 62% in net root production (Perry 1988). In Cuba, a new species of Pyralidae (Lepidoptera) larvae is a root borer of red mangrove roots (Padrón et al. 1993).

In the Colombian Caribbean, a mollusk borer (Teredinidae) that can reach one meter long is causing extensive damage to *Rhizophora* forests in many sites along the coast. The animal penetrates at root level, accending through the trunk and causing the death and fall of trees (Sánchez-Páez et al. 1997a).

The mangrove tree crab *Aratus pisonii*, although omnivorous (Díaz and Conde 1988) can have up to 42% of its diet based on mangrove leaves in some localities. In Pine Island (Florida, USA), Beever et al. (1979) found that *Aratus pisonii* can damage up to 80.6% (mean = 46.8%) of the leaves of the red mangrove and up to 7.1% (mean = 3.3) of the leaf area; those numbers translate in a mean consumption of 35.3 cm^3 of leaf area per month. *A. pisonii* is an important biomass exporter; Warner (1967) estimated that the average output of *A. pisoni* is 207 eggs per day per m^2. Most of this effort is consumed by aquatic organisms, given that the survival rate from egg to adult is only 0.041% (Warner 1967). Another contribution of *A. pisoni* as a biomass exporter is through frass. Beever et al. (1979) estimated that an adult *A. pisoni* would introduce 8.8 cm^3 of frass per month into the aquatic system.

In Australian mangroves, Smith (1987a,b) found that crabs belonging to the family grapsid could shape mangrove species distribution and abundance through selective predation of propagules. In some Neotropical mangrove forests, the results do not support the dominance-predation model (Smith et al. 1989). In Florida, no *R. mangle* propagules were eaten (Smith et al. 1989). However, in Panama, more propagules were consumed in the low intertidal, *Rhizophora*-dominated forest than the high intertidal, *Avicennia*-dominated forest, but there seems to be less predation on *Rhizophora* in Florida and Panama than in Malaysia and Australia (Smith et al. 1989). The predators responsible for consumption of propagules appear to be the grapsids *A. pisonii* and *Sesarma curacoense*; unidentified burrowing crabs; unidentified water-borne organisms (probably fishes); and the snails *Melampus coffeus* and *Cerithidea scalariformis* (Smith et al. 1989). The grapsid *Goniopsis cruentata* has also been observed to frolic with *Rhizophora* propagules in Venezuela (J. E. Conde, personal observation). Other important herbivores are the Scolitidae (Coleoptera) which attack the seedling of red mangrove (Padrón et al. 1993).

1.6.1.2
Hurricanes and Tropical Storms

The consequences of hurricanes Hugo, Gilbert, and Joan that struck the Caribbean in 1988–1989 were studied for many terrestrial plants, communities and populations (Walker et al. 1991). Nevertheless, mangrove forests were not specifically evaluated for that issue.

Tides and waves produced by hurricanes and storms, which may cause excessive siltation or erosion are considered natural stressors of mangroves (Lugo 1980). In the model of Cintrón-Molero et al. (1978), hurricane cycles are one of the most important external factors in mangrove function. Hurricanes may act as catalysts of succession or may retard or stop it in any of its cyclic stages (Lugo 1980). Also, they could prevent mangrove forests from reaching structural complexity, as could be the case in Florida (Lugo 1980). Periodic exposure of Caribbean mangroves to hurricanes and storms has been suggested as one of the mechanisms responsible for the low structural complexity and lack of climax elements in those forests (Roth 1992).

Although there is extensive literature where accounts of mangrove damage by hurricanes are included, most of this literature is restricted to descriptions of the immediate consequences, and does not consider the long-term effects and responses of mangroves to these climatic disruptions (Roth 1992). The damages can include defoliation, shearing of branches and trunks, and uprooting of trees (Roth 1992). The process of recuperation is slow (Lugo and Snedaker 1974) and depends on species and topographic characteristics, sedimentation and drainage patterns, and proximity to the hurricane pathway (review in Roth 1992). The regeneration of the mangrove forests usually occurs immediately, thanks to the surviving seedlings and saplings (Alexander 1967; Wunderle et al. 1992).

Roth (1992) carried out a detailed study of hurricane Joan's impact and regeneration on a mangrove forest in Nicaragua. Thirty-six percent of the trees died, representing 68% of the basal area of the pre-hurricane stand. The suppression was not random; larger trees were the most affected. As a consequence, the complexity index was diminished. Abundant regeneration by all the original mangrove species appears to have been favored by the hurricane. Seventeen months after Hurricane Joan struck the mangrove stands, seedlings of *Rhizophora mangle*, *Avicennia germinans* and *Laguncularia racemosa* were growing in great numbers throughout the stand, except on those places where the fern *Acrostichum* was present.

The responses of *R. mangle*, *A. germinans* and *L. racemosa* to the damages produced by the hurricane were dissimilar (Roth 1992). Sprouting capacity and initial seedling density were the lowest in *R. mangle*. This species, however, showed the highest seedling survival rate and an intermediate seedling growth rate. *L. racemosa* showed the highest sprouting capacity, although its seedling survival rate was the lowest. Initial seedling density was highest in *A. germinans*, but seedlings grew slowly.

Roth (1992) points out that that post-hurricane mangrove stand development is relevant to its management. It would appear that periodic, small-scale harvests could offer a combination of protection and profitable use in those places where hurricanes are frequent.

The deleterious effects of hurricanes can be compounded by other agents. Defoliation of *Avicennia germinans* caused by insects following hurricanes has been reported (Saur et al. 1999). However, the mechanisms underlying this phenomenon have not been elucidated.

1.6.1.3
Relative Sea-Level Rise

It has been suggested that the eustatic rise in sea level – an aftermath of global warming – might be the most important hazard to mangroves in the near future (Field 1995a). The specific impacts of marine transgressions on mangroves may include increases in the risks of flooding, erosion of the coastline, and saline intrusions. Field (1995a) points out, however, that the effect will depend on the local relative sea-level rise rate and the availability of sediment to support the reestablishment of mangroves, while Bacon (1994) emphasises the importance of variability of wetland type and geomorphological settings on a local scale. Potential changes could also be mitigated by gradual inland migration of mangroves and turnover in species composition (Bacon 1994). Semeniuk (1994) has compiled an exhaustive list of factors that could intervene in the fate of mangroves. Clearly, it is not the eustatic sea level change which is most important locally, but rather the relative sea level change, which includes the effects of changing sea level and also changing land elevation (Stewart et al. 1990). The Atlantic north coast of Brazil, for example, has extensive expanses of mangrove systems because of the relative fall in sea level during the late Holocene (Fig. 1.11; Kjerfve et al. 2001).

In any case, the net balance between sea-level rise and sedimentary accretion or peat accumulation will be instrumental in the fate of mangrove woodlands. Caribbean mangroves could survive a rise rate of 8–9 cm/100 yr, but they can not endure a rate in excess of 12–13 cm/100 yr (Ellison and Stoddart 1991; Parkinson et al. 1994). This latter figure seems to be very common for most of the Caribbean (Rull et al. 1999). For instance, in Playa Medina, Venezuela, the average sea-level rise rate has been estimated in 13.2 cm/100 yr (Rull et al. 1999), while in Hungry Bay, Bermuda, it has been estimated in 14.3 cm/100 yr over the last few centuries (Ellison 1993b). This rate probably will be exacerbated during the next century, due to thermal expansion of the oceans and increased melting of glaciers, brought about by temperature rises ensuing the greenhouse effect. Projected rates of relative sea-level rise have been estimated from 10 to 200 cm/100 yr for the next century, ranging from conservative to pessimistic scenarios (Nilsson 1992; Wigley and Raper 1992; Twilley 1998). The bests guesses range from 45 to 66 cm/100 yr (Stewart et al. 1990), which easily surpass the safety threshold; thus mangroves probably will be eradicated at many localities of the Caribbean Basin. Instances of mangrove degrada-

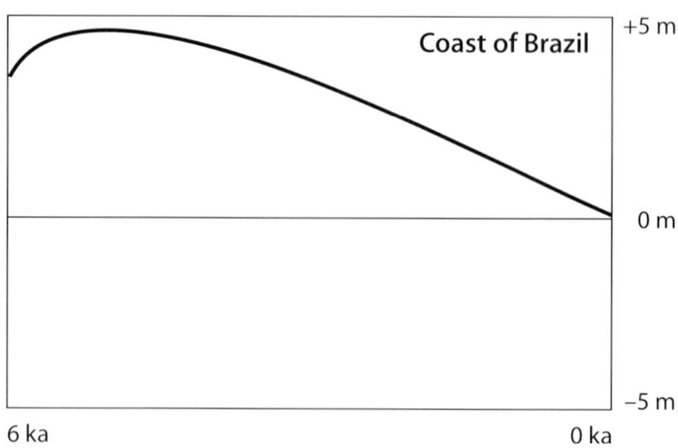

Fig. 1.11. Schematic view of the typical Holocene relative sea level change along most of the coast of Brazil, including the northern mangrove coast (Kjerfve et al. 2001)

tion caused by accelerated sea-level rise are available for Bermuda and southern Florida (Ellison 1993b; Wanless et al. 1994). In Bermuda, mangrove diebacks have been ascribed to a lesser rate of peat accumulation in mangroves (8.5–10.6 cm/100 yr), as compared to the rate of sea level rise during the same period (Ellison 1993b).

In areas of active sedimentation or peat formation, coastal progradation may curtail the effects of sea-level rise, as shown by archeological and palynological records. They show that in some areas, mangroves were able to keep pace with sea-level rise. For instance, cyclical shifts in sea level during the Holocene did not obliterate mangroves in southeast Brazil and northeastern Venezuela, but lead to their gradual displacement (Fairbridge 1976; Sanoja 1992), as inferred from the gradient of shell middens. In Venezuela the rising of sea level, some 5 300 B.P., gave rise to the formation of an ancient mangrove-bordered estuary some 100 km upstream, where shell mounds include remains of *Ostrea* sp., *Melongena* sp. and *Modiolus* sp. (Sanoja 1992).

The enhancing effect that sea-level rise could have on mangrove expansion has been indirectly illustrated by Colonnello and Medina (1998) at Caño Manamo, one of the distributary channels of the Orinoco River (Venezuela). In 1965, to reclaim flooded grounds for agriculture, a cofferdam was built at Tucupita, some 150 km upriver from the marine front. Among many other effects listed by Colonnello (1996) (including emersion of new banks and obstruction of waterways due to sedimentation), periodic intrusions of salt wedges up to the cofferdam have caused invasions of halophilic communities upstream. Since then, mangrove stands have expanded upriver into the oligohaline waters of Caño Mánamo, growing at a rate of 6–7 ha yr^{-1} at some sites; a remarkably higher rate than prior expansion ratios for the same locations (1 ha yr^{-1}) (Colonnello and Medina 1998).

Snedaker (1995) has examined a gamut of scenarios of mangrove affectation in the Caribbean basin due to climate change. He considers that changes in precipitation and catchment runoff may be more significant in shaping mangrove ecosystems at the regional level and may overshadow sea-level rise, which has been the focal point in global climate change.

Eventually, the survival of mangroves will probably depend on the interplay of several factors, co-acting in an intricate way. The prevailing factors will likely be geomorphological settings, processes that contribute to vertical accretion of mangroves and local rates in sea-level rise, although large-scale shifts in rainfall patterns can not be discarded. In general, mangrove forests established on highly urbanised carbonate platforms rimmed by coastal ridges, such as those in Florida (USA) and in many Caribbean Islands, will probably collapse, unlike mangrove ecosystems in deltaic or river-dominated settings, which will be able to migrate landward (Parkinson et al. 1994; Semeniuk 1994). In summary, a more complex response to sea level rise than originally suggested is highly probable.

1.6.2 Human Impacts

A detailed survey of anthropogenic impacts on Caribbean mangrove ecosystems has been conducted by Ellison and Farnsworth (1996). Their catalogue includes wood extraction, pollution by a variety of agents, dredge soil and solid waste dumping, and land reclamation. Most of these impacts are also present in the continental countries of the Americas.

1.6.2.1
Oil Contamination

Contamination of mangrove swamps by oil is commonplace in much of the tropics and has received much attention from scientists, in particular those working in threatened areas such as the Caribbean and Gulf of Mexico coasts. Mangroves, due to their particular geographic and hydrographic settings, are particularly vulnerable to oil contamination (Gundlach 1987; Lamparelli et al. 1997).

Major causes of oil contamination in mangroves are spills of various sizes following tanker accidents and other shipping incidents, as well as damaged pipelines and blowouts. Other, generally small-scale contamination events occur during loading and unloading operations in ports and terminals, such as overloading of tanks, tank cleaning, malfunction of valves, and carelessness during connecting and disconnecting of hoses. Such "small" spills, however, can involve up to 5 t of oil (Baker 1982). Without a doubt, accidents involving large tankers cause the most impressive impacts on mangroves, since they involve large quantities of oil including up to a hundred thousand barrels, and may affect pristine mangrove forests along extensive coastlines with devastating impacts.

New World mangrove forests have witnessed many such accidents, in particular during the 1960–1970s, when regulation and safety procedures were not commonly in use. Between January 1974 and June 1990, 31 oil spills occurred near coastlines in the Caribbean, Florida, the Bahamas and the east coast of Mexico (Burns et al. 1993). Other spills occurred in the open Caribbean and Atlantic, in the Amazon and Orinoco Rivers and in unspecified localities in the Caribbean. One of the first accidents to receive attention from mangrove scientists was the spill of nearly 70 000 barrels of crude oil from the "Argea Prima" in Puerto Rico in 1962. Oil accumulated preferentially on mangrove roots and a large mortality rate of invertebrates, fish and turtles were reported (Díaz-Piferrer 1962). Along the Panama coast, Rützler and Sterrer (1970) reported a reduction of many invertebrate species after the soaking of mangrove sediments with diesel oil from a 15 000 barrel spill from the "Witwatter" in 1968. Jernelov et al. (1976) described short-term and long-term effects of a spill of crude oil from the "St Peters" along the Ecuador/Colombian coast. Short-term effects followed thick oil covering of roots and trunks, resulting in partial defoliation, large mortality of sessile animals and reduction in crab density. After the removal of oil by wave action, defoliated parts recuperated as well as invertebrate populations through migration from non-affected areas. Chan (1977), studying the effects of a small (1 500 to 3 000 barrels) spill of crude oil in Florida Keys, showed that *R. mangle* seedlings were killed when over 50% of their surfaces were covered with oil, and *A. germinans* trees were also killed when 50% of pneumatophores were coated with oil.

More recent accidents involving oil spills and mangroves received better attention and detailed studies. For example, the effects of the 37 000 barrels of crude oil spilled from the "Zoe Colocotroni" in a Puerto Rican mangrove were monitored for over 10 years after it occurred by many authors. A summary of their major findings showed that three years after the spill, *R. mangle* and *A. germinans* trees that were defoliated following the spill had died, and associated fouling communities were reduced to a great extent (Nadeau and Bergqvist 1977). After four years, the oil present in sediments had been highly weathered, even in the most affected areas. Recolonization by seedlings from unaffected adjacent mangrove forests was observed for the first time (Page et al. 1979). However, two years later, remaining oil in sediments was still affecting the infauna, although the colonization processes were already in progress (Gilfillan et al. 1981).

In Venezuela, Bastardo (1993) studied the dynamics of mangrove leaf decomposition under the influence of hydrocarbons in the Tacarigua coastal lagoon. The results suggested that decomposition rates were accelerated by the presence of hydrocarbons. However, after 14 days of exposure, there were no traces of hydrocarbons left.

Finally the best studied case of oil spilled on American mangroves followed the spill of nearly 8 million liters of medium weight crude oil on 27 April 1986, from a ruptured storage tank on the Caribbean coast of Panama (Jackson et al. 1989). After 20 days, the oil surpassed the coral reef's fringe and reached the mangroves.

Upon reaching the mangroves, the oil accumulated on prop roots and pneumatophores, immediately killing the fouling communities. When defoliation started just one month after the spill, branches lacking the weight of leaves flexed upward, lifting the roots and thus also killing the subtidal epibiota that had escaped from direct oiling.

A band of defoliated trees was apparent within two months of the spill and widened thereafter. After 10 months, a band of dead mangrove trees (18 to 100 m wide) had been formed and after 19 months, dead mangroves occurred along 27 km of coastline. After 15 months, over 60% of the roots were dead, broken or rotting. *Rhizophora mangle* seedlings planted in the area failed to produce leaves, therefore the mangrove fringe habitat was largely destroyed and will not recover until new trees are able to grow there.

The long-term biological and chemical consequences of this spill have been monitored through several years (Garrity et al. 1994; Burns et al. 1994; Levings et al. 1994; Burns and Yelle-Simmons 1994). Five years after the spill, the length of shore fringed by mangroves was reduced at oiled sites as compared to unoiled sites. Trees at oiled sites had fewer and shorter submerged roots, as well as a higher proportion of dead roots than trees at those sites that were not oiled during the spill. Chronic contamination of nearby coastal habitats due to the role of mangrove mud as long-term reservoirs of petroleum residues from the spill, was also observed. The epibiota on mangrove prop roots was significantly reduced in most oiled sites, as well as the available surface for colonization of sessile organisms. Within six months of the spill, weathering processes removed most of the volatile hydrocarbons, and marker alkanes in oil adsorbed to surface sediments; however, oil leaching out of heavily contaminated sediments was bioaccumulated by bivalves for five years, even though the most soluble and toxic hydrocarbons had been depleted by that time.

The results of some of these studies allowed for the proposition of a pattern of events occurring in mangroves affected by oil spills. Getter et al. (1981), compared the effects of 5 oil spills on mangrove areas of the Gulf of Mexico and the Caribbean Sea. Despite the different circumstances of these spills, there were a number of similarities between them, which included a suite of stress symptoms and chronology of effects. In general in all sites, defoliation immediately follows the spill, resulting in tree deaths, depending on the proportion of crown defoliated. Leaf deformation was also a common symptom among all sites. Mortality of seedlings generally occurred later due to availability of internal reserves. Figure 1.12 illustrates the long-term major effects on mangrove forest structure parameters, after Lamparelli et al. (1997).

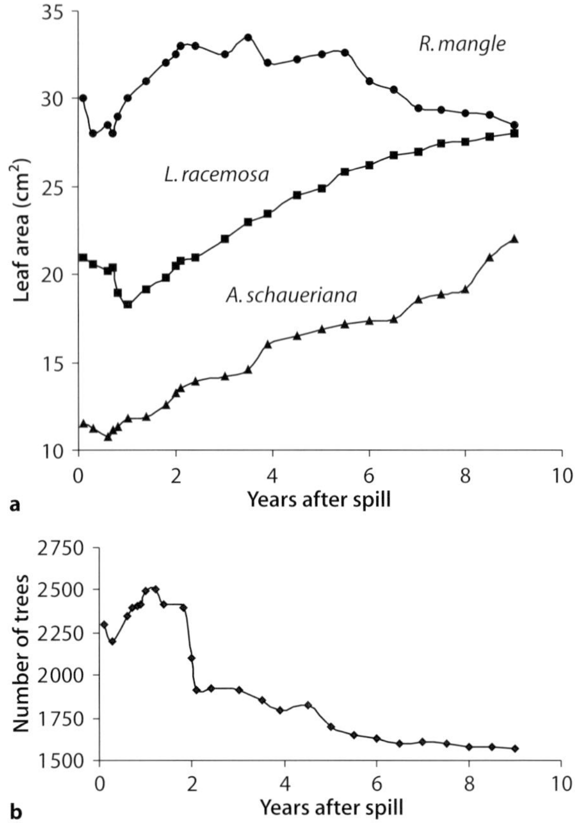

Fig. 1.12. Leaf area and tree density changes along the Iriri River Mangroves during 9 years after an oil spill (Lamparelli et al. 1997)

Among the many stress symptoms observed following an oil spill, such as root galls, leaf necrosis, root malformation and alterations in leaf morphology, defoliation has been identified as the major symptom which correlates well with most others, and always correlates with the degree of the stress imposed on the plant. Mangrove leaves also become significantly and proportionally shorter and narrower just after the onset of the stress (Getter et al. 1985).

Tree mortality due to oil contamination has been frequently related to the obvious blocking of gas-exchange surfaces by the oil coating. However, the waxy nature of mangrove surfaces, either leaves or seedlings, could minimise the effect of the coasting (Getter et al. 1985). Mortality seems to be more likely a result of the toxicity of the oil derivatives and/or dispersants that results in the degradation of plant metabolism. Toxic effects to roots decrease transpiration in mangrove plants. Physical damaging of stomata is also common due to coating affecting gas exchange and photosynthesis; an imbalance of the transpiration/ osmotic regulation system will immediately lead to permanent damage to the plant's metabolism (Getter et al. 1985; Getter and Ballou 1985).

Significant differences in response to oil by different mangrove species have also been reported and have been attributed to the differences in the physiology of the various mangrove species. For example, *R. mangle* seedlings resist better and recuperate faster than *A. geminans*, from chronically exposed sites. This seems to be related to the "filtering" capacity of *R. mangle* roots to cope with the saline environment. This mechanism which has been shown to be effective for salt exclusion and trace metals exclusion (Lacerda et al. 1988a) is also probably efficient in avoiding the uptake of toxic substances derived from oil. On the other hand, *A. germinans* that lacks this filtering mechanism and regulates osmotic balance through excretion of salt through glands in the leaves would more readily absorb toxic oil derivatives (Getter et al. 1985).

Another of the recently registered effects is the mutations caused at cytoplasmic and nuclear levels in the red mangrove *R. mangle*, which become evident to inclination in the metabolic and visible processes in the variation of the pattern of coloration of the leaves and of the viviparous embryos, apparently caused by the exhibition from the forests to the action of chemical pollutants, inside which the hydrocarbons stand out. (Klekowski et al. 1994b; Alvarez-León and Hernández-Avellaneda 1997). Another reported effect is the albinism presence-like consequence of the deficiency of chlorophylls, xanthophylls and carotenes in red mangrove *R. mangle*, as a consequence of recessive mutations, caused by the exhibition from the forests to hydrocarbons (Klekowski et al. 1994a; Corredor et al. 1995).

Site conditions and type of oil are the key determinants of mangrove recovery after a spill (Jacobi and Schaeffer-Novelli 1990). Removal of oil is dependent on the export of sediment particles and deposited litter to the sea. These fluxes are dependent on tidal amplitude and the consequent extension of flooded area during high tide (Ovalle et al. 1990). Removal time will be different along a given forest, resulting in a uneven distribution of the effects. For example, Burns et al. (1993) demonstrated that up to 20 years are necessary for deep mud coastal habitats to recover from the toxic effects of catastrophic oil spills. Defoliation and mortality are more intense in the inner parts of forests, where oil stays longer and weathering forces such as waves and currents are less effective. The outer fringe generally shows less serious effects, such as partial defoliation, root loss and loss of fouling communities. This difference, however, will be smaller at very protected sites, where there is a lack of intense waves or currents; the seaward fringe can also be strongly affected. Also, the use of dispersant greatly reduces the retention of petroleum hydrocarbons in mangrove sediments, reducing their availability for plant uptake and accelerating weathering processes, and therefore the forest recuperation.

1.6.2.2
Solid Waste Disposal

A chronic environmental problem in developing countries is the proper disposal of increasing amounts of sold waste generated by a growing urban population. Around the metropolitan areas of large coastal cities in the tropics and sub-tropics, mangroves are used as ideal sites for solid waste disposal, which is certainly one of the major anthropogenic impacts on mangroves in Latin America and the Caribbean region, causing habitat degradation and destruction. Among environmental oriented government actions, restoring mangroves along landfills has proved to be a large success. Mangroves drastically reduce the non-aesthetic impact of solid waste disposal sites, attracting a large number of animals, in particular birds, which also help to control rats and insect populations abundant in these areas. Recently, apart from obvious benefits, mangroves have proven to be efficient biogeochemical barriers to the transfer of pollutants, generated by landfills, to adjacent coastal areas. Research carried in Australia (Clarck et al. 1997, 1998) and Brazil (Lacerda 1998; Machado et al. 2000) are demonstrating this unsuspected benefit from mangroves, which, on the other hand, will be a strong argument in favour of their conservation and restoration especially at the periphery of urban areas.

Mangroves act as biogeochemical barriers to pollutants generated in solid waste disposal sites through different mechanisms occurring at the root level. Mangrove plants exude oxygen through their roots to cope with the anaerobic nature of the sedi-

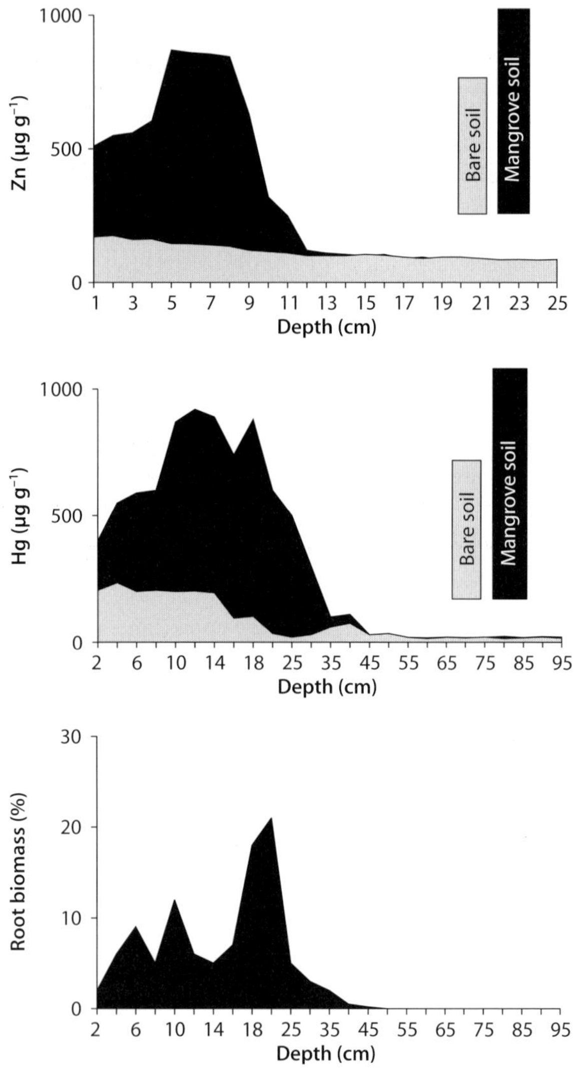

Fig. 1.13.
Root biomass and heavy metal distribution in sediment cores from mangroves and bare sediment areas, along the Gramacho landfill, Rio de Janeiro Metropolitan area, SE Brazil

ments. Thus, they create oxidised rhizospheres capable of fixating heavy metals under non-bioavailable forms. Moreover, large root systems developed by mangrove trees can retain and stabilise sediments more efficiently than bare sediments, avoiding the pollutant remobilization by physical disturbance.

These mechanisms have been successfully used at the Gramacho Metropolitan Landfill in Rio de Janeiro, Brazil, which receives about 7 000 t of solid wastes per day. The landfill leacheates carry high concentrations of heavy metals to Guanabara Bay. Mangroves around the landfill have been subjected to an intensive restoration programme by the landfill administration, and restoration is in full progress. The effect of mangroves on heavy metal retention can be seen in Fig. 1.13. The figure compares the distribution of root biomass, Hg and Zn in sediment cores collected inside a mangrove stand and in bare sediments adjacent to the landfill. The results from the bare sediment cores show lower heavy metal concentrations and a distribution profile suggesting remobilization and export. Under mangroves, much higher heavy metal concentrations are found around the rhizosphere depth, hampering their migration and eventual export to Guanabara Bay, as seem from the sharp decrease in heavy metal concentrations above this depth. Speciation studies (Machado et al. 2000) showed that at this depth heavy metals occur as refractory chemical species, unable to be absorbed by plant roots, which impede the intoxication of the mangroves. Thus minimization of pollution by heavy metals should also be considered when evaluating the amenities provided by mangrove ecosystems all over the word.

1.6.2.3
Global Warming: Temperature and CO_2 Levels Rise

The predicted global warming over the next century ranges from 1.5 and 4.5 °C (Field 1995a), which does not seem to be of significance for plants such as mangroves that can successfully tolerate diurnal oscillations of 20 °C or more in some localities. Although very little is known about the effect of high temperatures on the physiology and phenology of mangroves, the optimum leaf temperature for photosynthesis is 28–32 °C, a process that almost ceases at 38–40 °C (Clough et al. 1982; Andrews et al. 1984). However, interspecies variability should be taken into account when trying to forecast the effect of temperature. The majority of Australian mangrove species reach their maximum shoot growth at mean air temperatures of about 25 °C, with the exception of *Avicennia marina*, which continues to produce leaves even below 15 °C (Hutchings and Saenger 1987). This species is the only one classified in the cool-temperate thermal group by Hutchings and Saenger (1987), while the rest are considered warm-temperate, with the other exception being *Xylocarpus granatum*, which is regarded as tropical-subtropical. Additionally, it is known that mangroves growing in the discharge areas of power stations in Australia show no visible consequences, although some thermal effects may be masked by other factors (Hutchings and Saenger 1987). Although it is difficult to forecast results given the entangled web of intervening factors unchained by global warming, temperature rises will probably induce shifts in the guild structure and expansion of some mangrove species into previously subtropical and temperate regions.

In a similar manner, effects on mangroves brought about by increases of carbon dioxide levels in the atmosphere are difficult to appraise due to the shortage of specific information and diverging results. Kimball (1983) has collated a wealth of information on the effect of high concentrations of CO_2 on plant growth, which, in general, seems to be stimulated when concentrations of this gas increase. In aquatic communities, Drake (1992) reported substantial impacts on wetlands of sedge and grasses, which in general reacted like plants in greenhouse and controlled-environment experiments. Direct effects on photosynthesis dominated the effects of elevated CO_2, but other processes relevant to the ecosystem carbon budget were also altered. Some results did diverge from expected, though: increase in photosynthetic capacity, reduction of water loss, reduction in dark respiration in one of the species, increase in nitrogen fixation in C_3 species and a decrease in herbivory by insects and fungi (Drake 1992). The reports on mangroves are at odds. Several mangrove species were studied in Australia by Ball and Farquahar (1984a,b); their results suggest that photosynthesis in some species would be enhanced by increased levels of CO_2 in the atmosphere. On the contrary, Cheeseman et al. (1991), working with other species, did not observe any improvement in the photosynthetic performance due to increased levels of CO_2. The results available show a variety of responses to elevated concentrations of CO_2 that seem to depend on the species under consideration and the environmental setting in which the plant grows (Ball and Munns 1992). Factors that influence the outcome caused by elevated CO_2 include salinity, water, carbon, nitrogen and other nutrients, and ion toxicity (Ball and Munns 1992). For instance, elevated CO_2 can boost water uptake efficiency in mangroves, but this effect does not always translate into enhanced growth. Although it is too early to draw definitive conclusions given the shortage of information, it seems that interspecific variability will render a gamut of responses to the increase in atmospheric CO_2 levels. In turn, this will probably produce changes in guild composition.

1.7 Mangrove Uses

Mangroves play an important role in tropical coastal economies, providing many goods and services for the human population. These include: coastline protection and stabilization, nursery for a variety of economically important shellfish and finfish, and a source of important products to coastal human populations in the form of timber, firewood and charcoal, chemicals, medicine and waterways for transport, enrichment of marine coastal waters by nutrients, and an environment for aquaculture. Yet some of these benefits are presently poorly understood or unrecognised in Latin America countries. Examples of the importance of such direct and indirect benefits provided by mangrove are significant for the rational management of the Latin America coastal belt.

Waterway protection using mangroves are common in Ecuador and Colombia. In Brazil, mangroves have recently been included in the management plans of marinas and coastal condominiums. In Panama, up to 60% of total shrimp fisheries is based on 5 species that depend on mangroves for completing their development. Along the Maranhão coast, northern Brazil, the huge shrimp production includes two species of shrimps that develop inside the local mangroves. Apart from these indirect benefits, mangrove products themselves are particularly important for many coastal populations. Firewood and charcoal seems to be the major use of mangroves in Latin America. In countries like Nicaragua, where nearly 80% of households uses wood for cooking, mangroves provide a significant percentage of firewood. In this country, annual firewood extraction reaches up to 9 000 m^3. In Honduras, the use of firewood may range from 80 000 m^3 to 120 000 m^3, while in El Salvador, with only 350 km^2 of mangroves, up to 30 000 m^3 of firewood are extracted annually. In Brazil, mangroves are a regular source of firewood for bakeries and pottery, even along the most developed areas of the southeastern coast (Araújo and Maciel 1979).

Charcoal production is another major use of mangrove wood, although only a fraction of the total yield is collected due to inefficient extraction techniques. In Costa Rica up to 1 300 m^3 of mangrove charcoal is produced annually in the Terraba-Sierpe forests, while in Panama this may reach up to 7 400 m^3. Mangrove bark is still a source of tannin in most Latin America countries. Bark yields range from 1 840 to 4 490 kg ha^{-1} in Costa Rica, and bark production in Panama may reach over 400 t yr^{-1}. At the Parnaíba River Estuary, Piauí, Brazil, mangroves support over 10 000 people who depend on artisanal crab fishing.

Despite their importance for most coastal tropical countries in Latin America, mangrove ecosystems have been witnessing an accelerated rush for their resources, most of the time without the necessary care to maintain their integrity, thus threatening their sustainable utilization. Estimates of deforestation in mangrove areas of Latin America are scarce. Central America has annual cover losses estimated for Nicaragua (385 ha); for Guatemala (560 ha) and for Costa Rica (45 ha), mostly for conversion into rice fields, salt ponds and mariculture. In Ecuador, nearly half of the mangrove area (ca. 80 000 ha) has been deforested for various purposes, in particular for shrimp ponds, during the last two decades. In the Ilha Grande Bay, southeastern Brazil, which harbored nearly 600 ha of mangrove forests in the early 1980s, nearly 80% of it has been reclaimed to build condominiums and marinas.

Apart from deforestation itself, degradation of large mangrove areas is taking place in many Latin America countries due to misuse of coastal resources. Diversion of freshwater for irrigation and land reclamation purposes has been one of the major actions leading to mangrove degradation (Conde and Alarcón 1993). The mangroves of Guanabara Bay, Rio de Janeiro, which reached nearly 50 km^2 in the beginning of the century, is presently nearly totally degraded with less than 15 km^2 of pristine forests, mostly due to clear cutting of creek and river banks, oil spills, solid wastes dumping and decreased freshwater inputs. In French Guyana, over 20 000 ha of mangroves have been cut for rice culture, and large amounts of fungicides, pesticides and fertilizers are being used. In the insular Caribbean and the Caribbean coast of Venezuela, tourism development in coastal regions has been the major destroyer of mangrove areas (Hudson 1983; Bacon 1987).

One of the few mangrove silviculture experiences in Latin America was carried out in Venezuela from 1969 until the mid 1980s. In 1969, a vast area was granted to TAMAVENCA (Taninos y Maderas de Venezuela), in concession for 30 years to exploit timber in the Guarapiche Forest Reserve, San Juan River. This mangrove use has been considered a well-planned forestry enterprise. However, the results were not satisfactory. In 1983 only 20% of the proposed area was actually exploited and with a low yield of 150 $m^3 ha^{-1}$ of timber compared to over 500 $m^3 ha^{-1}$ expected. Recovery from this damage is still incomplete today (Medina E, personal communication).

In the 1990s, the government of Venezuela authorised diverse trans-national oil companies to prospect and explore in specific areas of the Orinoco Delta. Although previous environmental impact assessment studies were done according to Venezuelan laws (Geohidra 1996, 1997), a subsequent impact evaluation of these activities upon the extensive mangrove forests of the regions is yet to be done.

Pressures and menaces on mangrove forests vary from country to country, and even within a country, nearby localities can suffer different threats. In some places, the damage is inflicted slowly, and the degradation rate escapes the eye. In other instances, degradation occurs instantly and can reach dramatic proportions. These deleterious actions have not gone through uncontested. Rich legislation has been developed in many countries. Since D. José's Edict, in the eighteenth century, in Brazil, up to the '70s, when Carlos Andrés Pérez promulgated a presidential decree to specifically protect the Venezuelan mangroves, efforts have been made to face mangrove degradation from within a legal frame. The results have been varied and not always successful. Another route to mangrove preservation has been that of non-governmental organisations which develop programmes in mangrove conservation awareness, such as FUFENA in Venezuela, which is in charge of the conservation and management of a vast tract of mangroves on the western coast. The success of this approach, however, remains to be evaluated and should be a long-term task. Finally, rational sustainable development of mangroves in Latin America and the Caribbean is an endeavour where scientists play an important, but not exclusive role. Politicians, entrepreneurs and villagers should also be involved in the conservation of a resource whose ecological, scenic, medicinal and silvicultural values, among others, are unique and a heritage to be perpetuated.

1.8 Mangrove Management in the Neotropics

1.8.1 Mangroves and Man in Pre-Columbian and Colonial America

1.8.1.1 *Mangroves and Man in Pre-Columbian America*

In various countries on the American continent, there is strong archaeological evidence of mangrove utilization by pre-Columbian and even prehistorical human groups. The pre-Columbian man of the Americas started mangrove utilization during the transition period from nomad to fixed habits, between 9 000 to 3 000 B.P. Pre-Columbian inhabitants traditionally used mangroves for many purposes, particularly wood for building houses and fishing gear and for energy production. The use of mangroves varied from site to site, depending on the particular characteristics of a given human group. However, several common features are evident (Alvarez-León 1993).

The expansion of mangrove forests during the late Holocene along the Atlantic coast of South America was in particular triggered by the fall of relative sea level beginning 5 500 B.P. (see Fig. 1.10). This lead to important social changes among itinerant human groups along the coast of the tropical Americas, specially inducing an initial sedentarisation process in gathering, fishing, and hunting communities. In Venezuela 5 000 or 6 000 years ago, human populations, possibly coming from the inner lands, occupied the mouths of the rivers that descend from the Paria Sierra, and other coastal areas, many of them covered by mangroves (Sanoja 1992). This process is similar to the one hypothesised by Widmer (1988) for the southern coast of Florida, USA.

Mangroves, as a resource, including wood, resins, fibers and dyes, and also proteins of animal origin, provided a spectrum of resources for diverse forms of extraction and encouraged some incipient forms of management of native mangrove species. The present-day "cultivation" of the wood-boring teredinid mollusk *Teredo* spp., locally called "turu", by felling mangrove trees and leaving the trunk to rot in the forest floor, typically occurs among mangrove dwelling inhabitants of northern Brazilian mangroves, who may have been witnesses to such a process (Andrade 1984). Extraction of mangrove oysters by the indigenous populations of the coastal areas could have induced a kind of semi-nomad exploitation, which can be compared to the use of soil by slash-and-burn farmers. Major evidence of plant distribution manipulation, however, as seen in some areas of the tropical rain forest of the Amazon region, are lacking.

The nomadic human groups frequently formed semi-permanent settlements along the coast, close to lagoons and bays, where an abundant and easy to collect protein-rich diet was provided by mollusks (Reichel-Dolmatoff 1965). In these areas, they left large amounts of shells and organic and "cultural" debris, called "conchales" or "concheros" in Spanish speaking countries and "sambaquis" in Brazil. These shell middens provide important clues to understanding the characteristics of these populations, including food habits and utilization of natural resources (Perdomo-Rojas 1978; Prahl et al. 1990).

Castaño-Uribe (1989) suggests that in Colombia the process of sedentarisation of the population was a continuous chronological sequence, with great implications for the cultural development of the entire continent. An example of this was the finding in mangroves and adjacent areas along the Atlantic coast of Colombia of the most ancient ceramics of the American continent (ca. 2 400 B.P.), a discovery of pivotal cultural importance (Reichel-Dolmatoff 1985). Along the extensive mangrove fringes of the Pacific and the Caribbean coasts of northern South America, the development of human groups resulted in distinct patterns of cultural adaptation, which can still be recognised today. Their technology was continuously being changed and optimised for the exploitation of these resource-rich, non-seasonal wetlands, to provide the necessary infrastructure for the establishment of large human groups, whose economy was based on the collection of mollusks and on fisheries. The experience diversified and expanded throughout the continent (Castaño-Uribe 1989).

In Panama, there is evidence of human settlements in mangrove areas dating back to approximately 5 000 B.P., where dwellers made tools out of mangrove wood (D'Croz 1993). The importance of fishing of euryhaline species from mangrove-bordered estuaries and channels for pre-Columbian human settlements located in the central coast of Panama at that time has also been annotated (Cook and Ranere 1992). Utilising fences, harpoons, fishhooks and other primitive tools, many of them made of mangrove wood, people from these settlements harvested estuarine and coastal species belonging to the Carangidae, Batrachoididae, Ariidae, and Clupeidae.

In Brazil, this pre-Columbian record also extends from 8 000 to 1 000 B.P., when deposits of shells and fish bones from mangrove species were accumulated in mounds by nomad populations of fishermen and gatherers (Beltrão 1976; Kneip and Palestrini 1984). These groups formed semi-permanent camps visited for generations, resulting in the deposition of layers of debris not infrequently covering 2 to 4 thousand years (Kneip et al. 1984).

In Peru, the Tumpis culture, which settled at the Tumbes region, had an important role in Pre-Columbian America. They formed a large coastal population dedicated to agriculture, fishing and commerce, and were considered the best navigators of the Peruvian coast. They developed to a high degree the art of carving mangrove bivalves, including *Spondylus* sp., *Crassostrea* sp., and *Anadara grandis*. To the *Spondylus* shell, known to the Andean people as "Mullu", they attributed magical powers, and after carving zoomorph or phytomorph figures on them, they were offered in religious rituals (Echevarría and Sarabia 1993).

The migration of human populations to the coastal areas is also testified to in Venezuela by the presence of archaeological pre-ceramic places of shellfish gatherers and fishermen in the Tucacas area, Falcón State, where currently there are extensive forests of mangroves (Cruxent and Rouse 1958). In the south Pacific coast of Costa Rica, huge "concheros" testify to the early settlements of indigenous communities. The indigenous people of Puerto Rico (probably the Arawaks) occasionally ingested the red mangrove seedlings. This is a custom still observed among the fishermen, who claim that it helps them to quench their thirst during their prolonged fishing journeys (Carrera 1975). Archeological evidence suggests that mangroves were used considerably by the Arawak Indians in Puerto Rico. Many of the "concheros" are found in the coastal zone of the island. Among the mollusks ingested by the indigenous people are two mangrove species commonly found in the region: *Crassostrea rhizophorae* and *Isognomon alatus* (Carrera 1975). In the Los Roques Archipelago, 100 km north off the Venezuelan coast, indigenous people also used mangroves considerably. As a consequence of this exploitation, some mangrove species almost totally disappeared on some of the islands of the archipelago (Antczak and Antczak 1987). Almost five centuries after colonization, mangrove products still occupy an important position in the Venezuelan indigenous economy. In the Delta Amacuro State, indigenous populations use mangrove wood in the construction of their dwellings (Flores 1977).

1.8.1.2
Mangroves and Man in Colonial America

The origin of the word mangrove – literally a grove made of "*mangue*" – seems to rest on the term "*mangue*" (Portuguese for mangrove), which, in turn, originated during the colonial endeavours of Portugal and Spain during the sixteenth century. The word "*mangue*", presumably learned by the Portuguese somewhere along the central Atlantic coast of Africa, first appeared in a Portuguese chart of Angra dos Reis Bay, in southeastern Brazil, and elaborated by Lopo-Homem-Reineis, in 1519 (Vannucci 1989); this author comments that the word was adopted by the Portuguese, and later modified by the Spanish. Some sources, however, consider the term "*mangle*" (Spanish for mangrove) to have been used for the first time in 1519 by the Spanish (Vannucci 1989), as originated in the Antilles or Bahamas from a Taino vocable (e.g., WWWebster Dictionary On Line 1999; Diccionario de Uso del Español Actual 1997). Other possible origins and the philological track of these terms, that include the Moluccas, are discussed with due detail by Vannucci (1989).

Since their so-called discovery, mangroves have always been the subject of curiosity in Latin America and the Caribbean. Spanish colonisers were much attracted to mangroves, as can be inferred from the words of historian Gonzalo Fernandez de

Oviedo y Valdés, who wrote that the mangrove is " ... a tree of the best that in these places exists, and it is common in these islands ... rare and admirable trees to sight, because of its forms it is not known another being similar". This author, who added an inventory of mangrove uses, is allegedly, at the beginning of the sixteenth century, the first to describe American mangroves (Schaeffer-Novelli and Cintrón-Molero 1990; Rodríguez 1984). Additionally, in 1595, Sir Walter Raleigh pointed out some trees surrounding the channels at the mouth of the Orinoco Delta that were capable of living in brackish water. Tejera (1977) has catalogued other historical references to mangroves in American literature.

During colonization, the indigenous people's opposition lead to the practical extinction of native populations along the coast of the entire continent. By the time of the European conquest, mangroves represented such a nuisance to troops and horses that Cristobal de Molina, for example, described the mangroves in 1552 in Peru as the "most difficult land of these kingdoms". However, this did not hamper the discovery of the quality of mangrove timber for construction, and their exploitation was immediately started, mostly for poles and boats, and later for tannin. The first fortress the Portuguese built in northeastern Brazil was erected in Natal, in 1568, and was entirely made of mangrove wood from nearby forests. A particularly important role of mangrove wood was to supply fuel for the numerous sugar mills and the associated populations that were becoming established near the coast. On the arid northeastern coast of Brazil, salt extraction had been made as early as 1605 and resulted in some of the earlier mangrove wetland losses (Schaeffer-Novelli and Cintrón-Molero 1999). Since the sixteenth century, mangrove timber was exported from Colombia to Peru and reached 6 000 poles a year in the seventeenth century. In 1677 for instance, thousands of mangrove poles were exported to Cuba from Colombia (Prahl et al. 1990). Bark commerce was also intense. Ecuador, for example, exported nearly 600 t of bark per year to neighbouring countries from 1879 to 1906 (Bodero 1993).

During the colonial period, the commerce of mangrove products was so important that the colonial authorities produced specific legislation to protect and manage the use of this resource. In 1760, the King of Portugal and Brazil, D. José, issued one of the first, if not the very first, law to protect and adequately manage mangrove resources in Brazil. The law prescribed penalties for cutting trees, which had not been previously debarked. This is an example of environmentally sound lobbying by leather merchants who wanted to guarantee abundant, high quality material for their tanneries. In fact, as early as the seventeenth century, the Portuguese made methodical studies of the properties of the tannin from the bark of different species of *Rhizophora*, *Avicennia*, *Sonneratia* and *Xylocarpus* from many places in South India, Molucca, and Sri Lanka. These studies enabled them to select the best sources of raw materials for their tanneries, usually located in Portugal. The interest in the Brazilian mangroves grew only after the Portuguese supremacy in Asia was on a downward trend; the bark of the tree species of the mangroves bordering the Atlantic Ocean had been recognised as producing tan-bark of good quality, in particular *R. mangle* from Brazil (Vannucci 1999).

The Edict of the King of Portugal also indicates concern for sustainability of production, hence the banning of the practice of "ringing" the trunks to extract the bark. The Edict also reflects the accurate observation concerning the growth and regeneration pattern of mangroves south of Cabo de São Roque. The species of that coastal area grow slowly and do not reach the height and importance of the species of the northern Atlantic coast of South America or of the Indian Ocean and Southeast Asia.

Other examples of sound management legislation occurred in other parts of the Colonial Americas. For example, the exploitation of mangroves in Puerto Rico was so intense, that in 1839 a Royal Edict set up a "Junta" that should protect, among others, *R. mangle* and *C. erecta*; their woods could not be used in the naval industry (Carrera 1975).

During the second half of the nineteenth century, the inhabitants of Curaçao exploited the mangroves – especially *R. mangle*, of the Roques Archipelago, off the coast of Venezuela, using them as firewood for the furnaces of steamships; as a residual product, tannins were extracted from bark (Amend 1992). During the same period, mangroves also played a key, but unsuspected role in the development of the nineteenth century Brazilian chemical industry. On 25 January 1812, Prince D. João VI of Portugal and Brazil founded the first Practical Chemistry Laboratory in Rio de Janeiro, with the objective of discovering applications for the different natural materials of the colony for industry and commerce. The first invention of the recently established laboratory was the utilization of mangrove ashes, particularly rich in NaOH, to produce a good quality soap, a process developed by the Brazilian engineer Francisco Vieira Goulart, which resulted in reduced prices of soap relative to the then imported product from England. This "mangrove" soap was probably the first industrial product exported from Brazil (Schubbart 1955).

This panorama shows that resources derived from mangroves have been utilised from pre-Columbian times to the present. Presently, some forest products from mangroves are widely utilised in rural constructions, also as sticks in horticultural crops, and they are still used for the extraction of tannin. However, only recently have mangroves been considered as an ecosystem or conglomerate of ecosystems and microhabitats. This change of perception in evaluating mangroves has in many cases been due to the negative experiences in attempting to manage their resources based on mere pragmatic foundations.

1.8.2 Management and Conservation of Mangrove Areas

It is considered that in most insular Caribbean countries, legislation specifically covering mangroves is lacking, and where such legal frame is available, as in many continental countries of Latin America, regulation is more nominal than practical (Ellison and Farnsworth 1996). One of the main reasons for the failure in enforcing laws and norms is the scarcity of resources. Thus, the enactment of new protective legislation, although desirable, probably will not produce the sought-after objectives of conservation and management. A meaningful step to partly remedy this problem is to prioritise conservation needs of mangrove stands according to the degree of intervention and degradation. This approach would also take into account the diversity in structure, function and biodiversity of mangroves. An attempt in this direction – classification and systematization of mangrove stands – is presented ensuingly.

A summary of management units for the mangroves of Latin America and the Caribbean is presented in Table 1.9. It is important to note that New World mangroves are quite diverse in terms of structure and biodiversity, varying from site to site. The site-specific variability should be taken into account when analyzing the data in Table 1.4, and should always be accounted for whenever management policies are to be applied. Notwithstanding, a general classification of management categories for New World mangroves can be attempted. However, this scheme is far from being the ultimate classification scheme. Considering management guidelines, we propose the following four general categories: (A) Mangrove Conservation Reserves, (B) Mangrove Forest Reserves, (C) Mangrove Fisheries Reserves, and (D) Damaged Mangrove Areas (Table 1.9).

Table 1.9. Management categories and sub-categories and examples of mangrove areas in Latin America and the Caribbean

Management category and major objectives	Sub-category	Type example
A: Mangrove Conservation Reserves (conservation and scientific studies)	Fragile, of high scientific interest	– El Nino dominated mangroves of the Peruvian-Ecuadorian coast; – Inland mangroves of some Caribbean islands
	High biodiversity	– Mangroves along the Colombian-Ecuadorian border; – Amazon River estuary mangroves of Northern Brazil
	Special ecological role or significance	– Mangroves in industrialized bays of SE Brazil; – Storm threatened Caribbean islands and coastline
B: Mangrove Forest Reserves (sustainable utilization of forest products)	Large timber volumes	– Mangroves of Amapa, North Brazil; – Mangrove forests along the south coast of Cuba; – Mangroves in the Pacific coast of North Colombia and of Panama
C: Mangrove Fisheries Reserves (preserve and maximize fisheries)	Of actual mangrove species	– Crab fishing mangrove areas of north-eastern Brazil; – Oyster fishing mangrove areas of
	Open water species	– Shrimp fisheries along the Maranhão Coast of north Brazil; – Anchovy and shrimp fishing in the Gulf of Panama
D: Damaged Mangrove Areas (recuperation of mangrove ecological attributes)	Converted to other uses	– Abandoned shrimp farms in the Gulf of Guyaquil, Ecuador; – Abandoned resort areas in southeastern Brazil; – Abandoned silviculture area along the San Juan River, Venezuela; – Abandoned salt pond areas in north-eastern Brazil
	Impacted mangroves	– Oil impacted mangroves of the Caribbean coast of Panama; – Oil impacted areas in the Gulf; – Lake Maracibo, Venezuela; – Industrialized areas along southeastern Brazil and at Cartagena Bay, Colombia

Mangrove Conservation Reserves (MCRs) include those forests set aside for conservation, tourism, recreation, and scientific studies. A sub-category of this unit includes fragile areas of significant scientific interest. As an example, the 5 000 ha of mangroves of the Tumbes River estuary on the north coast of Peru are highly dependent on the El Niño phenomena. Inland mangroves of some Caribbean Islands also fit in this sub-category. High biodiversity sites comprise another sub-category of MCRs, because of their inherent biological significance and their potential as a refuge for endangered species. Type examples of this sub-category include the extensive mangroves along the Ecuador-Colombia border, various sites along the Pacific coast of Central America, and the unique Amazon River estuary mangroves. Also as a MCR unit, another sub-category includes those mangroves of particular importance in providing indirect benefits to coastal areas, such as maintaining water quality, providing protection against erosion and storms, and stabilising and building coastlines and islands. Examples of this sub-category include the mangroves of some southeastern Brazilian sites, such as Guanabara Bay, Santos-Bertioga region, and Cartagena Bay in Colombia, where intensive industrialization has lead to a decrease in coastal water quality. Also, the fringe mangroves of small Caribbean Islands, which provide effective shoreline protection, e.g., Corales del Rosario National Park, Mac Bean Lagoon Old Province National Park in the Colombian Caribbean and the Belize barrier reef mangroves are examples of mangroves of particular importance in providing indirect benefits.

Mangrove Forest Reserves (MFRs) is a category including those mangroves designated for sustainable production of timber and other forest products. MFRs include extensive mangrove stands, which allow for economically viable exploitation but still with conservation of functional aspects of mangrove ecology. Multiple sustainable uses of mangrove resources should be encouraged in MFRs. Examples of mangroves in this category are the extensive forests along the coast of Amapá State in northern Brazil, the Pacific coast mangroves of northern Colombia, the Pacific coast mangroves of Chiriquí and the Gulf of Panama, and the extensive mangrove forests on the south coast of Cuba.

Mangrove Fisheries Reserves (MFRs) is a category which includes those mangroves associated with extensive waterways with high production and utilization of fish and shellfish species that at least spend part of their life cycle in mangrove areas. We consider two sub-categories of MFRs. The first includes those systems where catches mostly consist of mangrove species proper, and which do not necessary involve the restriction of activities in adjacent coastal waterways. Examples are the mangrove crab fisheries in Piauí and other sites along the coast of northeastern Brazil, where over 30 000 people depend on crab harvesting; and oyster collecting along the Caribbean coast of Colombia and Venezuela. A second sub-category includes those areas where large catches of mangrove-dependent estuarine species are harvested, and protection and restriction of activities in waterways and adjacent coastal waters are necessary. Examples are the Panama Gulf shrimp fishing areas and the extensive mangrove waterways along the coast of Pará and Maranhão along the northern coast of Brazil, where there exist direct, well-established relationships between catch and mangroves areas.

A fourth category consists of Damaged Mangrove Areas (DMAs), those which have been converted to non-sustainable uses of mangroves or subjected to ecologically unsound management practices at the expense of mangroves, and those areas submitted to anthropogenic impacts with negative impacts on mangrove ecosystem functioning. Examples include the abandoned shrimp farms along the coast of the Gulf of Guayaquil, Ecuador, the abandoned large-scale silviculture sites on the San Juan River, Venezuela, the non-productive salt ponds in converted mangrove wetlands along the northeastern Brazilian coast, and the bankrupted tourism projects that caused deforestation of mangrove forests along Ilha Grande Bay in southeastern Brazil. Mangroves that have been impacted by oil spills include mangroves of the Caribbean coast of Panama, Lake Maracaibo in Venezuela, and the Santos-Bertioga coast in São Paulo, Brazil. Also included are the chronically impacted mangroves of Cartagena Bay, Colombia and the industrialised southern Brazilian coast, where a soup of pollutants is constantly been released into local mangroves.

In conclusion, although redoubled efforts have been invested in research during recent years, a detailed inventory of mangrove areas of Latin America and the Caribbean is still far from complete. With high probability, it will be necessary to designate new and different management categories, as more data are synthesised, due to the high diversity of mangrove habitats on the continent. Therefore, in implementing plans, it is always necessary to consider the limitations of our present understanding both of how mangrove systems function and how well management strategies will work.

1.9 Concluding Remarks

In recent years, mangrove researchers throughout the world have targeted their attention on a wide-ranging number of topics, on which an array of updates, reviews and breakthrough articles is now available. Although some of those themes have been under consideration for decades, lately they have been scrutinised from alternative standpoints or using state-of-the-art tools, such as those provided by molecular systematics and remote sensing. In this section, we have compiled a brief list of some of those publications, without the pretentiousness to be exhaustive, but mostly to offer a sample. Several of those articles are listed under additional references, but others have been already cited in one or another section of this chapter.

Many of the articles aim at restoration, management, conservation and impacts (Ong 1995; Ellison and Farnsworth 1996; Farnsworth and Ellison 1997; Ellison 1998; Field 1998, 1999; Lugo 1998; Twilley et al. 1998) and climate change and sea level rise (reviews in Field 1995; Snedaker 1995). Other areas of interest are fisheries, fish assemblages, nursery areas and aquaculture (Laegdsgaard and Johnson 1995; Louis et al. 1995; Robertson and Phillips 1995; Baran and Hambrey 1998), mapping and area assessment (Aschbacher et al. 1995; Green et al. 1998; Pasqualini et al. 1999), geographical patterns in litter fall and biomass (Saenger and Snedaker 1993; Bunt 1995), historical ecology (Schaeffer-Novelli and Cintrón-Molero 1999), evolution, genetics and ecological genetics (Duke 1995; Abeysinghe et al. 1999; Azfal-Rafii et al. 1999), demography and silviculture and population dynamics (Clarke 1995; Gong and Ong 1995), spatial dynamics at the mangrove fringe (Lebigre 1999), fungi (Hyde and Lee 1995), soils as sinks and as determinants of distribution (McKee 1993; Tam and Wong 1995; Lacerda 1998), interaction with other marine ecosystems (Patterson et al. 1993; Saintilan and Hashimoto 1999), nutrients and dwarf mangroves (Feller 1995), and outwelling, tidal modelling and sediment transport in mangrove areas (Lee 1995; Mazda et al. 1995; Wolanski 1995) among many others.

Unfortunately, and in spite of the this diversity of studies, the social dimension and man as a focal point have been subdued in this multifarious panorama. Even though mangroves are the basis of local small-scale economies in many areas, social studies are strikingly lacking. Cormier-Salem (1999) has discussed the role of social scientists in integrated approaches to mangrove studies and has pleaded for a better attention to the relations between mangroves and men, and particularly to the coevolution between mangrove natural and social systems.

Snedaker (1999) agrees with this emphasis on the human dimension, but he also points out that there is a lack of global consensus or focus on the future of mangrove research; a necessary condition to promote interdisciplinary initiatives. Besides social and anthropological studies, other areas in which efforts should be intensified are mangrove genetics and evolution, plant-plant and plant-animal coevolution, and mangrove forest dynamics – studied by means of advanced techniques in remote sensing.

Chapter 2

Mangroves of Africa

E. S. Diop · C. Gordon · A. K. Semesi · A. Soumaré · N. Diallo · A. Guissé
M. Diouf · J. S. Ayivor

2.1 Introduction

Mangroves are the dominant vegetation of over 70% of tropical and sub-tropical coastlines around the world (Fig. 2.1). Mangrove ecosystems are estimated to cover 150 000 km^2 world-wide, of which Africa has about 35 000 km^2 (Diop 1992, 1993). They constitute complex forests with high wood biomass and structural complexity. A significant feature of mangroves is their ability to develop morphological, physiological and reproductive adaptations, which enable them to colonise salty, waterlogged and erratic soil conditions. The mangroves are characterised by high primary production and play vital roles in the transfer of nutrients (Fig. 2.2), especially of carbon from the land to the sea, and serve either as sources or sinks in the global cycle of such substances (Lacerda and Novelli 1992). They are found in all countries along the western coast of the African continent, from Mauritania in the north, (Iles Tidra 19°50' N) to Angola in the south (Angola estuary of the Rio Longo, 10°18' S (Spalding et al. 1997).

Several authors including Chapman (1975), Lawson (1986), and Saenger (1995), have documented on the mangroves of the region.

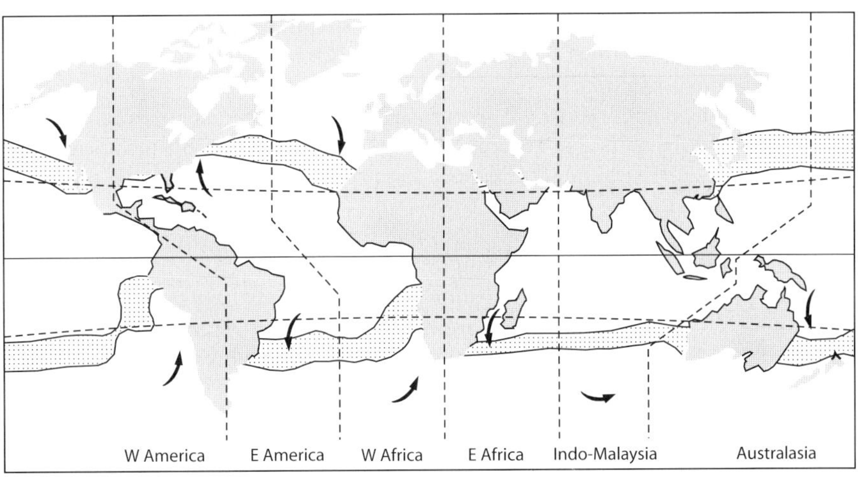

Fig. 2.1. Worldwide distribution of mangrove ecosystems

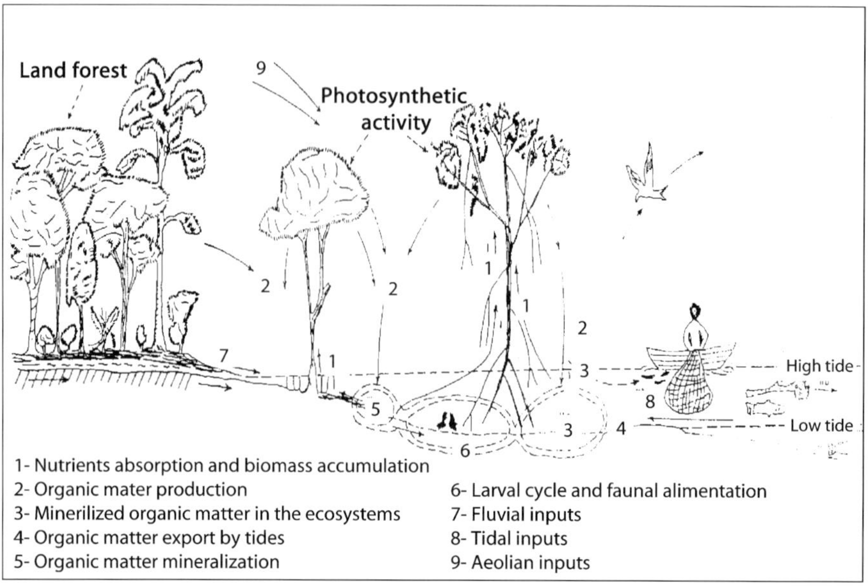

Fig. 2.2. Schematic representation of the nutrient cycling in mangroves

Until recently, the value of West African mangrove ecosystems was not of much relevance to many except for local populations. However, with the current recognition by governments of the relevance of environmental conservation in sustainable development, interest in the study of mangroves has surged up remarkably in the region. Studies on mangroves in Africa have so far covered an extensive area of 80%, underscoring the degree of interest. Most of these studies were descriptive; nevertheless, they recapitulate the importance of mangroves as firewood species, and as breeding grounds for fish, shrimps and prawns, as well as for recreation, tourism, and coastline and wildlife protection.

In spite of their importance, mangroves in West Africa are subjected to considerable stress, either by natural or anthropogenic factors. Natural factors include climatic perturbations, with its attendant effects on salinity and acidity, and coastal erosion or coastal subsidence, as the case of the Niger Delta.

Anthropogenic factors contributing to the degradation of mangrove ecosystems include: extensive exploitation of mangroves for fuel wood, charcoal and constructional purposes; badly designed irrigation dams, which may have negative repercussions on the soils; pollution resulting from industrial wastes, mining, oil exploration and eutrophication; and extensive exploitation of coastal resources for shells, oysters, clams etc.

About two decades ago, some countries adopted measures to promote the conservation, protection and sustainable development of mangrove ecosystems. Consequently, replanting of mangroves on an experimental basis was carried out in several countries such as Sierra Leone, Senegal, and Ghana. It is, however, pertinent to note that for any appropriate mangrove ecosystems and other coastal resource management

Table 2.1. General profile of countries with mangroves in the West African region (from Spalding et al. 1997)

Country	Land area (km^2)	Length of coastline (km)	Total forest extent (km^2)	Estimate mangrove areas (km^2)	Number of protected areas with mangrove
Mauritania	1 025 520	754	5 540	10	2
Senegal	196 720	700	75 440	1 853	2
Gambia	11 300	80	970	497	5
Guinea Bissau	36 120	300	20 210	2 464	0
Guinea	245 860	436	66 920	2 963	0
Sierra Leone	71 740	500	18 890	1 838	0
Liberia	96 320	579	47 900	190	0
Côte d'Ivoire	322 460	520	109 040	150	2
Ghana	238 540	540	95 550	100	5
Togo	56 790	56	13 530	26	0
Benin	110 620	121	35 200	69	0
Nigeria	923 770	850	156 340	10 515	0
Cameroon	475 440	402	203 500	2 434	1
Equatorial Guinea	28 050	296	18 260	257	0
Gabon	267 670	890	182 350	2 500	1
Congo	342 000	169	198 650	120	1
D.R. Congo	2 345 410	37	1 132 750	226	0
Angola	1 246 700	1 600	52 950	1 250	2

strategy to be worthwhile, policymakers and other stake holders need to require relevant information on physical, environmental and socioeconomic parameters of interest, as well as possible solutions to the emerging problems.

The East African region considered here includes Comoro, Kenya, Mauritius, Madagascar, Mozambique, Seychelles, Somalia and United Republic of Tanzania (Table 2.1). The status and future issues on coastal management in Eastern Africa focus on economic activities and population growth (Linden and Lunden 1996, 1997; Shah et al. 1997). Some of the key issues identified as affecting resource management relate to limited human resources, inappropriate policies, limited knowledge, insufficient public involvement, funding, and institutional issues. Implementation of sectoral policies and projects are poor, and the lack of adequate institutional arrangement remains as one of the most widespread constraints to effective coastal zone management. Lack of, or poor, enforcement of rules and regulations are widespread, and large parts of the coast are deteriorating due to habitat degradation, loss of biodiversity, erosion, unplanned development and human competition for coastal resources.

This chapter will give an overview of the current status of mangroves in Africa, describe their uses, examine their threats and present some of the issues related to their management.

2.2 Extent and Distribution of Mangroves

2.2.1 West Africa

The extent of mangroves in West Africa varies from one area to another depending on rainfall regime, freshwater supply, temperature and the substrate of the area (Fig. 2.3). The most extensive mangrove forests in the region are found between Senegal and the south of Sierra Leone; between Ghana and Cameroon; and the Niger Delta where they are most widespread.

Saenger et al. (1983) estimated the area covered by mangrove vegetation of Western Africa (from Senegal to Angola) to be 27 110 km^2. The principal areas along the African coast with extensive distribution of vegetation are those characterised by quaternary formation and active recent sedimentation.

Mangroves of Western Africa considered here include Mauritania, Senegal, Gambia, Guinea-Bissau, Guinea, Sierra Leone Liberia, Côte d'Ivoire, Ghana, Togo, Benin, Nigeria, Cameroon, Equatorial Guinea, Gabon, Congo D.R., Congo and Angola. Table 2.1 summarises the general profile of countries of western Africa that have mangrove forests.

Mauritania. In Mauritania, mangrove swamps are not seen in a continuum at this time. Rather, they occur as relics and patches of pure stands of stunted *Avicennia africana* in Cape Timirist, Tidra, Nair and Niroumi (Fig. 2.4). In the West of the Iwik headland, some dwarf plants may also be seen.

Senegal and Gambia. The mangrove forest of Senegal, located between latitudes 12°20' and 16°20', are the most significant stands in the northern part of West Africa (Fig. 2.5). The mangroves, *sensu stricto*, occupy an area of 1 853 km^2 (Spalding et al. 1997). They are estuarine mangroves located in the lower part of the rivers Senegal, Gambia, Saloum and Casamance. The estuary together with the mangrove swamps of Saloum consist of three important groups of islands that extend for over 800.00 km^2. The Saloum estuary, which may best be described as a ria (a submerged river valley), experiences

Fig. 2.3. The extent of mangrove ecosystems in West Africa

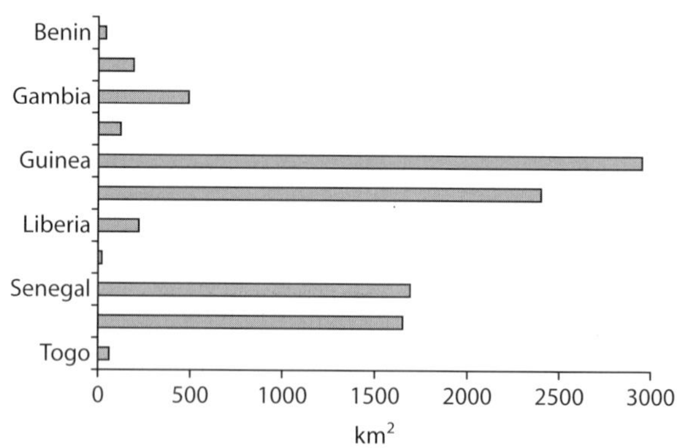

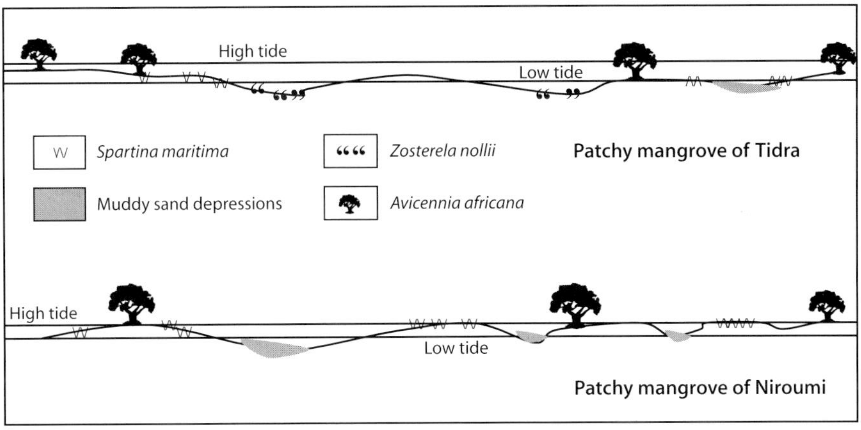

Fig. 2.4. Transects in the patchy mangrove of Tidra and Niroumi, mauritania

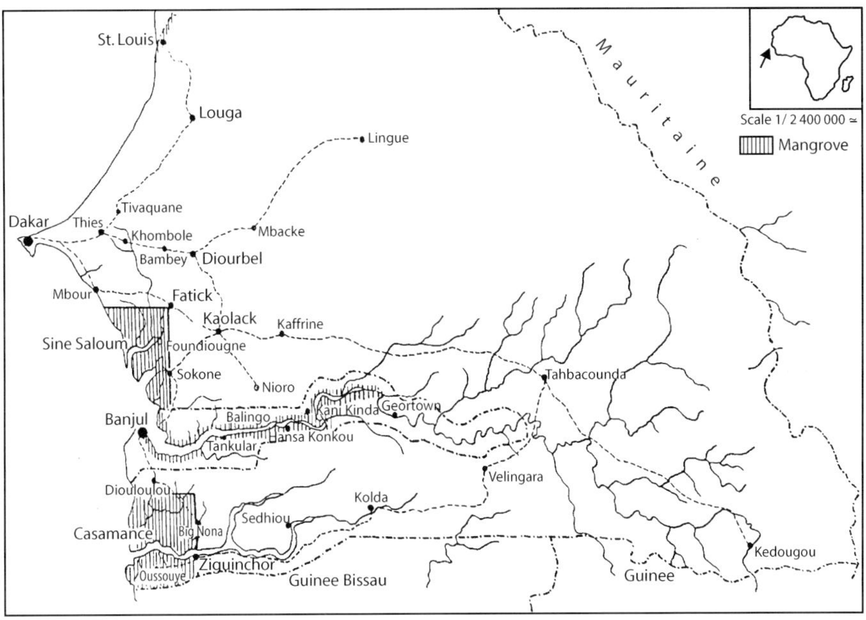

Fig. 2.5. Map of location of mangroves in Senegal and Gambia

regular influx of seawater. As a result, the tides are the main causal agents of water flow in the river in the absence of fresh water discharge.

In Gambia, there is an almost continuous belt of mangroves stretching inland for about 160 km, from the mouth of the Gambia River (Bintang Bolong and Bao Bolong). The mangroves occupy an approximate area of 497 km^2.

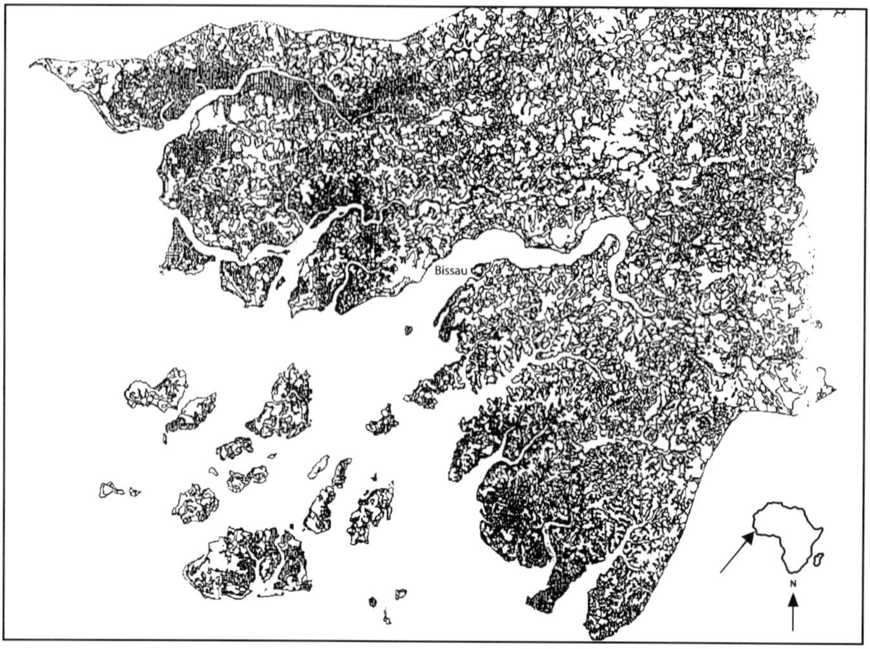

Fig. 2.6. Mangroves of Guinea Bissau

Guinea Bissau. The area occupied by mangroves has decreased from 4760 to 2464 km^2 over time (Spalding et al. 1997). The impact of tides is felt as far as 150 km inland thus inhibiting the rate of growth of the mangroves. Analysis of data on mangrove swamp distribution for each region of the country shows that they are most developed in the northern part of the country, followed by the southern parts including the archipelago, and less developed in the central part (Fig. 2.6).

Guinea. Mangroves are found all along the Guinean Atlantic shore, extending for about 300 km. This excludes the dolerite ranges of Cape Vergas and the central part of Kaloum peninsula. The total surface area of mangrove swamps is about 2963 km^2 (Spalding et al. 1997). The submergence of the Guinean Coast line favours deposition of alluvia and inundation at the mouth of rivers. Tides often go far upstream as a result of which several rias have been created. Notable among the rias are Rio Nunez, Rio Kapatchez, Rio Pongo and Rio Forecariah.

Mangroves specifically stretch along the valleys of these rias as well as along the bays such as those of Sangareah, in muddy areas and islands such as Tristao and Couffing, along the insular zone between Rio Pongo and Wassa-Wassa in Boffa, and on islands such as Quito and Marara.

In the Kaloum peninsula, mangroves occupy the muddy shorelines of north and south Conakry. They are particularly abundant in the sector between Gbessia Airport and the Soumbouya River. In all places the mangroves extend for about 10 km inland, sometimes extending to 40 km along major river channels.

Sierra Leone. The mangrove forest of Sierra Leone occupies 47% of the coastline and covers a total area of 1 838 km^2 (Chong 1979; Spalding et al. 1997) (Table 2.2). The mangroves are concentrated in four major areas namely the Scarcies River, the Sierra Leone River, Yawri Bay and Sherbo River. These are concentrated along the fringes of the Mahel, Sasiyek and Kambiadi creeks (Fig. 2.7) (ITTO/ISME 1993).

Table 2.2. Distribution of forest types in Sierra Leone (*source:* FAO 1979; Spalding et al. 1997)

Forest type	Area covered (km^2)
Closed high forest	3652
Secondary forest	3610
Forest regrowth	37744
Savannah	16192
Mangrove	1838

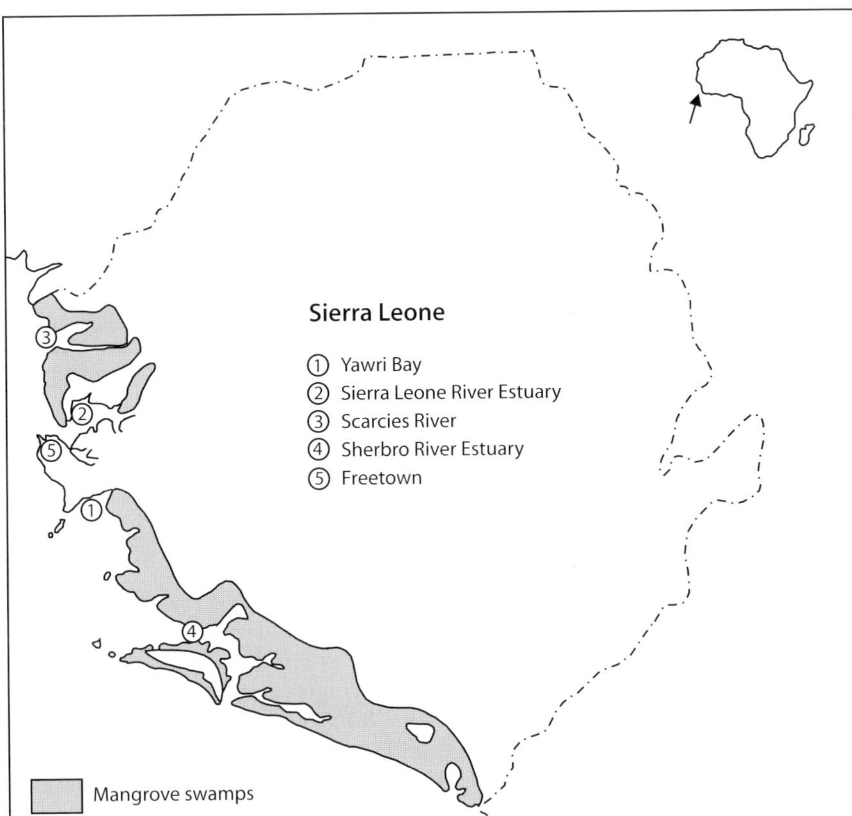

Fig. 2.7. Mangrove swamps in Sierra Leone

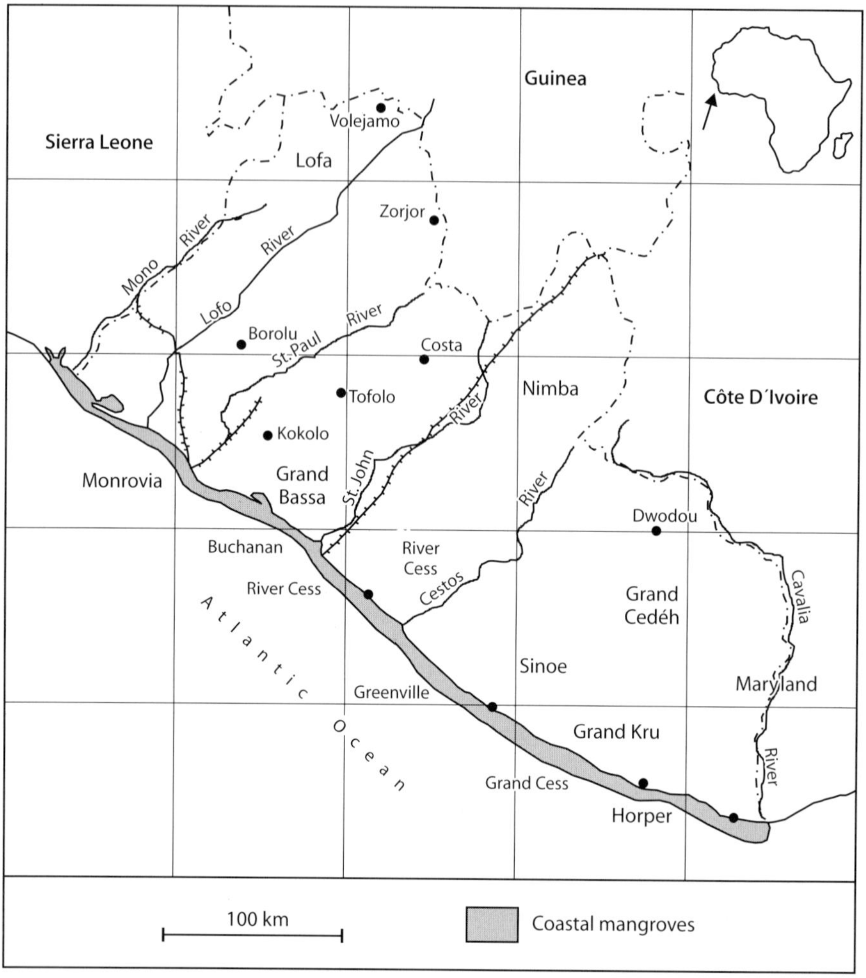

Fig. 2.8. Mangrove swamp areas in Liberia

Liberia. The area of Liberian mangrove forest is estimated at 190 km^2 (Spalding et al. 1997) and is found in the regosoils within the narrow coastal belt (Fig. 2.8).

Côte d'Ivoire. The extent of the 150 km^2 Ivorian mangroves is defined along a narrow strip almost parallel to the Equator. It stretches for about 500 km along the coast. Its distribution is explained by marine influence. There are two distinct clusters of mangroves: in the country of the Assinie and Fresco region where rivers empty into series of lagoons, and the region between Fresco and the border with Liberia at the Calvally River, which consists of a series of deltaic river mouth systems (Egnankou 1985; Spalding et al. 1997; Diop 1993).

Ghana. Mangroves in Ghana are estimated at 100 km^2. (Spalding et al. 1997). They are limited to a very narrow, non-continuous coastal area around lagoons on the west of the country; and to the east, on the fringes of the lower reaches and delta of the Volta River. They are extensive in the stretch between Cape Three Points and Côte d'Ivoire to the west, especially in areas around Half Assini, Amanzure lagoon, Axim, Princes Town and Shama, among others. To the east of the country, it is best developed at Apam, Muni lagoon, Winneba, Sakumo-1 lagoon, Botwiano, Korle lagoon, Teshie, Sakumo-2 lagoon, Ada, Sroegbe and Keta lagoon (Diop 1993).

Togo. The mangroves of Togo extend for only a limited area of 26 km^2. (Spalding et al. 1997). The most important area in terms of distribution is the Gbaga River and its tributaries. A few stands are also found along the banks of the Mono River and at the entrance of the Aheno lagoon.

Benin. The western littoral area of Benin where the influence of tides is most felt serves as the limit of mangroves in that country. They grow in zones around lagoons and lakes, extending over an area of 69 km^2. They extend along the mouths of the rivers Mono, Sazué and Ahé. The best-developed zones are those with species laid out in belts parallel to the shoreline (Spalding et al. 1997; Diop 1993).

Nigeria. Mangroves characterise the entire coastline in the southern part of Nigeria. They extend over an area of 10 515 km^2, constituting approximately 1% of the country's area (Spalding et al. 1997; Diop 1993). The mangrove forest of Nigeria, the largest in Africa, is most developed in the Niger Delta region where it attains a maximum width of 30 to 40 km on the flanks of the delta. The mangrove area of the delta region is bounded to the west by the Benin and the Calabar Rivers, and to the east by Rio del Ray estuary. To the west of the country, extensive mangroves flank the Lagos and Lekki lagoons. Another extensive area is the delta/estuary system associated with the Cross River to the Far East. The mangroves extend for about 7–8 km astride the estuary, and continue up to 26 km in the deltaic zone at the head of the estuary (Spalding et al. 1997).

Cameroon. Mangrove swamps occupy about 30% of the Cameroonian 350 km coastline. The country has a total of 2 434 km^2 of mangroves distributed between two main units: the "Cameroon Mouth" and the Rio del Rey, all situated at the opposite side of mount Cameroon (Spalding et al. 1997; Diop 1993). The "Cameroon Mouth" covers about 60 km of coastline from the mouth of the Sanaga to that of the Bimbia Rivers, extending to about 30 km inland. Rio del Rey stretches 50 km along the coast and extends 30 km inland. Pockets of mangroves are also located elsewhere around the Nyong mouth, the Campo River and the Man o' War bay (Diop 1993).

Equatorial Guinea. The 257 km^2 of mangroves of Equatorial Guinea are most extensively developed in the estuaries of three rivers: Mbini, Muni, and Ntem Rivers (Spalding et al. 1997). The Muni estuary, which is fed by several rivers, is 21 km wide. Mangroves extend for up to 17 km inland, from the mouth to the head of the estuary (Spalding et al. 1997).

Gabon. The mangrove forests of Gabon cover an estimated area of 2 500 km^2 (Spalding et al. 1997). They occur in the estuaries, bays and lagoons of the coast and are among the richest in species diversity in West Africa. The main areas of concentration are the mouths of the Como and Ogooué Rivers (Spalding et al. 1997). Most of the stands show poor development except for the narrow belt and areas limited to daily tidal influence. Growth seems hampered by chemical and/or physical parameters of the soil. Oil exploration poses a potential threat in terms of pollution (Diop 1993).

Congo. In Congo, the mangroves cover an area of 120 km^2 and are limited to the Kouilou and Noumbi estuaries, as well as the Malonda and Conkouati lagoon area, where they form a narrow strip of 10 m wide along the coast. The presence of the cold Benguela current near the coast coupled with a small tidal amplitude in the region seem to limit the development of mangroves in that country (Spalding et al. 1997; Diop 1993).

D.R. Congo. The Democratic Republic of Congo has an estimated mangrove forest area of 226 km^2, mainly occurring in the estuary of the Zaïre River (Spalding et al. 1997). The 40 km coastline is characterised by high cliffs, which had been breached by coastal rivers and streams, creating corridors where mangrove stands occur.

Angola. Angola's estimated mangroves of 1 250 km^2 are most extensive in the estuaries of rivers Lubinda and Zaïre, and the mouths of the Chiluago, Bambongo, Longo and Cuanza Rivers (Spalding et al. 1997). The mangroves are most diverse in flora in the enclave of Cabinda in the north, but with decreasing floral diversity to the south.

2.2.2 East Africa

In Eastern Africa, mangroves are estimated to cover 120 000 km^2. They are most extensive in Mozambique and Madagascar (Table 2.3). Therefore, this is much lower than in West Africa and it is correlated to the fewer large rivers emptying in the east coast

Table 2.3. General profile of countries in the East African region (*sources:* Ruwa 1995; Hatton and Massinga 1994; Gaudian et al. 1995; Semesi 1991a; Saket and Mattusse 1994; Rosolofo 1997a; Ahamada 1997)

Country	Land area (km^2)	Estimated shelf area-depth range (0–200) (km^2)	Length of coastline (km)	Mangrove area (km^2)
Comoros	2236	900	350	1.08
Kenya	582650	6500	500	5.00
Madagascar	595790	135000	4000	4200
Mauritius	1865	1600	200	1.00–2.00
Mozambique	738030	120000	2500	4000
Seychelles	443	48000	600	?
Somalia	637657	32500	3000	?
Tanzania	939703	30000	800	1335

than the west coast of Africa. Only small stands are found along the Red Sea, southern Somalia, South Africa and around the islands of Seychelles, Comoro and Mauritius (Steinke 1999; Semesi and Howell 1992; Ruwa 1995; Hatton and Massinga 1994; Gaudian et al. 1995).

Eleven tree species are found (Table 2.4), but *Xylocarpus moluccensis*, *Ceriops boiviniana*, *Bruguiera cylindrica*, and *Avicennia officinalis* are not very common. *Avicennia marina*, *Rhizophora mucronata* and *Ceriops tagal*, on the other hand, cover larger areas in the region. The most common species found in the Mida creek in Kenya, for example, are Rhizophora mucronata and Ceriops tagal, which together make up about 58% of that total area (Kairo and Gwada 1997). The Mida creek showed large differences in densities between plots. Of the total number of trees per ha (1 585), 56% were approximately 6 cm in diameter; only 8% had a diameter greater than 35 cm while less than 1% had diameter greater than 46 cm.

In the Zambezi delta and Mozambique, a total of 8 mangrove species are found (Doddema-de-Guia 1997). In the Zambezi delta mature climax communities contain large trees of *R. mucronata*, *B. gymnorrhiza* and *H. littoralis* of about 25 m tall and with a diameter at breast height of about 60 cm. In Mozambique, the diversity of mangroves decreases southward; the presence of *Heritiera littoralis*, *Lumnitzera racemosa* and

Table 2.4. Mangrove tree species and their local names in Eastern Africa

Scientific name	Local names in the various countries				
	Kenya	Tanzania	Mozambique	Madagascar	Seychelles
Avicennia marina	Mchu	Mchu	Invede	Afiafy or Mosotry	Mangliye blan or Manglier Blanc
Rhizophora mucronata	Mkoko	Mkoko or Makaka	Nhantazera	Anabovahatra or Tanga	Mangliye Rouz or Mangliye gro poumon or Mangliye hauba
Ceriops tagal	Mkandaa	Mkandaa	Nhacandale	Honko or Tangapbavy	Mangliye Zonn or Mangliye Zerof
Ceriops boiviniana	–	–	–	–	–
Bruguiera gymnorrhiza	Msinzi or Mshinzi	Msinzi or Muia	Mfinse	Tangapoly	Mangliye lat or Manglier latte
Bruguiera cylindrica					
Heritiera littoralis	Msikundazi or Mkungu	Msikundazi or Mkungu	Ncolongo		–
Lumnitzera racemosa	Mkandaa dume	Kikandaa	Mpiripito		Mangliye pti or Mangliye Fey
Sonneratia alba	Mpira or Mililana	Mililana	M'pia	Farafaka	Mangliye fler or Manglier Fleur
Xylocarpus granatum	Mkomafi	Mkomafi	InruboM'rubo		Mangliye Ponm or Mangliye pasyans
Xylocarpus molluccensi	–	–	–	–	–

Table 2.5.
Comparison of mangrove cover along the coast of Kenya as found in 1981 and 1990 (*source:* E. Martens, personal communication, Preliminary Mangrove Reports of Kenya)

District	1981 survey (km²)	1990 survey (km²)
South Coast	63.65	54.90
Vanga	42.65	34.21
Shimoni	21.00	20.69
Mombasa	35.65	27.49
Tudor	14.65	13.28
Port Reitz	15.75	10.20
Mtwapa	5.25	4.01
North Coast	418.15	–
Kilifi	3.60	3.75
Mida	16.00	16.58
Ngomeni	21.75	22.91
Tana Delta	18.45	Not surveyed
Kiunga	30.25	Not surveyed
Mto Kilifi	23.5	Not surveyed
Lamu	304.75	235.25

Xylocarpus granatum declines in the vicinity of Inhambane province and *Ceriops tagal* and *Lumnitzera racemosa* reach their southern limit in this province. In Somalia only two species, *Avicennia marina* and *Ceriops tagal*, are found close to the border with Kenya. Common mangrove associates in Eastern Africa are *Acrostichum aureum* (fern), *Barringtonia racemosa*, *Hibiscus tiliaceus*, *Thespesia populnea*, and *Phoenix reclinata* (palm).

Kenya. In Kenya, mangroves cover 630.28 km^2 under pure forest cover and 850.03 km^2 when the bare tidal flats and the enclosed dunes are included (Kairo 1997). Mangroves in Lamu alone represent 50% of all the mangroves in Kenya (Table 2.5).

Tanzanian. The areal extent of Tanzanian mangroves is double that of Kenya and covers a total area of about 1 335 km^2 (Table 2.6), 1 155 km^2 in mainland and 180 km^2 in Zanzibar (Semesi 1991b). If water in the creeks, clear-cut areas and saltpans are included, mangrove reserves cover an area of 1 728.89 km^2 (Table 2.7). The largest area and best-developed mangroves in Tanzania is found in the Rufiji delta.

Mozambique. In Mozambique, the main mangroves are in the part of the country where estuaries are frequent and large rivers such as the Zambezi, Pungui, Buzi and Save flow into the ocean (Table 2.8). About 20% of the total mangrove area is in the provinces of Sofala and Zambezi (Saket and Mattusse 1994). In the northern province of Nampula and Cabo Delgado, mangroves are well conserved in some sites. The annual rate of deforestation has been 0.2% and Maputo province with the highest deforestation of 15% from 1972 to 1990.

Table 2.6. Mangrove areas in the administrative blocks in mainland Tanzania (Semesi 1991a)

District	Forested area (km^2)	Non-forested area (creeks, salt pans, bare saline area) (km^2)
Tanga and Muheza	94.03	35.28
Pangani	17.56	12.79
Bagamoyo	56.36	35.48
Dar es Salaam	21.68	10.45
Kisarawe	38.58	21.93
Mafia	34.73	8.92
Rufiji	532.55	143.57
Kilwa	224.29	143.08
Lindi	45.47	27.54
Mtwara	89.42	44.08
Unguja island	60	–
Pemba island	120	–
Total	1 334.67	483.12

Table 2.7. Species composition and area covered by mangroves in mainland Tanzania (Semesi 1991a)

Classification	Area (km^2)	% of the total area
Rhizophora dominant, with *Avicennia, Ceriops, Sonneratia, Bruguiera, Heritiera* and/or *Xylocarpus*	555.499	49
Sonneratia – almost pure stands	12.233	1
Sonneratia dominant, with *Avicennia, Bruguiera* and/or *Rhizophora*	61.232	5
Heritiera – almost pure stands	0.912	0
Heritiera dominant, with *Avicennia, Bruguiera* and/or *Rhizophora*	81.884	7
Avicennia dominant, with *Rhizophora, Bruguiera, Heritiera, Ceriops* and/or *Xylocarpus*	171.416	15
Avicennia – almost pure stands	16.874	1
Mixture of *Avicennia* and *Ceriops*	174.327	15
Ceriops dominant, with *Rhizophora, Avicennia* and/or *Bruguiera*	80.379	7
Total Mangrove	1 154.756	100
Water in creeks	240.760	
Clear-cut areas	44.350	
Bare, saline areas	207.400	
Salt pans	30.930	
Non-mangrove forest inside the reserve	50.693	
Total reserve area	1 728.889	

Table 2.8. Mangrove areas and changes that occurred between 1972 and 1990 in Mozambique (Saket and Matusse 1994)

Province	Mangrove area (km^2) 1972	1990	Area depleted (km^2)	New area of mangroves (km^2)	Deforestation rate in 18 years (%)
Maputo	146.05	125.99	22.17	2.11	15.5
Gaza	3.87	3.87	0.0	0.0	0.0
I'bane	200.94	198.48	2.46	0.0	1.2
Sofala	1 299.97	1 253.17	63.34	16.54	4.9
Zambezia	1 594.17	1 557.57	37.66	1.06	2.4
Nampula	558.49	543.36	20.06	4.93	3.6
Gabo Del.	277.30	278.36	0	1.06	0.0
Total	4 080.79	3 960.80	145.69	25.70	3.6

Madagascar. Madagascar has about 3 270 km^2 of mangroves, but if the bare saline areas ("tannes") are included, the area is about 4 200 km^2 (Rosolofo 1997a). Almost all the mangroves in Madagascar are on the low wave energy western coast. The largest area is concentrated in Mahajamba bay (390 km^2), Betsiboka estuary (460 km^2), Tambohorano-Besalampy (457 km^2), Tsiribihima delta (280 km^2), Loza estuary (180 km^2) Manambolo delta (90 km^2), and Mangoky delta (232 km^2) (Rosolofo 1997a; ITTO/ISME 1993). The northwestern mangroves grow to large sizes, but the size and density decreases southward.

Comoros. The mangroves of Comoros occupy 1.08 km^2. The southern coast of Moheli Island has 0.91 km^2; 0.08 km^2 is found in Grande Comore and 0.09 km^2 in Anjouna (Ahmada 1997).

Seychelles. In Seychelles, the total area is not known but the largest mangrove areas are on Aldabra, Cosmoled and Astove Islands where some ten species can be found (Shah 1995). A continuous belt still remains in the port Launay and Baie Ternay, a Marine National Park on the west coast of Mahe also in Curieuse Marine National Park.

Mauritius. In Mauritius, very narrow stands are found along the east, southeast and northern coast totalling 1.00–2.00 km^2 (Gaudian et al. 1995).

South Africa. The total area of mangroves in S. Africa is around 10.0 km^2 and extends from Kosi system at 26° S to Nahoon River at 33° S with less than 0.005 km^2 (Steinke 1999). The Kosi system, which is near the Mozambique border has 0.59 km^2 of mangroves and has 6 tree species, but Richards bay with the largest mangrove area, 4.26 km^2, has only *A. marina*, *B. gymnorrhiza*, and *Rhizophora mucronata*. Other sites are St. Lucia 1.6 km^2, Mgeni 44 ha, Mngazana 1.50 km^2, Mlalazi 0.30 km^2, Mtolozi 0.26 km^2 and Mtata 0.34 km^2. Therefore, the area of mangroves, the number of species and the size of trees decrease southwards.

2.3 Physical Environment

2.3.1 West Africa

According to John and Lawson (1990), from the physical point of view, mainland water systems exert considerable influence on coastal water ecosystems. The distribution of vegetation is related to complex local factors among which climate and hydrology are dominant. This is particularly due to the competitive ability of each species to be linked to its climatic requirements (Fig. 2.9).

The climate also affects the soil substrate, not only because low rainfall and high evaporation rates produce hypersaline conditions, but also because the main mineralogical constituents and the pedogenetic processes are related to the prevailing climatic conditions (Blasco 1984; Diop et al. 1991, 1995, 1996a,b,c, 1997).

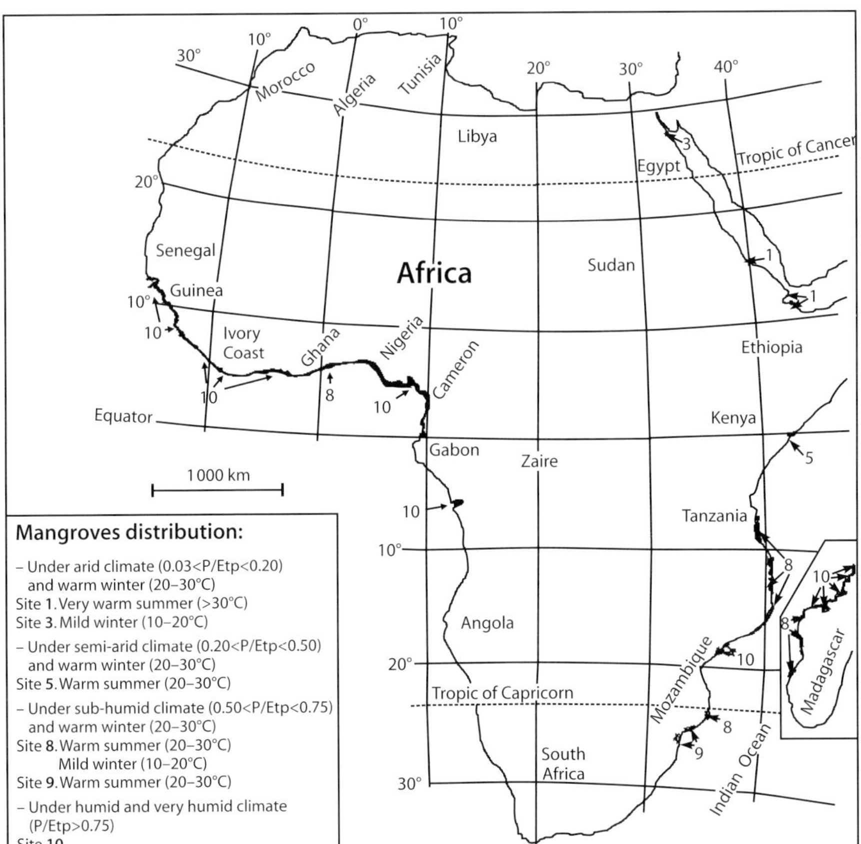

Fig. 2.9. Distribution of mangroves in Africa

2.3.1.1
Climate

Along the West African region, climate varies significantly according to latitude. It ranges from arid and semi-arid type in the north, to humid tropical climate of the Guinean type in the south. There are two distinct seasons: namely the rainy season that lasts from three months in the north of Senegal to six months in Sierra Leone and Liberia, and the dry season. The mean annual rainfall is between 139 mm in Mauritania and 5 000 mm in Sierra Leone (Table 2.9). Guinea, Sierra Leone and Liberia receive heavy rainfall (4500 to 5000 mm) from May to November.

Since 1968, there has been a general decrease in total number of rainy days and total amount of rainfall in the whole sahelian region with pronounced drought in 1968, 1972, 1973, 1974, 1977, 1979, 1983, and 1984. There has also been a simultaneous reduction in the duration of the rainy season. Table 2.10 illustrates the annual variation of the rainfall from 1980–1989 in Senegal.

With regard to temperature, the mean monthly temperature varies from 15 to 32 °C (Table 2.9).

As shown in Table 2.9, Guinea, Sierra Leone and Liberia are characterised by high rainfall regimes with rainfall receipts ranging between 4 500 and 5 000 mm during the major rainy season (May to November). It is pertinent to note that the coastal areas of

Table 2.9. Physical environmental parameters in the West Africa region (Spalding et al. 1997)

Country	Mean monthly temperature range (°C)	Mean annual rainfall (mm)	Spring tidal amplitude (m)	
Mauritania	21 – 30	139	0.5 – 1.5	(Port Etienne)
Senegal	15 – 28	700	0.3 – 1.6	(Dakar)
Gambia	17 – 28	800	0.2 – 1.8	(Banjul)
Guinea Bissau	24 – 27	1 750	2.3	(Cairo)
Guinea	23 – 32	4 500	0.5 – 3.7	(Conakry)
Sierra Leone	28 – 32	5 000	0.4 – 3.0	(Freetown)
Liberia	24 – 29	4 600	0.2 – 1.3	(Monrovia)
Côte d'Ivoire	24 – 27	2 800	0.2 – 1.2	(Abidjan)
Ghana	24 – 30	1 250	0.3 – 1.6	(Takoradi)
Togo	24 – 29	850	0.3 – 1.7	(Lome)
Benin	22 – 34	1 307	0.3 – 1.3	(Cotonou)
Nigeria	24 – 27	2 000	0.3 – 1.6	(Escravos)
Cameroon	21 – 31	5 000	0.4 – 2.3	(Point Olga)
Equatorial Guinea	26	3 000	0.3 – 1.7	
Gabon	23 – 30	2 645	0.4 – 2.4	(Libreville)
Congo	21 – 27	1 500	0.4 – 1.3	
D.R. Congo	25	800	0.3 – 1.6	
Angola	20 – 26	750	0.5 – 1.7	(Luanda)

Table 2.10. The annual variation of the rainfall from 1980–1989 in Senegal

Station	The most humid year		The driest year	
	Year	Rainfall (mm)	Year	Rainfall (mm)
St. Louis (Senegal)	1987	342	1983	95
Kaolack (Saloum)	1988	705	1983	285
Djibelor (Casamance)	1985	1 371	1980	691

Sierra Leone experience more rainfall than the rest of the country due to the coastal relief. Liberia is located at the heart of the wet tropical rain forest zone, hence the high rainfall receipts.

2.3.1.2
Marine Hydrology

Tides in the western African region are of semi-diurnal type with an average range of 1 m. There is a large-scale system of superficial and sub-surface marine water circulation. The main elements of this system are the North and South trade winds, the Canary current and the inter-trade wind counter-current.

In the month of January, temperatures of shallow waters in the region drop due to the influence of the cool Canary current. To the northeast, a layer of relatively warm water body from the inter trade wind counter-current is formed.

In November, there is a general anticyclonic circulation within the entire shelf. In a vertical plan, the circulation causes the transport of heat and salt as well as of biogenic elements. The prominent up-welling period between January and May weakens from June to November.

2.3.1.3
Hydrological Networks

There is in West Africa an increasing river flow from North to the South reflecting an increase of rainfall.

Three types of regimes can be distinguished:

- *Transitional tropical humid regime.* It is characterised by the rivers of Sierra Leone, Guinea and southern parts of Guinea Bissau, whose basins have high relief. Notable among the rivers are Scarcies, Rockel, Jong, Sewa, Kolente, Konkoure, and Cacine/Corubal.
- *Pure tropical regime.* This comprises rivers such as those located at the northern parts of Rio Geba where the topography is of less significance. The climate is of the sub-humid to dry type (Cacheu or Gambia after Gouloumbo). There is marked seasonality in river flow with characteristic dry season deficiencies.
- *Sahelian regime.* The Senegal and Saloum Rivers fall under this category. Before the closing of the Diama dam, the estuary of the Senegal was subjected to two forces:

the tidal currents and the river flow. The saline water intrusion could go as far inland as Podor, about 300 km away. In July 1982 salinity levels, higher than that of the sea, were recorded 70 km upstream from the mouth (Gac et al. 1986). Since the closing of the Diama dam 50 km upstream from the mouth, the estuary is now under the exclusive influence of the sea for 8 months. During the flood season (from July to November) the floodgates of the dam are opened, and this enhances freshwater inflow into the residual river downstream. The influence of the flow regime has modified the mangrove system (Soumare 1996) as well as the fauna of the area (Monteillet, pers. com.).

The salinity of the estuarine systems of the Saloum and Casamance increases as one advances inland, due to the rather poor freshwater supplies (Fig. 2.10).

In the Saloum, for example, the salinity level was 55% during the dry season and 50% during the rainy season about 100 km from the mouth (Gningue 1991).

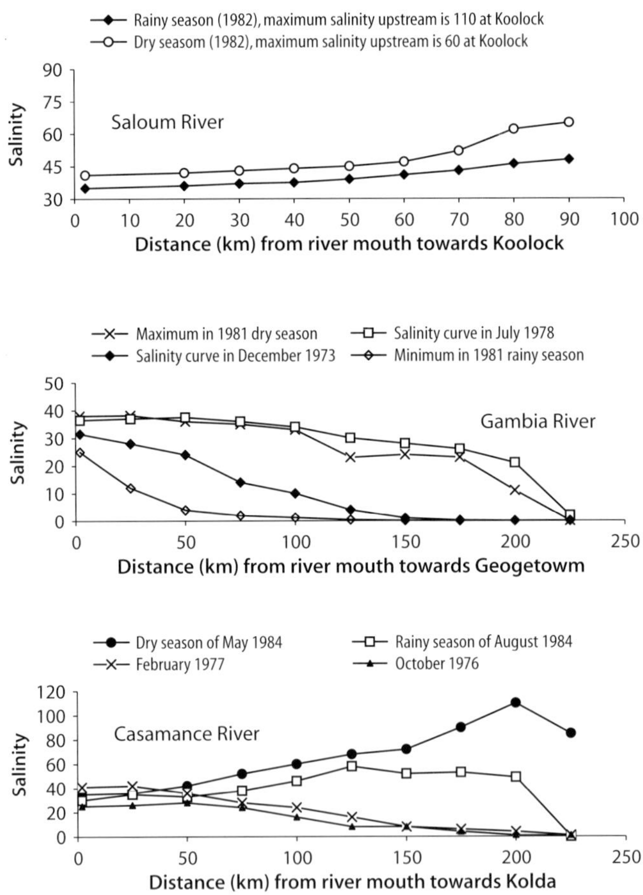

Fig. 2.10. Salinity regime on the surface waters of the Saloum, Gambia and Casamance rivers

Hypersalinity plays an important role in the metabolic efficiency of plants species. In mangrove areas, such conditions tend to reduce the productivity.

2.3.1.4
Soils

The soils of the West African mangrove area are essentially composed of quartzitic sand and kaolinite clay, with high organic matter component derived from decomposed mangrove litter. Within the mangrove swamps, silty clay or sandy silt layers are usually not very thick. The soils are generally clayey on the surface and sandy beneath. They are formed in a confined hot and dry environment and have two main characteristics:

- moderate to high concentrations of salts; NaCl being the dominant, followed by $MgSO_4$
- high accumulation of iron sulphide in the root system mainly in the form of pyrite, owing to the presence of decomposed organic matter. The iron sulphides are produced through the reduction of sulphates by sulphate reducing bacteria. Reduced amounts of sulphides are non-toxic and are in balance with conditions within the mangrove swamps. When aerated, the sulphides become oxidised to acid sulphates due to the action of *Thiobacillus* type bacteria, which thrives in areas of high acidity (pH 3 to 2).

There is thus an initial production of elementary sulphur, followed by a sudden corresponding decrease in pH, which can be as low as 2, and then the formation of salty acid sulphate soils.

The sulphates may also be partially neutralised and be transformed into basic sulphates of potassium jarosite or sodium jarosite. Under severe acidic conditions, aluminium sulphates may be associated, whereas in the presence of calcium, gypsium may be formed (Loyer 1983). It should be noted that these processes are common to all West African mangrove ecosystems even though their intensity may vary from place to place.

With regard to salinity, a seasonal and spatial gradient, which opposes the hypersaline conditions in the sahelian zone and the watershed areas is noticeable. The littoral plains of Guinea and Sierra Leone exemplify this.

In general, the soils of the Saloum estuary in Senegal are less acidic than those of the estuary of Casamance. This is due to the very high salinity of the Saloum estuary soils that neutralises their normal acidity (Fig. 2.11).

In Guinea Bissau, traditional techniques have been adopted by specialised rice farmers who encroach on mangrove soils to avoid the oxidation of the rice fields (*bolanhas*). This is done by allowing seawater to enter the fields during the dry season.

There are two kinds of clayey soils associated with *Avicennia* and *Rhizophora* respectively (Table 2.11). The coastal swamps of Sierra Leone consist of alternate bands of silt, sand, gravel and clay. Silt is dominant in the northwest, while in the south, coarse sand over which are waterlogged areas interspersed with bands of dry terrain is a common feature. In Liberia, the regosoils (peat soil) on which mangroves grow are found within the narrow coastal belt (Hasselman 1979).

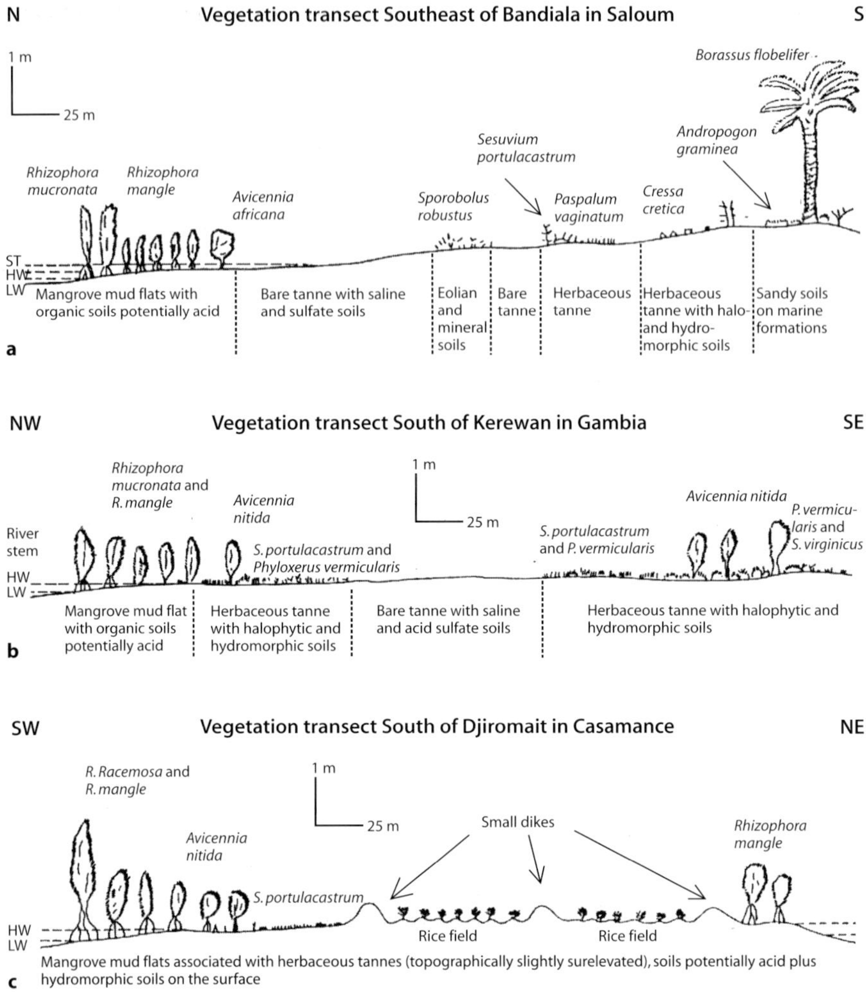

Fig. 2.11. Vegetation transects SE of Bandiala in Saloum (**a**), S of Kerewan in Gambia (**b**), and S of Djiromait in Casamance (**c**)

Table 2.11. Major differences between *Avicennia* and *Rhizophora* soils (Simao 1993)

Type of soil	*Avicennia*	*Rhizophora*
Un-mature and non-oxydized soils	Rare brown spots	–
Semi-mature soils	Predominance of brown spots (iron oxides)	Predominance of jarosite spots (yellow and few brown spots)
Well mature soils	Predominance of brown spots and red spots (goethite)	Predominance of red to dark spots (hematite)

2.3.2 East Africa

2.3.2.1 *Climate*

The climate of Eastern Africa is dominated by monsoon winds, which change their direction from south west in April–August to north east in October–March. During the SW Monsoon there is much mixing and upwelling along northeastern coast of Somalia, between 5° N and 12° N about 46 km offshore. Sea-surface temperatures very rarely fall below 20 °C, except off the coast of northern Somalia during upwelling events. The maximum temperature is about 29 °C for offshore waters, but shallow inshore lagoons with limited water-exchange may experience very high maxima of up to 39 °C. Humidity is high throughout the year, reaching up to 90% during the rainy season. The average salinity is 34.5‰. Most parts of the Kenyan and Tanzanian coast experience mean annual temperatures in excess of 27 °C, and high relative humidity year-round and mean annual rainfall is between 760 mm and 1300 mm. The weather pattern along the eastern coast is dominated by eight to thirteen-year alternating wet and dry periods (Wright et al. 1997).

In Madagascar and Mozambique, the hot, wet season extends from November to April and the cooler, drier season from May to October. July is the coolest month, with mean monthly temperatures ranging from 10 °C to 26 °C. December is the hottest month, with temperatures between 16 °C and 29 °C. In Madagascar and Mauritius, tropical cyclones are an important climatic feature especially from December to March, bringing torrential rains and destructive floods. The highest monthly rainfall occurs in January 275–375 mm, and annual rainfall 1100 and 2900 mm, being highest on the northeast sides of the islands.

2.3.2.2 *Marine Hydrology*

Ocean currents influence the distribution of organisms and nutrients. Sheltered deltas are the most usual places where mangroves establish themselves. Mangroves establish in low energy coasts and are restricted to the intertidal zone. The South Equatorial current when reaching Africa splits into two main currents: one flowing south comprising the Mozambique and Madagascar currents and the other flowing north as the East African Coastal Current (EACC). EACC flows as far as southern Somalia from where it is deflected eastward, becoming the Equatorial Counter current during the NE monsoon period. During the SE monsoon period, the EACC flow increases to 4 knots (120 cm s^{-1}), and its range extends further north to become the Somali current. EACC is weaker during NE monsoon; less than 1 knot. Except for Somalia, which experiences upwelling, most parts of the region experience down welling, and the surface waters are generally nutrient poor.

There are two high tides and two low tides of differing magnitudes each day. Springs tides occurring at full and new moon have large tidal ranges varying from 0.5 m in Mauritius to 5.7 m in the area near Beira in Maputo. The tidal range of the entire portion of eastern Africa, including the islands of Comoros and west coast of Madagascar is 2–4 m. The east coast of Madagascar, Seychelles and Mauritius all experience much smaller tidal ranges of about a meter or less.

Mangroves are restricted to the area between the mean tide level and the extreme high water springs. Tides export nutrients and protect the ecosystem from toxic accumulation of salts. Tides are also responsible for dispersal of plant propagules and of planktonic stages of animals. Tidal movement also affects activities in mangroves; usually mangrove cutting and collection of mollusks is carried out at low tide. However, the transport of poles and bulky firewood is transported from mangroves by boat during high tides.

2.3.2.3
Hydrological Networks

In contrast to West Africa, the east coast of East Africa is drained by only a few large rivers. The Zambezi is the largest river. Apart from the Tana River, most of the rivers of Kenya are short or ephemeral, disappearing in dry seasons. Tanzania's major rivers are the Ruvuma, the Rufiji, the Wami, and the Pangani. The Rufiji is the largest river whose drainage system extends over most of southern Tanzania. The Zambezi River and its tributaries dominate the central region of Mozambique, and the rivers Lúrio, Save, and Messalo systems follow in size. The Ruvuma River defines most of Mozambique's northern border with Tanzania, and the Maputo River forms part of the southernmost boundary with Swaziland and South Africa. Others include the Ligonha, Changane, Komati (Incomati), Púngoè, Revuè, and Búzi rivers. In Madagascar the western coast is crossed by many rivers, including the Onilahy, the Mangoky, the Tsiribihina, and the Betsiboka, which bring huge deposits of fertile alluvium down into the vast plains and many-channeled estuaries. The river mouths, while not completely blocked by this sediment, have numerous sandbanks.

2.3.2.4
Soils

Mangrove sediments, which are often waterlogged, have pH values that range from 3.5 to 8.3 due to the limited buffer capacity of these sediments and intense acidifying processes such as aerobic degradation of organic matter, oxidation of reduced components, ammonium uptake by roots, and root respiration. The sediments are mostly firm to soft mud, but in other sites they can be sandy or high in organic matter. The oxygen diffusion is greatly reduced, and in the mud there is little water movement except near the creeks. Therefore, better growth of the mangroves takes place near the creeks and as an adaptation to anaerobic condition trees have developed pneumaptophores. Near the roots there is a higher level of oxygen due to aeration by the roots. Crab holes also form interconnected tubes that help in draining and flushing salts. Nutrients and salinity also vary with season and tidal water movements. For example in Mida creek mangroves, Kenya, higher concentrations of dissolved nutrients are found in the rainy season than in the dry season (Kazungu et al. 1986; Kitheka 1999). The nutrient concentrations decline with ebb tides and are lowest during low water. The concentration of NH_4^+-N ranged from 0.002 to 5.45 μM, and those for $(NO_2 + NO_3)$-N, PO_4^{3}-P and SiO_3^{2}-Si were 0.12–5.63, 0.10–0.58 and 1.31–81.36 μM, respectively.

Mangrove sediments are nitrogen-rich compared with mangrove litter, as a result of microbial nitrogen retention, uptake and fixation, and the import of nitrogen-rich material (Middelburg et al. 1996). In mangroves at Gazi in Kenya, Woitchik et al. (1997) showed the maximum rates of nitrogen fixation in a *C. tagal* zone to be 380 nmol N_2 h^{-1} g^{-1} dw in the rainy season but 78 nmol N_2 h^{-1} g^{-1} dw in the dry season. In the *R. mucronata* zone the rates were 390 nmol N_2 h^{-1} g^{-1} dw in the rainy season and 189 nmol N_2 h^{-1} g^{-1} dw in the dry season. They also reported that although N_2 fixation rates were highest during rainy season, total nitrogen immobilised in the leaves was highest during the dry season. It was estimated that biological nitrogen fixation could account for between 13 to 21% of the maximum nitrogen immobilised in the decaying mangrove leaves.

Ground water inflow to mangrove areas induces stratification of the water column, limits salinity during the dry season, supplies nutrients, and is an important factor governing plant zonation. Groundwater outflow sustains the mangroves during periods of severe salinity stress and nutrient deficiency in dry seasons (Wolanski et al. 1992; Thibodeau et al. 1998; Kitheka 1999). Contribution of groundwater seepage to the net nutrient flux (particularly on nitrite-nitrates) is largest during the dry seasons. The freshwater influx via rivers and direct rainfall is also responsible for lower salinities in riverine mangroves and those mangroves occurring in areas with high rainfall, contrary to those, for example, found in the Red Sea where limited rainfall occurs and where lack of large rivers result in very high salinities in these mangroves.

Cutting mangroves, on the other hand, alters benthic nitrogen dynamic because the abundance of burrowing animals decrease (Stromberg et al. 1998).

Mangrove leaves represent a major source of organic carbon to the mangrove sediments in eastern Africa (Rao et al. 1994). Rates of leaf decomposition are highest in *Sonneratia alba*, followed by *Avicennia marina*, then *R. mucronata*, and least of all for *C. tagal*. Rates of decomposition eare highest during the rainy season. High rainfall and diurnal tidal inundation appear to enhance the leaf decomposition process.

Furthermore, many micro-organisms play a role in the sulphur and nitrogen cycles in mangroves. The microorganisms are involved in the reduction and oxidation of sulphur compounds. For example, Methanogens are the most important dimethylsulphide (DMS) utilised in mangrove sediments as compared to sulphate reducing bacteria. The most probable numbers of methanogens from sediments are high (up to 10^5 cells g^{-1} of fresh sediment). Recently, a new obligatory methlylotrophic metahnogenic archaea, *Metahnosarcina semesii* MDI^T was isolated in mangroves of Mtoni, Tanzania (Lyimo et al., in press).

Studies done by Steinke and Jones (1993) on marine and mangrove fungi from the Indian Ocean Coast of South Africa identified 93 species of marine fungi from three habitats, and the highest number (55 species) were obtained from the mangrove woods, while Singh and Steinke (1992) found various phylloplane fungi in the decomposing leaves of *Bruguiera gymnorrhiza* including *Fusarium* spp., *Penicillium* spp., *Trichoderma* spp., *Verticillium* spp., *Phoma* spp. and *Mucor* spp.

2.4 Mangrove Vegetation

Floristic composition generally decreases from north to south in response to the moisture gradient.

The distribution of herbaceous species that accompany woody formations is equally ordered by the moisture gradient.

- In the northern mangroves, halophytes are exclusively dominant;
- In the south where the rainfall is higher, hydrophytes, which are particularly sustained by rapid sedimentation, are most common.

Vegetation formation is therefore dependent on hydro-sedimentation that determines the establishment of the pioneer species, the succession of the different populations and the evolution towards equilibrium. The determining parameters are the speed of sedimentation, the rainfall and the temperature.

2.4.1 Flora Composition and Distribution (Mangrove and Associated Plants)

2.4.1.1 *West Africa*

The floristic composition of West African mangroves is very limited compared to that of Eastern African mangroves. The species are typically the same from Senegal to Angola. The woody plants belong to about 9 species and 4 families (Table 2.12).

Table 2.12. Distribution of mangrove species in West and East Africa (Field 1996; Semesi 1992)

Family	Genus	Species	West Africa	East Africa	Structure
Pteridaceae	*Acrostichum*	*aureum*	x	x	Fern
Sterculiaceae	*Heriteria*	*littoralis*		x	Tree
Combretaceae	*Conocarpus*	*erectus*	x		Tree
	Laguncularia	*racemosa*	x		Tree
Sonneratiaceae	*Sonneratia*	*alba*		x	Tree
Rhizophoraceae	*Rhizophora*	*harrisonii*	x		Tree
	Rhizophora	*mangle*	x		Tree
	Rhizophora	*racemosa*	x		Tree
	Rhizophora	*mucronata*		x	Tree
	Bruguiera	*gymnorrhiza*		x	Tree
	Bruguiera	*cylindrica*		x	Tree
	Ceriops	*tagal*		x	Tree
	Ceriops	*boiviniana*		x	Tree
Meliaceae	*Xylocarpus*	*granatum*		x	Tree
	Xylocarpus	*molluccensis*			Tree
Avicenniaceae	*Avicennia*	*germinans*	x		Tree
	Avicennia	*marina*		x	Tree
Combretaceae	*Lumnitzera*	*racemosa*		x	Tree

- The Rhizophoraceae are dominant with three species: *Rhizophora harrisonii*, *Rhizophora mangle*, *Rhizophora racemosa*
- The Avicenniaceae are represented by *Avicennia africana* known as *Avicennia nitida* in Sierra Leone
- The Combretaceae are represented by two species: *Conocarpus erecta* and *Laguncularia racemosa*
- In Guinea, the Malphighiaceae are represented by three species: *Banisteria leona*, *Heteropteris leona* and *Stigmaphyllum ovatum*

Other non-mangrove plants species that grow in association with the mangrove include: *Dalbergia ecastophyllum*, *Drepanocarpus lumatus*, *Dodonaea viscosa*, *Terminalia scutifera*, *Sophora occidentalis*, and *Thespesia populnea*.

Herbaceous species characteristic of marshlands are: *Ipomea pes-caprae*, *Ipomea cairica*, *Paspalum vaginatum*, *Scirpus littoralis*, *Philoxerus vermicularis*, *Sesuvium portulacastrum*, and *Canavalia rosa*.

2.4.1.2
East Africa

The mangroves display considerable variety in structure and floristic character, which ranges from dwarf shrubs in areas that are dry and receive less ground water, to 20 m tall in areas where the rainfall, salinity, temperature and nutrients are favourable, such as the Rufiji and Zambezi delta. Some mangrove areas are composed of one tree species, others two or three, and some with up to ten tree species. Most mangroves occur as dense patches, but those in marginal sites occur in scattered form. Enclosed within mangroves are also bare hypersaline patches. Red Sea mangroves, for example, do not grow as luxuriously as most other tropical mangroves due to conditions of high salinity, poor soil texture, low precipitation and low nutrients. In the Red Sea, only *Avicennia marina*, *Rhizophora mucronata* and *B. gymnorrhiza* are found (Mandura 1997). However, *A. marina* is the dominant species, and most of the Red Sea mangrove stands consist only of monospecific stands of it. Besides, most mangroves in the Red Sea do not have a clear tree form, but are bushes. In south Africa the dominant species are *Avicennia marina* and *Bruguiera gymnorrhiza*. Apart from physical factors that affect species distribution and zonation, biotic factors also influence the partial distribution because of herbivory, predation and competition for space, light and nutrients. For example, shading by vegetation lowers salinities and favours the growth of those species that prefer low salinity and can tolerate shade.

In eastern Africa, eleven different mangrove tree species are recorded (Table 2.12) (Graham 1929; Walter and Steiner 1936; Semesi 1986, 1998; Steinke 1999) and these normally occupy specific habitats in the forest. In the Red Sea there are three species; in Kenya, eight species; in Tanzania, Mozambique, and Seychelles there are ten species; in Madagascar there are 11 species, and in South Africa there are six species. Among the common mangrove associated species are the climber *Derris trifoliata* (used for fish poisoning and rope making), the fern *Acrostichum aureum*, a herb *Sesuvium portulacastrum*, *Barringtonia racemosa*, *Phoenix reclinata*, *Thespesia populnea*, *Hibiscus tiliaceus*, and *Pemphis acidula*. Two common lichens on dead branches of mangroves are *Ramalina verulosa* and *Ramalina fecinda*. Many algal mats are found on sediments and attached on mangrove roots e.g., *Bostrichia* sp., *Centrocerus* sp., *Hypnea musciformis*, etc. In some parts in the *Sonneratia* zone, macrophytic algae such as *Acanthophora spicifera*, *Gracilaria salicornia*, *Hypnea* sp. and seagrass *Halodule* sp. are found, as well as many algal mats and cyanobacteria, which all contribute to the primary production of the mangrove's ecosystem.

2.4.2 Mangrove Zonation

2.4.2.1 *West Africa*

Observations along the West African littoral zone reveal three distinct types of mangrove swamps:

- Primary mangrove swamps characterised by minimal disturbance;
- Secondary mangrove swamps resulting from over-exploitation of primary swamps that are characterised by signs of degradation with increasing mortality and salinity towards upstream; and
- Mangrove swamps converted to rice fields (Fig. 2.12).

The vegetation distribution is related to topography and the nature of the substratum. As a result of recent droughts, mangroves suffer high salinity effects. The zonation of mangroves along a transect shows:

- An internal zone dominated by *Rhizophora racemosa* over silty coastline soils. Here, the prop root systems resist the action of the sea waves. In Saloum and Casamance, tall *Rhizophora racemosa* are replaced by groups of relatively shorter *Rhizophora harrisonii* and *Rhizophora mangle*. In Guinea Bissau, Guinea and Sierra Leone, the *Rhizophora* occupy narrow coastal fringes, attaining heights of 15 to 30 m.
- An external zone of *Avicennia africana*, located in the higher limits of the silty zone. This zone is irregularly submerged by tidal waters and maintains a shallow, highly saline water level. Serious pedological changes have occurred here as a result of drought.

Associated with mangroves are halophyte marshes locally referred to as "herbaceous tannes". The species include *Sesuvium portulacastrum*, *Philoxerus vermicularis*, *Scirpus maritimus*, and *Sporobolus robustus*.

2.4.2.2 *East Africa*

The occurrence of individual mangrove species in the mangrove forest is reliant on environmental factors such as salinity, nutrient availability, level of oxygen in soil and wave energy. As mangrove species are variable in their tolerance of these factors, a pattern of species distribution known as zonation occurs (Walter and Steiner 1936; Macnae 1968; Chapman 1977; Semesi 1986; Kairo 1997). For example, the upper limit

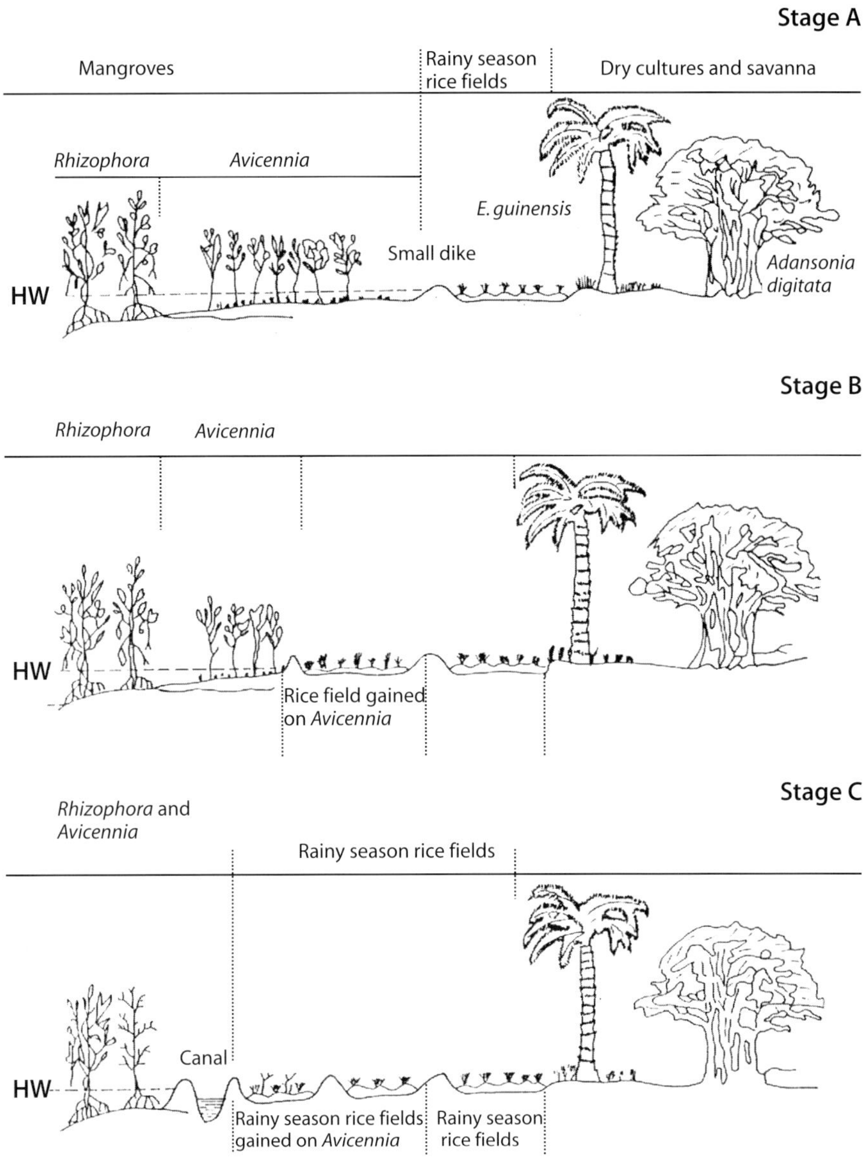

Fig. 2.12. Conversion of mangrove soils for rice culture in Casamance, Senegal (adapted from Blasco 1983)

of tolerance of mangrove trees in terms of soil osmotic potential (atm) for the Kunduchi mangroves in Tanzania are: *S. alba* (38.8 atm), *B. gymnorrhiza* (40.7 atm), *X. granatum* (44.0 atm), *R. mucronata* (46.7 atm), *C. tagal* (50.1 atm) and *A. marina* (97.8 atm) (McCusker 1977). Therefore, usually *Sonneratia alba* occurs in areas where the salinity is almost constant, close to that of seawater where tidal water reaches daily.

Avicennia marina is the most widely distributed because it can tolerate high ranges of salinity, varied flooding regimes, compacted substrate, sand flats, and newly deposited sediments. On the seaward side, the species attains large sizes but on the landward margin it is present only as bushes. However, it does poorly on muddy soils. *A. marina* possesses salt glands and has a high photosynthetic rate at full sunlight. *Rhizophora mucronata* forests are dominant on muddy soils and often form extensive pure stands. On sandy soils, however, the species fails to compete with others. *Bruguiera gymnorrhiza* is often found as a zone between *Rhizophora mucronata* and *Ceriops tagal* zones or mixed with them. *Xylocarpus granatum* is most often found mixed with *Avicennia marina*, and grows on raised portions where flooding takes place only for a few days a month and where there is fresh water influence. As is the case for *Heritiera littoralis*, *Xylocarpus granatum* is an important element of the riverine mangroves but does not form pure stands, and in South Africa only one specimen is found in Kosi bay (Steinke 1999). *Heritiera littoralis*, common in riverine areas, grows mainly in habitats with low salinity and is restricted to areas usually only flooded by spring high tides. In Gazi Kenya, the distribution of *Rhizophora mucronata* was correlated with the soil redox state, sulphide concentration, and salinity, and was absent from the less-reduced zone with high salinity. *Bruguiera gymnorrhiza* and *Heritiera littoralis* are restricted to sulphide-poor and less-reduced substrates (Matthijs et al. 1999). Where mangroves run along riverbanks or on coral islands, there is insufficient space for zonation to develop. Normally, it is the extreme factors that limit the distribution of a species (Macnae 1968). Therefore, different mangrove stands in Eastern Africa have differing dominant tree species, and the total number also varies. The most vigorous mangrove stands are often associated with riverine conditions, and the enhanced growth may be controlled by groundwater.

2.4.3 Productivity

2.4.3.1 *West Africa*

In West Africa, scientific information on primary productivity of coastal areas is limited. Since mangrove species are the same on both sides of the Atlantic it seems the productivity values for much of West African mangroves are comparable to those of Florida (John and Lawson 1990). In Mauritania, the northern limits for West African mangroves, lower mangrove productivity values are to be expected.

In an experiment carried out on *Rhizophora racemosa* at sites in the eastern part of Saloum, Toubacouta (Fig. 2.13), which involved ecological analysis and measurements of biomass density and total net primary productivity (Doyen and Agobba 1985), the following results were obtained:

- Total maximum height of dominant species: 10.60 m
- Estimated density: 3 400 stems ha^{-1}
- Average circumference at the base of the stem: 42.4 cm
- Average age of the stock: 35 yr

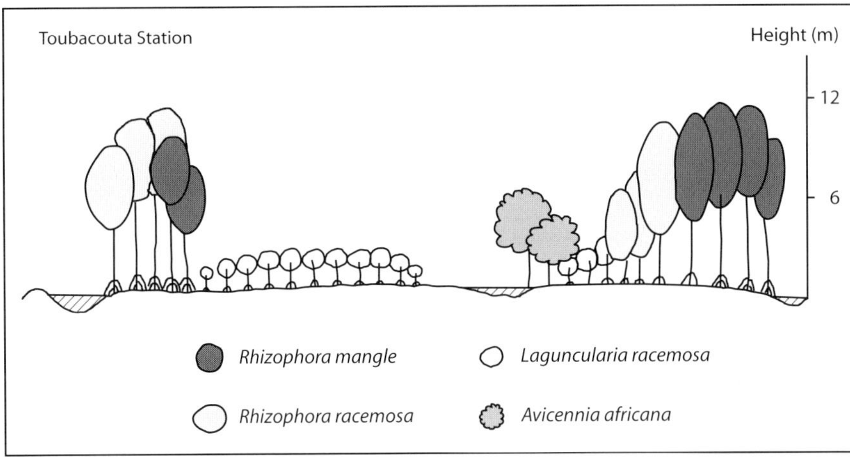

Fig. 2.13. Presentation of the station 1 of Toubacouta, Senegal

- Total aerial biomass: 61 730 kg ha^{-1} yr^{-1} of dry matter or 92 595 kg ha^{-1} yr^{-1} of fresh matter
- Net primary productivity: 1764 kg ha^{-1} yr^{-1} of dry matter or 2 646 kg ha^{-1} yr^{-1} of fresh matter
- Ligneous biomass: 53 475 kg ha^{-1} of dry matter or 80 213 kg ha^{-1} of fresh matter
- Net ligneous productivity: 1 529 kg ha^{-1} yr^{-1} of dry matter or 2, 292 kg ha^{-1} yr^{-1} of fresh matter

Data obtained on *Rhizophora racemosa* at the Toubacouta mangrove swamps were compared to what was obtained elsewhere, and the results revealed low quantities of biomass coupled with low primary productivity for the mangroves of Toubacouta as against the much healthier and more productive mangroves of the Niger delta, East Africa and Asia.

2.4.3.2 *East Africa*

There is almost no information about the productivity of mangroves in Eastern Africa. Most studies have looked at species composition, as well as physical and chemical composition of sediments, and only few monitor rates of growth, and none monitor the overall productivity of the mangroves. Productivity of the mangroves is mostly inferred from the presence of many organisms, which derive food and shelter in the mangroves. As is the case of other ecosystems, the process of photosynthesis is the cornerstone of all life for mangroves. Therefore all those factors, which limit photosynthesis and delay recycling of nutrients or limit their uptake, will have an impact on productivity. In effect, water deficiency, excessive salinity and nutrient deficiency, particularly phosphorus and nitrogen and light will limit primary productivity. It can be therefore assumed that the mangroves in the large deltas are more productive than those which do not receive large fresh water quantities. This is because the river brings both nutrients from the catchment area and water that lowers the salinity and stimulates growth. Nitrogen fixation, as a source of allochthonous nitrogen, also sustains a nitrogen input to the mangrove ecosystem. The mangroves are limited in southern Africa because of lower temperatures and those in the Red Sea by too much salinity. The biomass of plants that accumulates is either taken by herbivores or by detritus feeders. Therefore, secondary production is dependent on primary production. Many studies show that the detritus route is the most important in the mangroves.

Naidoo et al. (1998) showed that the gas exchange of *Avicennia marina* and *Bruguiera gymnorrhiza* was strongly influenced by photosynthetic photon flux (PPFD), leaf temperature and the leaf to air water deficit. Maximal CO_2 exchange occurred between 12:00 and 14:00. At saturating light intensities, optimal temperature and leaf to air water deficit, CO_2 exchange was consistently higher in *A. marina* (880 μmol $m^{-2} s^{-1}$) than in *B. gymnorrhiza* (5.3 μmol $m^{-2} s^{-1}$). *A. marina* also exhibit higher leaf conductance and transpiration but show lower leaf water potential than *B. gymnorrhiza*.

The estimated wood volume at one site in the Rufiji delta had a mean of 286 $m^3 ha^{-1}$ (Matta 1997). For the size class $5 \leq DBH \leq 10$ cm (diameter class 1), the volume was 15 $m^3 ha^{-1}$ and for $10 \leq DBH \leq 20$ cm (diameter class 2), it was 64 $m^3 ha^{-1}$ and for DBH > 20 cm (diameter class 3), it was 189 $m^3 ha^{-1}$. The stocking (stems/ha) for the diameter classes 1, 2, and 3 was 754, 492 and 241, respectively, and the total number of stems was 1 488 ha^{-1}. *Rhizophora mucronata* contributed 42.3% followed by *Xylocarpus granatum* with 21.1%, *Ceriops tagal* 11.0%, *Sonneratia alba* 9.1%, *Avicennia marina* 8.8%, *Bruguiera gymnorrhiza* 4.5%, and *Heritiera littoralis* 3.2%. The mean basal area was 28 $m^2 ha^{-1}$. *Rhizophora mucronata* and *Bruguiera gymnorrhiza* had the largest bark percent, 12% and 11%, respectively. This is why these two species, especially *Rhizophora*, were historically exploited commercially for its tannin. *Avicennia marina* has the least bark of only 3%, *Xylocarpus granatum* has 6% but *Ceriops tagal* and *Heritiera littoralis* have around 8%.

Steinke et al. (1995) reported a mean aboveground living biomass of mangroves at Mgeni estuary in South Africa of 94.49 ±7.83 t dry matter ha^{-1},while dead wood contributed a mean mass of 7.63 ±0.89 t dry matter ha^{-1} and underground biomass of 9.67 t dry matter ha^{-1}. The underground biomass represented only 9.8% of the aboveground value. There was a mean density of 4 700 living stems ha^{-1} with plant heights ranging from 0.57 m to 5.80 m. Mean Leaf Area Index was 4.95±0.80. Mangrove leaf litter formed 74.6% of the diet of the crab *Sesarma meinerti* in South Africa (Steinke et al. 1993).

Machiwa (1999) found in a mangrove stand in Zanzibar that there is a significant export of mangrove litter to the adjacent marine environment during spring tides. Net organic carbon export from the entire forest was 79×10^6 g C yr^{-1}; dissolved organic carbon accounted for about 78% of the total export. Rates of import and export of particulate matter in the forest were not statistically different. A relatively high export of macrodetritus was recorded at the marine fringe (mainly colonised by *Sonneratia alba*). There was a low export of litter in the terrestrial fringe zone, a monospecific stand occupied by *Avicennia marina* (Forsk) Vierh. While Machiwa (1999) showed a mangrove stand in Zanzibar was a net exporter of carbon, Middelburg et al. (1996) showed that mangrove sediments in Gazi Bay act as a nutrient and carbon sink rather than as a source for adjacent seagrass and reef ecosystems. Other studies show that tidal asymmetry, characterised by stronger ebb flows than flood flows in the mangroves partly promote the net export of organic matter to the seagrass beds. The brackish and turbid water plume in the mangrove creeks is often trapped in the mangrove swamp and does not reach the coral reef.

The algae and cyanobacteria growing in the mangroves must also be contributing a lot in overall primary production. Species in the genera *Oscillatoria*, *Spirulina*, *Lyngbya*, *Richelia*, *Nostoc*, *Pleurosigma*, *Gyrosigma*, *Hantzschia*, and *Nitzschia* are found in channels and as algal mats. Julius (1997) found a total of 23 species of benthic microalgae in Mtoni mangrove sediments, Tanzania. Most of the species encountered were pennate diatoms such as *Navicula* spp., *Pleurosigma* spp., *Hantzschia* spp., *Nitzschia* spp., and *Amphora* spp. Cyanobacteria such as *Oscillatoria* spp. and *Nostoc* spp. were abundant. The mean average number of microalgae was found to be 91 856 cells per gram of soil sample while an average of 77 920 and 9 262 cells were diatoms and cyanobacteria, respectively. She found the biomass to range from 0.27 to 1.33 $mg\ g^{-1}$ of soil sample. In the tidal pools, the dominant microalgae were cyanobacteria, i.e., *Oscillatoria* spp., *Spirulina* spp. and *Nostoc* spp. On mangrove roots Julius (1997) reported that the epiphytic algae *Caloglossa leprieurii* and *Bostrychia murayella* were dominant on *Avicennia marina* roots. *Rhizoclonium grande* was found to colonise the *Rhizophora mucronata*, and on the roots of *S. alba* the algae species *Bostrychia radicans* were most common.

2.4.4 Ecophysiology

Growth curve analysis for successive *Laguncularia racemosa* leaves on an annual basis in Guinea showed uniformity in the curves (Diallo 1992). The curves showed more similarity where the leaves maintain equal distance between the internodes. Thus, an annual pause increases continuously but in an irregular way. In the dry season (December–February), plastochrone and duration of growth increase while the growth rate and the size of adult leaves decrease. This implies a positive correlation between plastochrone and duration of growth. In contrast, the growth rate and the duration of growth are negatively correlated (Fig. 2.14).

Trees of the mangrove swamps are characterised by an average transpiration. The minimal transpiration (about 3 $mg\ cm^{-2}\ h^{-1}$) occurs in *Dalbergia* and *Canavalia*, whereas the highest transpiration rate (20 $mg\ cm^{-2}\ h^{-1}$) is recorded for herbaceous species such as *Philoxerus* and *Sporobolus*.

A preliminary ecophysiological study on Rhizophora was conducted for two sites within the reforested area of Somone lagoon (first site close to the lagoon near a transect established in 1995, and the other at the outer fringes of the lagoon). Measurements, namely stomatic conductivity (Gs) and leaf transpiration (Tr) as well as other microclimatic elements such as humidity, radiation and temperature were determined using a porometer (Licor-1600) at three different hours of the day (10 A.M., 13 P.M. and 16 P.M.). For each plant, two leaves located opposite each other were measured in descending order (from the apex) for the first four orders.

The ecophysiological behaviour of *Rhizophora mangle* leaves in Somone lagoon was observed and compared to those of the degraded area on the Mar Lothie Islands (Saloum Islands, Senegal). The results are illustrated in Fig. 2.15, 2.16, and 2.17.

Fig. 2.14.
Successive leaf growth of **a** the annual *Rhizophora mangle* and **b** *Laguncularia racemosa* sprout

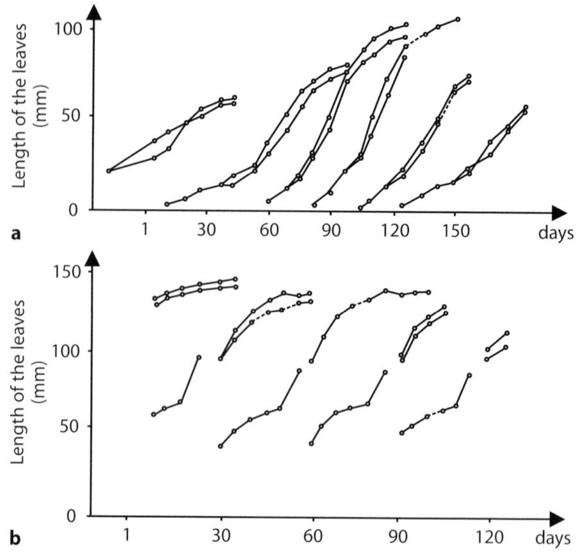

Fig. 2.15.
Kinetics of the daily stomatical conductivity and leaf transpiration of the 1st (*F1*), 2nd (*F2*), 3rd (*F3*) and 4th (*F4*) leaves of *Rhizophora mangle* at Somone Lagoon, Senegal, in relation to microclimatic factors (PAR and DPVS) (values are average of three measurements)

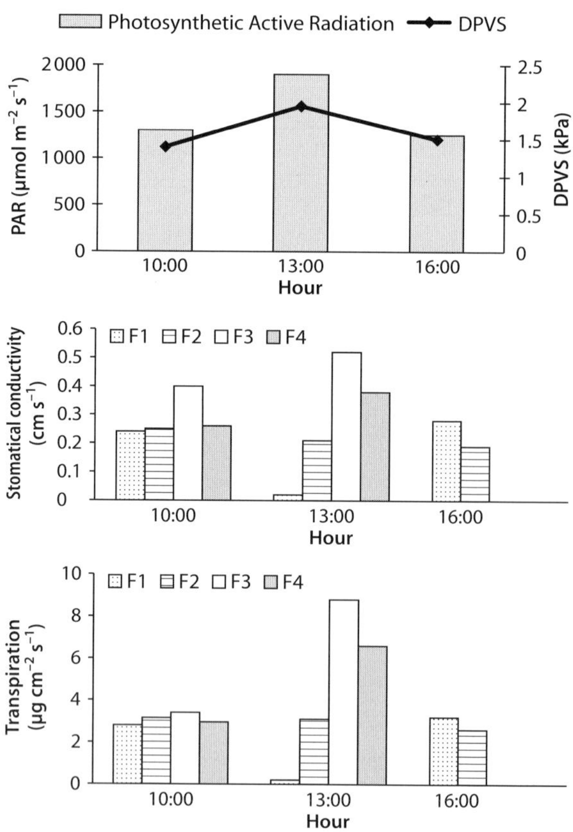

A strong relationship could be observed between Gs., Tr. and the microclimatic elements (Photosynthetic Active Radiation – PAR and Saturated Vapour Pressure Deficit – SVDP) on the one hand, and Gs. and Tr., which have similar evolution, on the other. The maximum values for Gs. and Tr. are generally obtained at the beginning of the day (10 A.M.), while PAR and SVDP do not reach their maxima which would lead to stomatic closing.

Figure 2.16 compares the evolution of Gs. and Tr. under different conditions and for the orders F1, F2, F3, and F4. It reveals a higher average Gs and Tr values for F3 leaf-order, and more so for the *Rhizophora* plant located along the transect.

For the *Rhizophora* located close to the lagoon, leaf-orders F2 and F3 remain active until the end of the day. With regard to plants located along the transect and along the lagoon fringe, leaf-orders F2, F3 and F4 maintain gaseous exchanges throughout the day. Gs and Tr also tend to decline at the end of the day because of the reduction in radiation (PAR) and evaporation (SVDP).

A comparison was also made between the behaviour of *Rhizophora* in Somone lagoon and those of degraded areas in Mar Lothie.

Data collected at 10 A.M. indicated that the values of Gs and Tr are more significant for the leaf-order F2 (Table 2.13). This suggests that the leaves were photosynthetically active. On the degraded site, however, the leaf-order F3 was never observed on the *Rhizophora*.

Fig. 2.16.
Kinetics of the daily stomatical conductivity and leaf transpiration of the 1st (*F1*), 2nd (*F2*), 3rd (*F3*) and 4th (*F4*) leaves of *Rhizophora mangle* at Somone transect, Senegal, in relation to microclimatic factors (PAR and DPVS) (values are average of three measurements)

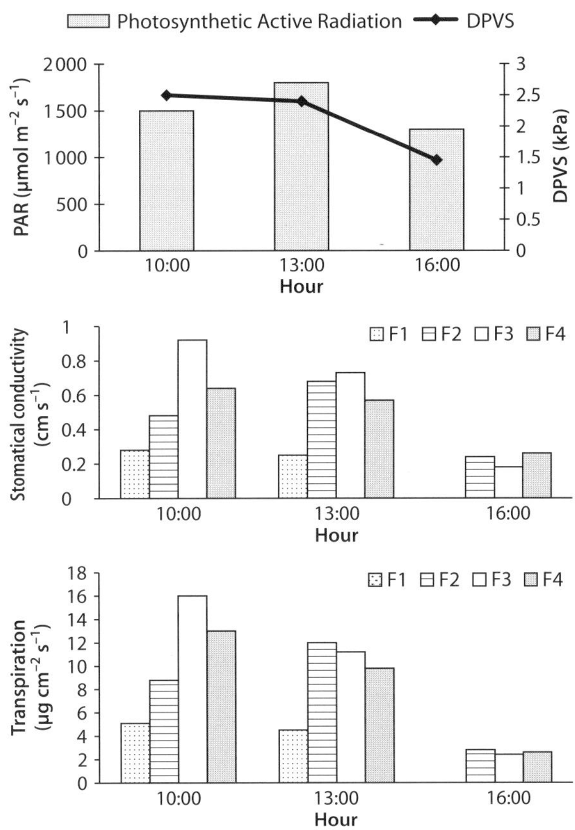

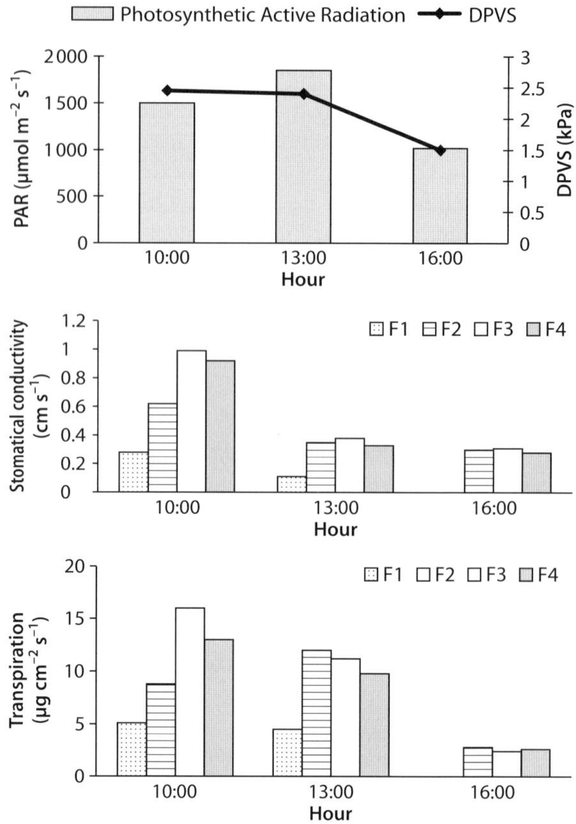

Fig. 2.17. Kinetics of the daily stomatical conductivity and leaf transpiration of the 1st (*F1*), 2nd (*F2*), 3rd (*F3*) and 4th (*F4*) leaves of *Rhizophora mangle* at the external fringe of Somone Lagoon, Senegal, in relation to microclimatic factors (PAR and DPVS) (values are average of three measurements)

Table 2.13. Stomatic conductivity (*Gs*) and leaf transpiration (*Tr*) values of *Rhizophora mangle* and some microclimatic parameters at Mar Lothie Island

Leaf order	PAR (µmol m^{-2} s^{-1})	HR (%)	Tf (°C)	DPVS (kPa)	Gs (cm s^{-1})	Tr (µg cm^{-2} s^{-1})
F1	1140	50.4	30.6	2.1	0	0
F2	924 – 21.9	43 – 11.6	31.9 – 1	2.5 – 0.5	0.20 – 0.05	3.99 – 1.4
F3	1160 – 56.6	52.6 – 3.7	30.3 – 1	2 – 0.2	0.03 – 0.04	0.44 – 0.6

The difference in behaviour observed between the *Rhizophora* for the regenerated site (Somone) and that of the degraded site (Mar Lothie) in respect to the functioning of hydric leaves according to their level of development presents important perspectives on:

- Adaptive response of *Rhizophora* to environmental conditions, and
- Identification of potential sites for mangrove rehabilitation.

2.5 Mangrove Associated Fauna

2.5.1 Invertebrates

There is scattered information on invertebrates living in mangroves of eastern Africa and this is mainly on macrobenthos. The insects, although the largest group of invertebrates, appears not to attract scientists to the region. Insects such as bees are visitors to the flower of the mangrove and honeydew, which also attract the weavers or tailor ants (Semesi 1992). Few studies have been done on the community structure and seasonal variation of mangrove invertebrates in eastern Africa. Prawns, mullusks, crustaceans, and polycheate worms live in the mangroves. The main shrimp species caught along the eastern African coast include *Penaeus indicus* (white shrimp), *P. monodon* (jumbo shrimp), *P. semisulcatus* (tiger shrimp), *P. japonicus* (flower shrimp) and *Metapenaeus monoceros* (brown shrimp). The potential for crab fishery is good, and the important species are *Scylla serrata* and *Portunus pelagicus*.

The most conspicuous animals in the mangroves are crabs, and the most common crab species are *Sersama* (marsh crab) and *Uca* (fiddler crabs) (Macnae 1963; Semesi 1992). They show characteristic zonation. Burrows of *Sersama* crabs inhabit the upper zone, when it is sandy. The zone of *Uca* spp. are dominated by the *Ceriops* and *Bruguiera* mangrove species. Mangrove snails like *Telebralia* spp., *Cerithidea* spp. and *Nerita* spp. crawl over the mud surface, sometimes climbing the trees and clustering on the shaded side of trunks. Barnacles and spring oysters are common on stems of the mangroves especially *Sonneratia* trees. On the leaves of mangroves especially those of *Bruguiera*, *Sonneratia* and *Ceriops*, species of *Littorina* are seen.

Mangrove oysters (*Crassostrea cuculata*) are very prolific on pop roots and lower branches of *Rhizophora mucronata* and of *Sonneratia alba*. Removal of fallen leaf litter in mangrove forests is effected by benthic communities dominated by crabs and litter feeding snails. One of the most conspicuous crabs are the fiddler crabs, which have an enlarged claw. Total densities of macrobenthic infauna of mangroves and surrounding beaches at Gazi Bay, Kenya ranged from 265 to 6 025 individuals m^{-2} (Schrijvers et al. 1995). They also found that virgin mangrove sediments rich in mud and organic material are characterised by high densities of macro-Oligochaeta and Mollusca, but denuded and less dense mangroves or mangroves higher in the tidal zone were much more sandy and had a high abundance of Polychaeta and Nemertini. Structural patterns in the macrobenthos were only vaguely associated with vegetation type.

Schrijvers et al. (1997) working on an *Avicennia marina* dominated mangrove in Kenya reported that *Terebralia palustris* and *Cerithidea decollata* are no-selective deposit feeding gastropods, but the crabs *Sesarma meinerti* and *Metopograpsus thukuhar* are almost exclusively vegetarian. The above species are represented by 57, 36, 0.25 and 1 individuals per m^2, respectively. They are also found on the same site as nematodes that were the dominant meiobenthos (93%), followed by Oligochaeta (2%), Copepoda (1%), Turbellaria (1%) and Halacaroidea (1%). Studies by Slima et al. (1997) in the same mangrove area showed that in the relatively elevated *Ceriops tagal* vegetation, which is only flooded during spring tides, the detritivorous snail *Terebralia palustris* (Linnaeus) was the major macrobenthic organism responsible for litter removal. While in the low lying *Rhizophora mucronata* stand, which is flooded by each high tide, the crab *Sesarma guttatum* (H. Milne Edwards) was responsible for most of the litter removal (consumption and burial). They estimated on average, 11.2% and 18.6% of the fallen litter is processed by macrobenthic animals in the *C. tagal* and in the *R. mucronata* vegetation, respectively. *Sesarma leptosoma* is the only species of crab inhabiting the mangrove swamps of East Africa that can climb to the top of tall mature trees, where it feeds on fresh leaves (Canniccia et al. 1996). Feeding-site fidelity seemed to be related more to the chance of finding suitable leaf-buds than finding leaves.

Schrijvers et al. (1998) reported that the macrobenthic density was 23 058 individ. m^{-2} under *Ceriops tagal* in Gazi mangroves in Kenya. Oligochaetes constituted 94% that was dominated by the family Tubificidae (98.4%). Almost half of the remaining 6% was composed of polychaetes (47%) with the family Terebellidae (33%) and the nereid genus *Namalycastis* (14%). These polychaetes were followed by macro-nematodes of the genus *Oncholaimus* (21.5%), gastropods (17%), and amphipods (13%) dominated by the genus *Grandidierella* (making up 11.5%) with a small fraction of the genus *Ampelisca*. The insect larvae made up only 0.08% of the total macrobenthic infauna. Resource competition for muddy detritus was found as the structuring force for the dominant oligochaetes in mangrove sediment (Schrijvers et al. 1998). The dominant tubificid population and the polychaete *Namalycastis* sp. that increased in cage-covered sediment indicated that polychaetes might be impacted by epibenthic predation.

Kamau (1997) noted at Ngomeni mangroves in Kenya that high macro-invertebrates species associated with moderate salinity and temperature had a high percentage of organic matter and water content. Highest infaunal densities (566.7 individuals m^{-2}) were found in mangroves. Epifauna commonly found are: *Sesarma guttatum* (A. M. Edwards), *Sesarma ortimani* (Crosnei), *Terebralia palustris* (Linnaeus), *Macrophthalamus depressus* (Ruppel), *Uca lactea* (de Haan), *Uca inversa* (Hoffman), *Uca vocans* (Linnaeus), *Uca urvellei* (A. M Edwards), *Eurycarcinus natalensis* (Krauss), and *Anadara* spp.

A survey by Olafsson (1995) within 5 mangrove areas on the west and east coast of Zanzibar showed that Meiofauna densities in surface sediments (0–5 cm) ranged from 205 to 5 263 individuals per 10 cm^2, being on average 1 493 individuals per 10 cm^2. Of the 17 major taxa recorded, nematodes dominated (64–99%) in all samples, while harpacticoid copepods were usually the second most abundant. Within all areas, the numbers of meiofauna were very variable, and significant differences among areas were only detected for oligochaetes and turbellarians. Densities of nematodes, harpacticoids, polychaetes and turbellarians were, however, significantly ($P < 0.001$) higher at low water stations compared with mid and high water stations. Harpacticoids were negatively correlated with the numbers of fiddler crab (*Uca* spp.) burrows. Other correlations between environmental factors (grain size, temperature, salinity, oxygen tension, prop root density, fiddler crab burrows) and major meiofaunal taxa were non-significant. A total of 94 nematode genera were recorded from four mangrove areas. The most abundant and frequent genera were *Microlaimus* and *Spirinia*, followed by *Desmodora* and *Metachromadora*. In a hypersaline area, diversity was much reduced and where salinity was over 100‰, the fauna was restricted to 3 nematode genera: *Microlaimus*, *Theristus* and *Bathylaimus*. Stromberg et al. (1998) found out Ocyponinae and Serminae comprise a large component of the intertidal macrofauna in Kisakasaka mangroves, Zanzibar.

Many studies show that there is a higher diversity of species in healthy mangroves than in degraded ones (Schrijvers et al. 1995; Stromberg et al. 1998). The density of Ologochaetes in uncut mangrove areas was found to be 3105 individuals m^{-2} but on 40 individuals m^{-2} in cut areas (Stromberg et al. 1998). In general, the macrofauna decreased in the cut area, except crab density was similar in the cut and uncut area because they can move around. However, (Kamau 1997) found that *Uca lactea* (de Haan) and *Uca inversa* (Hoffman) were found in large quantities in degraded mangroves. Fondo and Martens (1998) found that species of epifauna recorded were similar in both deforested and uncut areas, with crabs and mollusks dominating. Similar groups of infauna were also found in both areas with nemetoda, Nermetea and Errantia being dominant.

Osore et al. (1997) demonstrated that rainfall and tidal regime had substantial influence on the zooplankton community structure in the mangrove creeks.

Very little information exists for West African mangrove invertebrate fauna. Four main association types of mollusks are among the few dominant groups studied in association with mangroves of West Africa (Elouard and Rosso 1977). These include:

- Species associated with sandy-silty areas, characterised by low energy: *Tympanotonus fuscatus*, *Crassostrea gasar*, *Littornia angulifera*, *Tagelus angulatus* and *Tellina nymphalis*;
- Characteristic lagoon formation with euryhaline, eurythermic mollusks, which live in the medio-littoral and upper infra-littoral zones: *Anadara senilis*, *Dosinia isocardia*, *Loripes aberrans*, *Natica macroccensis*, *Semifusus morio*, and *Bullastriata adansonii*;
- Marine species: *Mactra glabrata*, *Natica collaria* and *Terebra senegalensis*; and
- Species association with the intermediate zone between the lagoon and the sea: *Anadara senilis* and *Natica fulminea.*

Crabs, *Uca tangeri* and mudskippers are also found in typical marine areas of these mangroves.

2.5.2 Vertebrates

Ecologically, the mangrove ecosystems are feeding grounds for a variety of aquatic and terrestrial animals including birds. Many fish species feed and breed in the mangroves the most common being marine catfish (*Arius* sp.), milk fish (*Chanos chanos*) and mudskippers. In Tudor creek Kenya, 83 species of fishes was reported by Little et al. (1988) and that 90% of 21 fish families were made of juveniles and larvae. Clupeid and gobiid fish species dominated both beach seining and plankton samples and the species, which use the creek as a nursery area, enter the system principally at a post larval/juvenile stage. Laroche et al. (1997) showed in Sarodrano mangrove creek (SW Madagascar) that tidal, lunar and diel changes have effects on the composition of the fish assemblage in mangroves. The species overlap between the Sarodrano mangrove fauna, and the adjacent coral reef fauna was weak with six species in common, and they concluded that the mangrove plays only a very limited nursery role for coral reef species. Sixty species of juvenile fishes were caught in Sarodrano mangroves, 44 species being of commercial interest. Catches were dominated by Gerreidae (27% of total abundance), Teraponidae (16%), Carangidae (13%) and Sparidae (12%). The temporary resident fishes in the mangrove zone represented 50% of the species and 97% of the total abundance. The species' richness, abundance and biomass per netting were low in the middle of the cool season (July–August).

Birds of various types are found in mangroves. These include: Mangrove Kingfisher, Pied Kingfisher, Little Egret, Dimorphic Heron, Common Sand Piper, White Pelican, Greater Flamingo, Black Winged Stilt, Little Stilt, etc. Also found are monkeys, baboons, rodents, duikers, fruit bats, pigs, snakes, hippopotamus and crocodiles.

Large vertebrates that are associated with the mangroves of West Africa include: *Tragelatus scriptus scriptus*, *Cercopiterus aethiops*, *Tricherus senegalensis*, *Hippopotamus amphibius*, *Crocodylus niloticus*, *Mellirora capensis*, *Tragelaphus spekei*, and *Soussa teuszii*.

The avifaune is represented by many species found in humid areas (Ramsar site), which are of international importance from an ornithological point of view, such as palearctic birds.

2.6 Uses of Mangroves

2.6.1 West Africa

Mangroves are of high economic value in the region. They are well noted for their wood products (tannin, firewood, charcoal, timber, and more recently, pulp) and medicinal value. Additionally, mangrove swamps have traditionally been used for agriculture, especially rice cultivation, and for fishing. Most of these uses have rather negative repercussions on the mangrove ecosystem.

2.6.1.1 *Rice Cultivation*

Local populations across the spectrum in the West African region depend on mangrove swamps for rice cultivation in order to enhance their livelihoods. A few such people are the Sereres, Niominka, Diola (Senegal), Balanta, Felupe (Baiote) and the Papel (Guinea Bissau). Rice cultivation, it must be noted, is one major activity that accounts for the destruction of the mangrove.

In Guinea, for instance, out of a tidal swam area of 3 900 km^2 potentially occupied by mangroves, 780 km^2 have more or less been converted into rice fields, of which only 40% are effectively productive (Diallo 1991).

To prevent saline water from entering into the rice fields, local techniques including the building of polders have often been adapted (Fig. 2.18).

It is important to note that the traditional way of converting the mangrove swamps into agricultural fields is only temporary, as natural vegetation is always re-established when human activities cease.

Fig. 2.18.
Example of rice fiel protection against tidal floods using mangroves in Casamance, Senegal

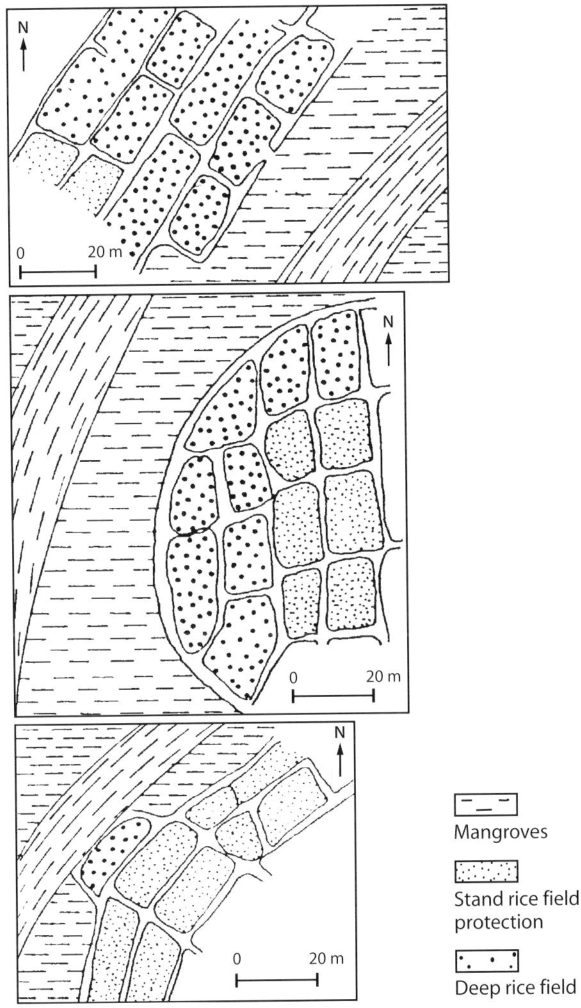

2.6.1.2
Fishing

Estuarine and inland fisheries play a very important role in the supply of local fish requirements. The most important fish species are mulets, ethmalosa and tilapia. In the Saloum estuary, the population of fishermen is estimated at 4000–6000. It is characterised by a high rate of migration, which can reach 95% in some islands.

In Guinea Bissau, the Papel are subsistence fishermen whose surplus production is marketed. The catch mainly includes Polynemidae, Cichlidae, Arriidae, and Scianidea (Simao 1993). Traditional fishing is also practised intensively along the Guinean coast, particularly in the region north of Conakry (Kamsar, Taboria) (Table 2.14).

Table 2.14. Characteristics of fishing activity in some countries of West Africa (Samb and Dioh 1996)

	Mauritania	Senegal	Gambia	Guinea Bissao	Guinea	Total
Period	1995	1993	1992	1993	1995	
No. of fisherman	11 581	35 000	1 568	8 219	11 000	113 268
No. of boats	1 565	6 055	1 653	2 000	2 595	15 338
No. of machines	1 565	9 593	15 701	6 000	2 306	35 165
Motorization (%)	96	90	36	20	47	

2.6.1.3 *Shell-Picking*

The picking of shells mainly by women and children is an age-long activity evident by the huge accumulation of shells in the mangrove areas. The shells are mainly those of *Crassostrea gasar* and *Anadara senilis* species. Some of the shell-accumulated sites are preserved as sacred places.

2.6.1.4 *Fuel Wood*

Rhizophora, an excellent firewood species, is exploited extensively within the environs of Conakry on a commercial basis because of its great value. The ports at Dixin in Conakry and Dubreka, 45 km away, are very important firewood marketing centres.

Besides its importance as a fuel wood species, *Rhizophora* serve numerous medicinal purposes. All parts of the *Rhizophora racemosa* plant, for instance, are known to cure various ailments: the yellow leaves are an efficient remedy for wounds; fresh leaves when tied on the head are believed to cure headaches; while the roots when taken as a beverage give relief to toothaches. The dried leaves are also used as a dyeing material and as a detergent.

2.6.1.5 *Aquaculture*

Oyster and shrimp culture in addition to the traditional pisciculture are the main forms of aquaculture practices in the region. In Senegal, oyster culture started in the 1940s in the lagoon of Joal-fadiouth. This was to be later turned into a co-operative venture. Production figures have, however, been on the negative trend since 1973, reducing from 57 000 dozen in 1981 to less than 20 000 dozen in 1988. This decline was mainly the result of degradation of the mangrove swamps emanating from droughty environmental conditions that have characterised the region of late, in addition to anthropogenic factors including over-exploitation of the mangroves for wood products.

Shrimp culture essentially is on an experimental basis with established sites at Katakalouse in Casamance (Senegal), Koba-Lamodia (Guinea) and Makali (North of

Sierra Leone). In Senegal, species that have been tried include *Penaeus notialis*, *P. kerethurus*, *P. vannamei*, *P. indicus*, *P. monodon*, and *P. japonicus*. The best results are obtained with *P. monodon*, *P. vannamei* and *P. stylirostris* from New Caledonia.

2.6.1.6
Constructional Materials

Wood products from mangrove forests are a very good source of constructional timber for local communities. In Senegal and Sierra Leone, the wood is used for fencing, building dykes, and making fish traps and poles. In the rural building industry, wood derived from mangroves is used at all stages of construction, particularly as supporting poles in the construction of huts, building houses and ovens, making frames and as roofing material. The *Anadara senilis* that is accumulated in mangrove areas is exploited and used as a substitute for gravel in the building industry.

2.6.1.7
Tourism

In Senegal the tourist industry, which started about 20 years ago is presently the second most important economic activity of that country. An issue of relevance is the fact that the industry is most developed along places where mangroves grow.

2.6.1.8
Other Uses

Most rivers of the region empty their waters into the sea. In countries where the rivers flow over mineral rich areas, some mineral particles are washed and transported into deltaic mangrove areas, thus making them potential mining sites. This is exemplified by Liberia, especially at the mouth of the Lofa River in the Grand Cape Mount.

2.6.2
East Africa

The direct uses of different mangrove species are summarised in Table 2.15. In coastal communities, mangroves are used to supplement local demands for fuel wood, fencing materials, house construction, boat building, fish traps and medicine (Semesi 1991a, Rosolofo 1997b). Most coastal village communities use small quantities of mangrove resources for local needs, and it is mainly commercial activities that over-exploit the mangrove resources. Selective cutting is the normal practice for subsistence use, and this practice usually allows for profuse natural regeneration of the mangroves. However, large-scale clear felling of mangroves prevent natural regeneration, and as such, artificial planting is needed. At a commercial level, mangroves are an important item of trade and a source of employment and income for the coastal communities. Mangrove poles are cut for both export and the local market. Revenue collected through the sale of mangrove poles in Mombasa, Kenya varies from US$300 to US$1100 per year (Kairo 1997). However, most pole cutting is done illegally because of ineffective controls.

Table 2.15. Direct uses of the mangrove tree species in Eastern Africa

Species	Uses
Avicennia marina	Inferior firewood, but used for boiling of brine, fish smoking and production of lime, building dugout canoes and beehives; leaves used as goat and cattle fodder; branches support beehives, construction of beds
Bruguiera gymnorrhiza	Good firewood; used for fish smoking, fishing stakes; poles for building
Ceriops tagal	Good firewood; poles; drying fish, fishing stakes; fence posts, poles for house building, timber for boat building, children use fruits to make whistle
Heritiera littoralis	Good firewood; timber for boat-building; furniture; dhow masts
Lumnitzera racemosa	Good firewood
Rhizophora mucronata	Good firewood; poles for building; fence posts; fish traps; fishing stakes barks used for dying nets
Sonneratia alba	Inferior firewood; commonly used in boat-building; pneumatophores used as fish net floats, floral nectar sucked by children and half open buds used by children to make earings and the fruit into spinning top
Xylocarpus granatum	Good firewood; used for fish smoking; boat-building; making furniture. The seeds are used to treat stomach problems and the fruit pulp to cure rashes fruit also used to induce abortion (high dosages are said to be lethal)

2.6.2.1
Rice Growing

Clearing mangroves for rice farms takes place in some riverine mangroves such as the Tana delta in Kenya, Zambezi in Mozambique and Rufiji delta in Tanzania (Kairo 1997; Hatton and Massinga 1994; Semesi 1991a). In the Rufiji delta, on farms cleared from mangroves after approximately the 7th year, yields fall dramatically and the farms are invaded by grasses, phragmites and sedges, and working the farms becomes a problem. The extent of this activity is unknown but it appears to have been started in late 1960s to early 1970s.

2.6.2.2
Fishing

Traditional fishing in the East African region is concentrated in shallow water habitats close to the shore, as the requisite capital and technology needed for deep-water fishing is lacking in the local communities. The catch, therefore, depends largely on the productivity of the inshore ecosystems including mangroves.

Mangroves also play a very important role in fish smoking especially for fishing communities located far away from urban centres. In some rural communities of Madagascar, for example, the rate at which mangroves are exploited for fish smoking far exceeds the natural rate of rejuvenation (Rosolofo 1997b).

Most of the marine species like shrimps need the mangrove habitat during part of their life (Macnae 1974; DeFontaubert et al. 1996; Odum and Heald 1975). Other species such as mud crabs spend most of their lives in the mangroves and move to the open sea to spawn. The degree of linkage between the mangroves and offshore habitats is dependent on the proximity of the habitats to one another.

In Mozambique, for example, the Sofala bank is the most productive fishing ground especially for shrimps, due to the large areal extent of the mangroves. The shrimp-fishing sub-sector contributes about 40% to the net foreign exchange earning in the country (Doddema-de-Guia 1997). Similarly, in Tanzania, the mangroves of the Rufiji delta make its surroundings the most important prawn fishing grounds in that country, contributing to about 80% of the total Tanzania's commercial catch (Fisheries Division annual report). About 3 000 fishermen are involved in prawn fishing and 6 000 in other forms of fishing.

Due to the extensive area of the mangroves of Madagascar, shrimp fishing yields as much as 10 500–11 000 t each year (Rosolofo 1997b). It could be noted from the above analysis that in Tanzania, Madagascar and Mozambique, there is a high correlation between the areal extent of nearby mangroves and prawn catches.

2.6.2.3 *Shell-Picking*

Most of the shellfish are in the intertidal mud flats and seagrasses close to mangroves. Edible species of oysters, mussels, cockles, and gastropods are collected extensively for local consumption, usually by women and children (Kayombo and Mainoya 1986; Semesi et al. 1999). In the Bagamoyo area of Tanzania, the important crab species are *Scylla serrata* locally known as "Kaa ungo" or "Mbuba" and *Portunus pelagicus* known as "Kaa upembe" (Mgaya et al. 1999). Women and children collect *Terebralia palustris* (locally known as "Suka") for food, and men collect it mainly as fish bait or to be used as feed in the aquaculture pond. In Kaole mangroves, "Kombe kichaa" (*Anadara* sp.), "Kombe kazungu" (*Anadara* sp.), "Nyaluale" or "Nyuwali" (*Strombus gibberulus*), "Kinyonga" (*Polinices mamilla*), "Kijino" (*Strombus* sp.), "Kikola" or "Kikopola kome" (*Pleuroploca* sp.), and "Suka" or "Tondo" (*Terebralia palustris*) are the common shellfishes collected (Semesi et al. 1999).

2.6.2.4 *Fuel wood*

Fuel wood demands are quite high in East Africa because it is largely utilised in commercial quantities for the production of salt, lime, extraction of coconut oil and the processing of fish, prawns and sea cucumbers. The ability to burn well and the high calorific value makes mangroves a good source of firewood. In several villages of Madagascar, the most common use of mangrove wood is for boiling prawns and smoking fish. In the village of Andavoanemboka, 65 m^3 of mangrove wood is burned every year to cook prawns, and 39 600 m^3 is burned for processing fish (Rosolofo 1997a). The estimated available wood is 260 000 m^3 yielding a renewable stock of 26 000 m^3 if the forest renews after every 10 years, and only 13 000 m^3 if it takes 20 years to grow. In the village of Andavoanemboka, *Avicennia marina* and *Ceriops tagal* are the preferred species for prawn and fish processing, respectively. *Ceriops tagal* has thus already become rare.

2.6.2.5
Aquaculture

In Kenya about 0.14 km^2 is used for prawn production and 0.45 km^2 have been cleared for development of aquaculture ponds (Ruwa 1995). In Madagascar, 1547 t of prawns were produced from 6.67 km^2 ponds in 1995, and sixty potential sites have been identified for shrimp aquaculture covering an estimated 500 km^2 (Rosolofo 1997a). At Mahajamba, for example, there is a potential of 40 km^2 of shrimp for farming. So far only three sites have been developed, and a semi-intensive rearing is being used. In Tanzania, three companies have been given licences to start shrimp farming. These are Cencor ltd, (1.20 km^2 in Ruvu Mangroves), Prawntan ltd (6.03 km^2 in Ruvu) and the African Fish Company (AFC), which has been allocated 100 km^2 land in the Rufiji Delta. Some AFC operations will be located within Mafia Marine Park. The proposed AFC prawn farm has raised many issues related to land, influence of structural adjustment policies, and institutional issues at local, national and international levels. It has also demonstrated the power of politicians in directing or misdirecting the "development" in Tanzania. Communities' views and interests are usually not given high consideration. Some of the concerns are:

- Lack of guidelines on aquaculture in Tanzania. It is being recommended that these be put in place before development can proceed.
- The proposed farm overlooked the regulations of the Forest Ordinance and of the Marine Parks and Reserve Act.
- The land policy puts a limit of area of land that can be owned by individuals, but this proposed farm passes the limit by far.
- There is externalisation of costs on expense of public goods. The beneficiaries of commercialised exploitation, for example of the mangrove poles or prawn fishery are outsiders, not the local people whose habitat is being destroyed (von Mitzlaff 1989).
- Market economy and the structural adjustment policies seem to favour the rich and marginalise the poor.

A global debate around this project has been on-going for about two years, and many useful recommendations for and alternatives to the massive project have been put forward. It is, however, sad that the majority of the people have lost to the powerful minority as the project has been approved as proposed. Through this process it became clear that local communities should be made aware of their rights as well as their responsibilities for effective participation to take place. The mangrove management plan of 1991 has been a very useful reference document in this connection. Academicians and several environmental Non Government Organisations are helping the villagers to appeal against the project.

2.6.2.6
Building Materials

In the rural areas, mangrove poles, coral stone and mud are still the main building materials (von Mitzlaff 1989). For building boats, *Xylocarpus granatum*, *Rhizophora mucronata* and *Heritiera littoralis* are taken for the mast and *Sonneratia alba* for the ribs (Kairo 1997; Ruwa 1995; Rosolofo 1997b). Large *Avicennia* trees are made into canoes. Kenya and Tanzania used to export mangrove poles of species *Rhizophora mucronata*, *Ceriops tagal* and *Bruguiera gymnorrhiza* to Somalia, Iran, Kuwait and Saudi Arabia, which were mainly used for building houses. This is because the poles are of high density, strong and durable and are resistant to termites and fungi. However, both countries suspended the export of mangrove poles because of the high rate of mangrove cutting.

2.6.2.7
Tourism

In Eastern Africa, there has been little development of ecotourism related directly to the mangroves, but elsewhere, boardwalks have been built, and mangrove tours are popular with many visitors. Few visitors from other countries have had the opportunity to see mangroves, and to walk and take boat tours through the mangrove forests. In South Africa, there is some tourism to mangrove areas as part of the mangroves that belong to national parks such as St. Lucia. Tourist fishing dependent on the mangrove and reef ecosystems is becoming more popular.

2.6.2.8
Other Uses

An herb, *Sesuvium portulacastrum*, which grows on sandy portions in the mangroves, is eaten as a vegetable in Tanzania. *Avicennia* species are used for animal fodder and *Xylocarpus* for local medicine. Orange and greenish sponges growing on the aerial roots of *Sonneratia alba* are used as fish bait (Semesi et al. 1999). The mangrove bark is a source of tannin which is used to treat leather. Extract from the climber *Derris trifoliata* found in the mangroves is used as fish poison and for rope making. Fish and prawn trapping fences are made mainly of *Ceriops*, *Rhizophora* and *Avicennia* samplings. The poles of mangroves are also used in seaweed farming in Tanzania. Only rarely are bees kept in the mangroves (Semesi 1991a).

2.7
Natural Threats to Mangrove Ecosystems

2.7.1
West Africa

Climatic perturbation is one of the major threats to the development of mangroves of West Africa. Since 1968, most of the countries in the sahelian zone have gone through a severe drought period, which lasted for 20 years. This phenomenon affected many coastal systems in the region and consequently reduced the area covered by mangroves (Table 2.16). The reduction in the size of mangrove swamps seems to be linked to the progression of the sahelian zone southwards, in response to the prevailing moisture conditions (Samb and Dioh 1996). Drought has further introduced changes in inundation patterns and river regimes, and also affected the fresh water supply and saline water influx to coastal river systems. Consequently, a record high salinity of surface

Table 2.16. Major coastal environmental issues in some coastal West African countries

Issue	Senegal	Gambia	Guinea Bissao	Guinea	Sierra Leone	Liberia
Coastal erosion	x	x		x	x	
Mangrove felling	x	x		x	x	x
Reef destruction				x		
Over fishing	x			x		
Eutrophication	x			x		
Pollution	x			x	x	x
Salinization	x			x	x	x
Flooding	x		x	x		

and groundwater has been experienced. This condition has negative effects on the mangroves of Gambia, particularly in Bitang Bolon and in the Tendaba swamps.

Marius (1984) has observed that the worsening climatic condition of the region has brought changes in the geochemistry of soils resulting in acidification, oxidation and neogenesis of gypsium.

An issue of critical concern is the continual exposure of the fragile mangrove ecosystem on the northern fringes of the region to climatic hazards, with its attendant dangers among which are the biological impoverishment of the aquatic and land areas; the adaptation of animal and plant populations to gradual hypersaline conditions; and the tendency towards confinement. Changes are also reflected on micro and macrofauna. The ecological stress in the Gambian mangrove swamps, for example, is the result of phytopathological effects (Teas and Mc Van 1982).

2.7.2 East Africa

Some natural phenomena known to kill mangroves in eastern Africa are storms or hurricanes, tidal waves, pests and diseases, sediment movement, and river floods (Semesi 1988; Kairo 1997; Steinke 1999), but these are not well quantified in the region. Poor drainage due to deep flooding or to poor tidal flushing also may cause considerable areas of mangroves to die as a result of alteration of water level, bank erosion and diversion of water courses. Mangroves cannot tolerate prolonged inundation by fresh or salt water and will die. Such flooding occurs regularly in some parts of the Rufiji delta. If the site is overwashed by strong tides, seedling establishment is poor. Sand deposition from sea and land, which can cut off portions of mangroves from salt water also causes them to die.

Weed competition, e.g., *Acrostichum aureum* and *Derris trifoliata* might overgrow the area and prevent the establishment of seedlings. Infestations of *Sonneratia alba* by a stem boring caterpillar of *Salagena discata* (Gaede) (Cossoidea-Metarbelidae: Lepidoptera) is serious in Lamu and Gazi mangroves of Kenya (Kairo 1997). Die back of *Sonneratia* is also noted in some mangroves of Tanzania, but no studies have been carried out to see if it is also caused by the same pest as in Kenya.

2.8 Human Threats to Mangrove Ecosystems

2.8.1 West Africa

Coastal areas are one of the most favoured sites for settlements because of the opportunities they offer. They provide extensive areas of complex and specialised ecosystems such as mangroves, coral reefs and seagrasses among others. Fertile lands, marine resources and access to external markets through river transport promote economic activities, which in turn attract a high concentration of population.

In West Africa, a high percentage of the urban population lives in coastal cities (Table 2.17). This tendency has a degradational effect on the resources of the coastal ecosystems due, *inter alia*, to population and demand pressures.

Reduction in number of mangrove species is often a sign of over-exploited resources due to excessive pressure. Mangrove swamps near village settlements are particularly disturbed and often completely destroyed. However, the destruction of mangroves by the activities of local people cannot be compared to the damage caused by giant agro-management enterprises. One such enterprise is ILACO, which has well-defined sterilised zones for rice cultivation in Tobor (Casamance).

The construction of anti-salt dams across rivers leads to reduction of exchanges from upstream to downstream and has negative repercussions on the estuarine productivity.

In Guinea, bauxite mining in Fria is the main cause of pollution on the Konkoure River and on the mangrove swamps located downstream. The level of pollution has, however, been reduced today owing to some innovations introduced in bauxite processing.

2.8.2 East Africa

In recent years, the rate and variety of human influences have increased to the point where a large proportion of the mangrove resource is threatened by destruction. Man-

Table 2.17. Population of coastal countries in West Africa in millions of inhabitants

Country	Total population (million)		Coastal urban (million)	
	1995	2025	1980	2000
Mauritania	2.34	5.00	0.23	1.18
Senegal	8.39	17.0	1.38	3.08
Gambia	0.98	1.88	0.11	0.29
Guinea Bissau	1.07	1.98	0.17	0.35
Guinea	6.70	15.0	0.69	2.03
Sierra Leone	2.74	9.80	0.45	1.18
Liberia	3.04	7.23	0.47	1.19

groves that are severely affected are either close to urban centres, large villages or those converted to aquaculture or saltpans. Activities that are destructive include: clearance for agriculture and aquaculture (Semesi 1991a; Hatton and Massinga 1994; Rosolofo 1997a); clearance for salt production (Karanja 1997; Doddema-de-Guia 1997; Kairo 1997; Semesi 1991a); clearance for urban and industrial development (Saket and Mattusse 1994; Shah 1995); diversion of freshwater due to dam construction; and over-exploitation of firewood, pole and charcoal production (Rosolofo 1997b; Semesi 1986). For example, 60% of the mangroves at Kunduchi in Tanzania have been cleared (Banyikwa and Semesi 1986). While in Mozambique, the Kariba and Cabora Bassa dams along the Zambezi River are causing mangroves to dry; accelerated erosion of the delta and salt water now ascends some 80 km up the Zambezi River (Kairo 1997). In Ngomeni, Kenya, 7% of the mangroves have been lost (Karanja 1997) to aquaculture and saltpans (Table 2.18).

During the civil war in Mozambique, some mangroves were not used because people fled to other areas (Saket and Mattusse 1994). However, areas such as those close to Maputo and Sofala are heavily used because of immigration of the population to the coast. The Maputo province shows the highest rate of deforestation, 15.5%, followed by Sofala with 4.9% and the northern province of Nampula with 3.6%. In addition, 0.05 km^2 in the Maputo province along Tembe and Umbeluzi rivers are very degraded because of firewood cutting and construction material (Doddema-de-Guia 1997). In Mozambique, depletion of mangroves is mainly due to four factors:

a Cutting for energy and building material which occurs all along the coast but with increased intensity around Maputo and Beira;
b Clearing for salt production especially in the northern province;
c Clearing for agriculture practised in the northern and central provinces; and
d Degradation caused by ecological changes, mainly damming of rivers such as the Zambezi, which is leading to desiccation of mangroves (Saket and Mattusse 1994).

Table 2.18. Intensity of major threats on mangroves in Kenyan coastal districts (*X*: nil, *L*: low, *H*: high) (Kairo 1997)

Key problem	Coastal District				
	Lamu	Tana River	Kilifi	Mombasa	Kwale
1. Conversion					
Agriculture	X	X	X	L	X
Salt-pans	X	L	H	X	X
Aquaculture	X	L	H	L	X
2. Diversion of fresh water	X	H	H	L	X
3. Improper land use pattern	X	H	H	L	L
4. Illegal felling	L	L	H	H	L
5. Pollution	L	X	X	H	X
6. Pests	H	L	L	L	H

The relative importance of these different activities in terms of mangrove destruction has varied greatly from country to country, and within countries. The threat to the mangroves of South Africa, for example, is mainly due to harbour development and closure of the river mouth by sand. Resort development and boat traffic increase erosive boat wakes. Of these different activities, clearance for salt production and clear cutting for charcoal are the most important factors in destroying the mangroves of the Bagamoyo area in Tanzania (Semesi et al. 1999). Polychaete worms mostly found in the *Sonneratia* and *Rhizophora* zones are dug up and used as fish bait, resulting in the ecological damage of some mangroves of Tanzania.

2.8.2.1 *Conversion to Agriculture*

Conversion to agriculture mainly for rice cultivation is practised in the rich riverine mangroves. However, productivity does not last for a long time. Therefore, farmers abandon these farms and move to other sites. The *Sesarma* crab (locally known as Koe) destroys rice seedlings in Rufiji. In order to kill these crabs, farmers use DDT which also affects other organisms such as fish and prawns. Similar observations are made in Chinde part of the Zambezi delta, where salt intrusion and crab infestation are also a problem in rice farms located in former mangrove areas (Hatton and Massinga 1994). Therefore, if not regulated, cultivation of mangroves will deplete the healthiest mangroves in the region.

2.8.2.2 *Conversion to Salt Pans*

Some solar saltpans are located on the saline bare areas behind mangroves, but others have been constructed on mangrove areas. For example in Nacala Bay, Mozambique, almost all former mangroves at the head of the bay have been cleared for salt pans (Hatton and Massinga 1994), and in Ngomeni Kenya, salt work has taken 2.43 km^2 of mangroves (Karanja 1997), and in Tanzania, salt pans cover about 0.031 km^2 (Semesi 1991a).

2.8.2.3 *Conversion to Aquaculture Ponds*

Shrimp farms are major contributors to the destruction of mangroves in South East Asia and Latin America (Primavera 1997). Developers are now moving to Eastern Africa, although the environmental and social problems related to shrimp farming are well-known. Prawn farming deprives local communities of fish and shellfish nurseries, storm buffering and other ecological services. Shrimp farms also contribute to the depletion of ground water and pollution of water with shrimp waste, chemicals and antibiotics (DeFontaubert et al. 1996).

2.8.2.4 *Coastal Development*

Only the mangroves found near large towns such as Dar es Salaam, Maputo and Mombasa are being filled in for house construction or for other activities (Semesi and Howell 1992; Ahamada 1997; Shah 1995). From nearby towns, neighbouring residents also use mangrove areas as dumping places, thus introducing untreated waste to those areas. Small mangrove stands near large towns are also being killed because of oil pollution, e.g., Mombasa, Dar es Salaam and Maputo (Doddema-de-Guia 1997; Kairo 1997; Semesi 1991a). Others are cleared to provide a good view of the sea, and this is causing beach erosion as can be seen in Nunge, Bagamoyo.

2.9 Management of Mangroves: Restoration and Rehabilitation

2.9.1 West Africa

Mangrove management is very important for the sustainability of mangrove ecosystems. However, one major factor inhibiting their effective management in the region is land tenure with its attendant problems. To address this issue, most countries of the sub-region have formulated forestry legislation by which the Forestry Departments are enjoined to play supervisory roles in the control of destructive activities on the mangrove systems. At the local front, some communities have also instituted mechanisms aimed at effective management of the mangroves. Such local initiatives, it is believed, could be streamlined and integrated into state management strategies for good results.

In Guinea Bissau, the Coastal Planning Project has begun a programme aimed at:

- Developing the mangroves of Rio Cacheu into the dos Tarafes National Park; and
- Preserving and rationalising the management of mangrove swamps with the view towards:
- Quantifying the use of resources and their economic values;
- Measuring the evolution and the causes of degeneration of the mangroves; and
- Proposing national legislation for the regulation of their use.

The programme is supposed to continue once the existing neighbouring conflicts are over.

In Guinea, CERESCOR and the University of Conakry have initiated studies into the biological, ecological, pedological and sedimentological characteristics of mangrove areas of the country.

Table 2.19. Protected areas with mangrove coastal states in West Africa (UNEP 1982)

Country	No. of proposed protected mangrove areas
Mauritania	2
Senegal	3
Gambia	2 – 3
Guinea Bissau	3
Guinea	6
Sierra Leone	3
Liberia	2

Box 2.1. Case Study of Mangrove Restoration in Mar Lothie, Saloum Islands, Senegal

Mar Lothie is located in the northern part of the Gandoul Islands, which are within the Saloum Islands. The mangrove ecosystems of the area have been plagued by severe drought, particularly over the last few decades. Anthropogenic activities, notably fuel wood extraction, are also taking a great toll on the mangroves. As a result, the area originally occupied by the mangroves has been drastically reduced to only a narrow strip along the river-banks. As an intervention measure, the International Society of Mangrove Ecosystem (ISME) and a UNESCO support-multidisciplinary research team from the University of Dakar, in collaboration with local groups (Pinda Lothie) has been constituted to execute a rehabilitation programme. This programme is divided into several phases viz.:

1. Educational Phase – Awareness Creation at the Community Level

Close communication links with all users of the mangrove resource is a key step towards any successful mangrove restoration programme. Mindful of this, several meetings were held at Mar Lothie to sensitise the people to the need to preserve the mangrove ecosystem of the area. The educational meetings were also aimed at popularising the techniques of mangrove restoration and replanting. Restoration techniques similar to those used in Asia and Latin America were to be adopted.

2. Collection of Propagules

Drawing from the experience of an earlier restoration programme at the Somone Lagoon-site, propagules of *Rhizophora mangle* of various lengths were collected for planting. The propagules were divided into two categories based on their lengths. The first category consisted of those ranging in length between 25 and 30 cm while the second category ranged between 10 and 20 cm.

3. Site Preparation

Two sites were selected for reforestation:

Site A. This site is located in the northern part of Mar Lothie Island along the Ndangane River (Saloum Islands). It initially covered an area of 1 177 m^2 (i.e., 47 × 25 m) and was later increased to 4 165 m^2. The area is characterised by a clayey substrate approximately 15 cm thick, good surface sedimentation enhanced by tidal waves, a salinity range of 46–47% and a pH of 7.33. The discovery of young shoots of *Rhizophora mangle* at the site also guided the decision to select the site.

Site B. The second site is to be found on the South of Mar Lothie Islands. It covers an area of 1 456 m^2 (i.e., 52 × 28 m), having been increased from an initial size of 150 m^2 (i.e., 25 × 6 m). The area is dominated by remnants of *Rhizophora mangle* along with a few other naturally regenerated species at the edges. The salinity of the water is maintained at 46% by the regular influx of tidal waters. The sandy-clay soils have a pH of about 8.

4. Direct Planting

This involves direct planting of the propagules without necessarily preparing nurseries. The local people, under the guidance of the scientific team, carried out the planting. A regular spacing of 1 × 1 m was maintained between the propagules. The exercise which is still on-going also targets the involvement of school children at the primary level to inculcate in them environmental consciousness, and to teach them techniques of mangrove restoration while they are still young.

5. Establishment of Nurseries

Conversional nurseries were established using poly-sacks. They were located on high ground areas, away from the effects of high tides. Again, this was done with high community participation. The nurseries were established to make available a stockpile of seedlings for transplanting in areas earmarked for the reforestation programme on the one hand, and on the other, for replenishing the existing plots where direct planting was carried out, should the need arise.

6. Observations on Sites A and B

For site A, in February 1998, the salinity was between 46–47% while the pH averaged about 7.3. In May and August of the same year, these parameters were 65% and 51%, respectively for the salinity, and 8.3 for the pH.

In site B, in February 1998, about 46% was recorded for salinity while the pH averaged about 8. In May and August, 60–55% and 46% respectively were recorded for salinity and 8 while the pH value of 8.28 was obtained for the period.

The decrease in salinity for the two sites (A and B) in August may be attributed to the onset of the rainy season.

For both sites, observations made in May, April and August 1998 indicated that:

- The young propagules looked very healthy with the appearance of small leaves;
- Mature propagules, especially those brought down from Somone lagoon, showed signs of necrosis and mortality;
- A good general preservation of the parcels from any anthropogenic or natural pressure,
- Besides, a new nursery ground of 100 propagules have been implemented on the site A in August 1998.

Mangrove vegetation in Sierra Leone is under the jurisdiction of the Ministry of Agriculture, Forestry and Fisheries. The Ministry is thus charged with the task of effective management of the mangroves.

In a nutshell, efforts still need to be made for the assessment of the mangrove resources, especially their development, by proposing strategies for the sustainable management and integrated use of the mangrove ecosystems in Africa.

Table 2.19 shows a summary of the protected mangrove areas proposed for the West African region, and an interesting case study of mangrove restoration is the Mar Lothie, Saloum Islands, Senegal, described in Box 2.1 and shown in Fig. 2.19.

Fig. 2.19. Reforestation activity at the Mar Lothie study area, Saloum Islands in Senegal

2.9.2 East Africa

Forest management encompasses the administrative, economic, legal, social, and technical measures involved in the conservation and use of natural forests and forest plantations (FAO 1997). It involves the various degrees of human intervention to safeguard the forest ecosystem, its functions and its resources for the sustained production of goods and provision of environmental services. The concept of sustainable use involves the sustainable harvest of economic products while at the same time maintaining the existing system as naturally, or closely to its original state as possible. Therefore, globally the protective roles of forests are being given more emphasis, and non-wood products are also receiving more attention. Similarly, the needs of the forest dependent people are being given greater weight. However, the complexity of the mangrove ecosystem makes attempts to manage it extremely challenging.

Legislation relating to mangroves is generally included in that for terrestrial forestry. In Kenya and Tanzania, with their long history of mangrove pole use and export, all mangroves were gazetted as Forest Reserves in 1928–1932. In Tanzania, mainland legislation governing mangrove forest reserves is included under the Forest Ordinance of 1957 (Semesi 1991a). In Zanzibar, forestry legislation is covered by the Wood Cutting (Amendment) Decree 1968, and mangroves are listed as protected. In Kenya, the mangroves are under forest reserve through proclamation N° 44 of 30 April 1932 and later by Legal notice N° 174 of 20 May 1964 (Kairo 1997). Theoretically, the legislation imposes many restrictions on access to and utilisation of the mangrove forests but is rarely enforced. In Mozambique, the National Directorate is finalising a set of new rules and regulations for the management of forestry and wildlife for Forestry and Wildlife (Doddema-de-Guia 1997). Within this new set of rules and regulations, the mangrove areas in the whole country will be considered as nature reserves. In Madagascar, mangrove forests are not exploited for timber or poles on a commercial scale, and there is no management or control. In Mauritius, Seychelles and Comoros, the mangroves are also protected, and many stands fall within marine reserves and parks. It was not possible to get information on management or uses of mangroves of Somalia.

Most planning in the region does not look at the wider national economic implications. Therefore, previous plans emphasised the harvesting of mangrove trees for commercial wood products, that is, building poles, firewood, and charcoal. The essential role of mangroves in supplying the basic needs of coastal communities or their important value to the environment such as supporting fisheries was downplayed. Management for non-wood resources in mangroves such as for prawns, crabs, fish bait and honey is not being done, although now awareness of these issues is taking place. There are no regulations regarding the type of harvesting, time of harvesting and allowable quantities of the faunal resources in the mangroves. Since there is no close follow-up on these resources, there is no information about their status.

Generally, no overall national authority exists that can effectively resolve conflicting issues related to conservation and development of the mangroves. For example in Mozambique, seven different institutions are involved in the management of mangroves: the Ministry of Industry and Energy, Ministry of State Administration, National Directorate of Tourism, National Marine Directorate, Directorate of Forestry and Wildlife, University of Eduardo Mondlane, and Ministry of Coordination of Environmental Affairs (Doddema-de-Guia 1997). Similarly in Tanzania, although the Forestry Division is charged with the managing of the mangroves at present, conflicts arise between those branches of government responsible for forests, fisheries, wildlife, agriculture, ports, surveying of land and issuing of titles, and mineral mining (Semesi 1991a).

In Kenya, according to the Forest Act and the Memorandum of understanding (MoU) between Kenya Wildlife Services (KWS) and the Forest Department (FD), the FD is the lead agency for managing the mangroves. KWS is supposed to work closely with the FD in planning and implementing its activities in the mangroves, which are included in national parks (Provisional Forest Officer, personal communication). Since the MoU between KWS and FD was drawn in the Head Offices in Nairobi, it was reported by the Provisional Forest officer that Field Officers tend to interpret the MoU differently, because they were not involved. Furthermore, he emphasised that MoUs are only effective if the partners have equal opportunities, resources and incentives, which is not the case between KWS and the FD staff in Kenya. Initially, the Wardens and Forest Guards carried out joint patrols, but after some time, the Forest Guards refused to patrol because they were not paid any field allowance while Wardens, on the other hand, were compensated. This is because KWS is a para-statal organisation (responsible for the management of all national parks of Kenya), which has support from many donors as well as the Government of Kenya, and it also generates its own revenue, while FD is solely supported by the Kenyan government. Therefore, differences in salaries, infrastructure, etc. between these two institutions prevent real co-operation in the field.

Trials of mangrove restorations are taking place in the region. In Gazi, Kenya about 15 ha have been replanted and that *Sonneratia alba* shows the highest annual increment (1.18 m yr^{-1}), (Table 2.20) (Kairo J. G., personal communication).

Planting of saplings (0.3–1 m) from nursery or those collected from the wild had the highest survival rate (80–100%) after one year compared to propagules. Survival rate was also dependent on abiotic factors, such as wave action, frequency of tidal inundation, etc. In exposed areas, survival was as low as 10%, but in protected areas it was more than 70% (Kairo, personal communication). In Tanzania, a Mangrove trial plot at Pangani shows that after seven years, *Rhizophora mucronata* attained 5.7 m and *Bruguiera gymnorrhiza* 4 m, but *Ceriops tagal* attained only 2.4 m. In all the trial experiments, it was found that the species have to be planted in the zone where they

Table 2.20. Maximum annual growth rate for five mangrove species planted in Kenya (J. G. Kairo, personal communication)

Species	Annual height increment (cm)	Diameter at first internode (cm)
Sonneratia alba	118.0	1.90
Avicennia marina	98.0	1.12
Rhizophora mucronata	45.3	1.01
Ceriops tagal	27.1	0.39
Bruguiera gymnorrhiza	70.1	0.93

normally grow, otherwise at an older stage if grown at the wrong site, the trees will usually die.

Because there has been little appreciation of the specialised nature of mangroves and their management, there is often a lack of adequately trained field and management personnel. Furthermore, the lack of transport and general infrastructure greatly hampers patrol and management efforts.

Tanzania was the first country to prepare the mangrove management plan, and currently Mozambique and Kenya are preparing theirs. In 1991 a National Mangrove Management Plan was prepared in Tanzania (Semesi 1991b, 1992). This plan emphasises the need to have close co-ordination among the various users of the mangrove ecosystem. The mangrove forests are divided into four management zones:

- Zone I, forests which will receive total protection
- Zone II, forests which are ready to be brought into production
- Zone III, degraded areas which will be closed to allow recovery
- Zone IV, areas that will be set aside for different developments

Although it has been eight years since the management plan of Tanzania was prepared, it is still not followed closely, and community involvement is minimal. Collaborative management is being practised in three pilot villages in Tanga: Kipumbwi, Sange and Kisa (Ngunyali 1997), and 4.22 km^2 of mangroves have been allocated for trial. The mangrove management project is collaborating with the Tanga Coastal Zone Development Project and local communities. However, the government through its Forest Division still assumes overall responsibility. The primary objective of this collaborative management is protection and wise use of mangroves for the benefit of the villagers. The main constraint has been limited trained personnel to carry out the implementation of the management plan and lack of experience of involving communities. However, the mangrove management plan of Tanzania has been useful, as its concept is widely used in the region. Within Tanzania, it is used to raise awareness about the mangroves, and the maps and information prepared in 1989–1991 provide a very good baseline material for future work.

Recently, in the region many activities have been carried out so as to raise public awareness such as through wildlife clubs, video, slides, posters, etc. A series of workshops supported by SAREC-Sida and the World Bank on Integrated Coastal Zone Management (ICZM) has raised awareness of policy and decision makers about the need for proper management of coastal resources (Shah et al. 1997; Linden and Lunden 1996; Ahamada 1997). Also, a SAREC-Sida Regional workshop on Mangrove ecology, Physiology and Management was organised in Zanzibar, Dec. 3–14 1997.

Each country has established mangrove projects that are implemented by various governments and non-governmental agencies (Ruwa 1995; Hatton and Massinga 1994; Doddema-de-Guia 1997; Rosolofo 1997a; Ahamada 1997; Shah 1995). Governments are starting to incorporate mangroves within the ICZM initiatives in such a way so as to enhance the contribution of mangrove ecosystems to the economy of the countries. Furthermore, the new forest policies have statements that emphasise the need for active participation of all the stakeholders in forest reserve management and conservation through joint management agreements, with appropriate user rights and benefits.

Therefore, current objectives of managing various mangroves are according to the Kenya Forestry Master Plan (1995–2020) (MENR 1994):

- Conservation of mangrove areas that serve protective functions such as windbreaks and barriers to erosion.
- To maintain a habitat for the fauna and flora, especially those enclosed within marine parks and nature reserves.
- The optimisation of a combination of direct and indirect uses of mangrove areas.
- Recognition of the needs of communities living in and depending on mangrove environment.
- Management of representative areas for biodiversity, tourism, research, and education.
- Improvement of institutional capabilities of the body responsible for mangrove management.
- Identification and planning of research activities.

Detailed ecological knowledge of the mangrove ecosystem is, however, still limited although it is accumulating. Information on the growth and yield of natural mangrove forests needed to determine optimum thinning and planting strategies is lacking. An understanding of the food-chain relationship and specific physicochemical processes within the coastal waters is scant. Research is being undertaken throughout the region on reforestation, forest structure, nutrient dynamics, microbial processes, crabs, ground water, fauna, and diseases. Also, local uses are being studied in much more detail. However, indigenous technical knowledge and traditional management systems need to be understood and evaluated.

The mangrove inventories, preparation of management plans and their implementation rely heavily on donor support. For example in Tanzania, funding is from NORAD, while in Kenya it is from USAID, the Netherlands government and FAO. Donor contributions to forestry projects and related activities is around 90% of the sectoral funding in Tanzania (Final Draft, Tanzania Forest Policy 1997). Therefore, in order to sustain these activities, mechanisms of insuring funding should be found.

2.10 Conclusion

The current rate of destruction of West African mangrove ecosystems for purposes of farming, domestic fuel demands, constructional timber, and fishing calls for pragmatic efforts at both local and supra-local levels for their effective management to reverse the trend. Paramount to this is the need for programmes that would ensure habitat conservation for the survival of mangrove-associated flora and fauna, biodiversity conservation and the enhancement of aesthetic quality of the environment, among others.

The success of such management programmes necessitates the adoption of strategies that embody education, local participation, training and continuous research and monitoring.

- Education: There is a need to intensify educational programmes for awareness creation on the conservation of mangroves and rehabilitation of degraded areas. Such programmes should aim at mobilising the local people to participate more actively in reforestation programmes such as involvement in propagules collection and the extension of the already rehabilitated mangrove areas.

- Training: Techniques of mangrove restoration used by Forestry Departments have to be transferred to target local groups who would in turn pass it on to other local people. The involvement of school children in the rehabilitation programmes should be reinforced to ensure early knowledge of management techniques among the youth. Final year university students in related fields should also be encouraged to research different aspects of mangrove ecosystems such as sedimentology, cartography, anthropogenic studies and monitoring.
- Establishment of Nurseries: Local involvement in nursery establishment as already in practice in some areas of the region should be intensified and spread to other areas to ensure constant availability of seedlings for reforestation. Appropriate guidelines for site selection should also be followed to avoid high mortality of propagules.
- Research, monitoring and information dissemination: Existing scientific data on coastal ecosystems in general and mangrove systems in particular dates back to two decades. These constitute the main reference point for current research on mangrove ecosystems and their management. In line with these two principles to be considered are:
  - Systems of mangrove management to incorporate inventories of all biophysical elements within the mangrove areas and the involvement of the local people in the management programmes; and
  - The spelling out of guidelines for delineating areas to be preserved and strategising both scientific and local approaches for the judicious management, restoration and preservation of mangrove ecosystems.

From a long-term perspective, applied research on mangrove ecosystems in West Africa will certainly facilitate linkages between protection and maintenance of ecological process and sustain multiple uses of mangrove resources. Furthermore, a better synergy between university research and the implementing agencies should be reinforced for the realisation of the set objectives.

In the Eastern Africa region, the demand for wood energy and other natural resources is expected to continue increasing due to population growth, poverty and the Structural Adjustment Policies (SAPs). The devaluation of currencies, retrenchment of workers, and removal of subsidies force people to revert to wood fuel, to farming and other activities based on natural resources for survival. Current globalisation trends will fuel exploitation pressure even further (Myers 1997). As long as problems with poverty, population growth and inequitable terms of trade are not solved, the pressure on mangrove resources will continue to grow.

Multiple approaches are needed to slow the pressure on the resources. For example, to reduce wood consumption used to process fish and prawns in Madagascar, improved technologies such as solar drying or cooling might reduce the wood needed and hence allow for the sustainable use of mangroves. Also, the use of improved cooking stoves or of solar cookers and heaters would most likely reduce the demand for fuel wood. In the case of solar saltpans, the improved utilisation of the pans and diversifying their uses, for example for aquaculture, would enable operators to generate more funds and employ more workers. In order to restore degraded areas, reforestation must be carried out using ecological knowledge of the trees, sediments and zonation.

In the medium term, evidently, harvesting of mangroves will continue. Therefore, greater knowledge of the resources as well as careful control and management is needed. However, the growing awareness of the importance and fragility of the mangroves' ecosystems in Eastern Africa is a good sign. It is recommended that management should aim at assessing the root causes that threaten the mangroves and trying to target these for a long-term solution. These should go hand and hand with short-term solutions.

Training foresters on community related matters and on mangrove ecology is needed, because their formal training is presently focused on terrestrial forests. The management of non-wood products such as shellfish is also important, therefore, close collaboration with fishery experts is needed. Proper valuation should be made before mangroves are converted to other land uses such as rice or prawn farms.

Dedication

Adelaida Kleti Semesi (1951–2001)

Adelaida Semesi; fortunate are those who had the opportunity to know her.

As a scientist, her first paper was published at age 25, in the Bulletin of the University of Dar-es-Salaam. The paper was on the agar content of three species of *Gracilaria*. She passed away on Tuesday, 6 February 2001 as a full Professor, not yet 50 years of age, and her last work was a synthesis review of the mangroves of Tanzania. She was buried on Saturday, 10 February 2001 in Dar-es-Salaam. In a short span of 25 years, she produced singly or as a co-author over one hundred important scientific research papers and valuable reports. From marine algae, phycocolloids and carragenan in particular, she advanced into marine biology and coastal management, especially mangrove ecosystems, research, use, and management. She proceeded along a straight path that she must have idealised from the very beginning; a path or self imposed ideal for the fulfilment of which she relentlessly worked. She demonstrated practically that good scientific research is and will always be useful when the ultimate goal is to protect, use and manage rationally the natural wealth of one's country. All her work, from strict experimental science to the most brilliant reports illustrate the need for marrying traditional wisdom to scientific research. Her analytical mind was particularly fit to study the complexities of coastal waters and apply the result of her studies to rational use and management of her country's natural biotic resources. The tropical coastal zone is particularly dynamic; hence the need to know the structure and function of the basic resource system and later knit the accrued knowledge into a comprehensive whole. The list of publications by Adelaida Semesi shows how she made the passage from the elementary to the more complex systems, always bearing in mind the benefits for her people and her country. Since 1990, she had a major involvement with mangroves and thanks to her, one of the first countries in the world to have a detailed plan in 10 volumes for the management of mangroves is Mainland Tanzania. She extended this to "The Role of Science in Conservation and Management of Natural Resources in Eastern Africa: The Case of Mangroves, Seagrasses and Seaweeds", presented at the 1997 AAAS Annual Meeting, Seattle. Throughout her career, she was alert to the challenge posed to the scientist whose duty is to convince governments, especially in so-called developing countries, that scientific research leads to responsible use of permanent natural wealth and guides towards management for sustainable production and increased productivity of ecosystems for the present and future generations. Adelaida Semesi was a council member of the International Society of Marine Ecosystems. As a Professional Biologist, she ensured the supervision of more than 46 MSc and 6 PhD students. She awards many fellowships, in particular with the Rockefeller Foundation, International Foundation for Science and Senior Fullbright in the USA. Her last award was with the Agricultural University in Norway in July 2000 when she was appointed Director of the Institute of Marine Science of Zanzibar in Tanzania, her home country.

What a rare person was Adelaida Kleti Semesi, and what a great loss for all of us. She faced the challenge and showed the way, especially to us from the so-called developing world. Others as great as she are rarely born, yet we should all follow in her footsteps: strengthen the marriage between science and traditional wisdom, then share our understanding with all people. She ran the gauntlet; who will be the next leader?

As a human being, as a woman, wife and mother, she lingers in the blood of her family and in our memory as an example to be followed, though difficult to reach. May she rest in peace, as she will live long with us. We thank her for the example and legacy she left with us, and with her dear ones, we mourn her demise.

Chapter 3

Indo-West Pacific Mangroves

M. Vannucci

3.1 Introduction

There are biogeographical, palaeontological, biological and historical reasons why it is useful to consider the Seas from the East coast of Africa eastwards towards the western Pacific as a major geographic domain, later to be conveniently sub-divided into Regions and Sub-regions for detailed studies. Though the mangroves are the go-between for land and sea, man is a *terra firma* species and as such, the habit prevails to describe continents and their margins, rather than the oceans and their margins: oceanography was recognised as a distinct science only in the nineteenth century. To write on function and management of the mangroves of Asia and the Pacific, it is preferable to consider the seas that border the thousands of kilometres of coasts that form the transition between land and sea. Mangroves are a gift of the tides; sea and brackish waters disperse their seeds and propagules. They are the living place of aquatic, terrestrial and aerial flora and fauna and they would not exist without the interaction of land and sea. Mangroves form a special domain of their own: the mangrove ecosystem that thrives only in the intertidal belt. It is from this angle that we will consider the mangroves of the Indo-West Pacific.

Mangrove ecosystems cover an immense part of the Indo-West Pacific realm, including Australia and New Zealand, which cannot be dissociated from it. As diverse as they may appear, and in fact are, it is not only area-wise that mangroves are a cohesive unit, therefore the question arises on how best to introduce the subject. The focus of this review is on the "Function and Management" of mangroves. Therefore, a logical subdivision of the text is based on the role that different types of mangrove ecosystems play in nature at large and in relation to man. The best way to manage them is to consider their intrinsic structure and dynamics. The structure and dynamics of mangrove ecosystems varies widely from place to place. Hence is the need to systematise the matter and to avoid lengthening the text unduly, though at present there is enough knowledge to write many volumes. To save space and time, I decided to avoid describing mangroves from the country-by-country point of view, for two reasons: First: most countries produced and are now releasing scientific publications, brochures and leaflets about their mangroves, either as scientific papers, ecotourism guides or school books. Second: it is more rational to subdivide description of function and management according to the different types of mangrove ecosystems. The word management in the title of this book inevitably also involves local administration and central governments of every sovereign nation with mangrove wetlands along its coastline. Laws and regulations are formulated and must be enforced at the local government level. On the other hand, the management of natural ecosystems, which extends beyond man-made country boundaries, has suffered and often continues to suffer from contrasting natural and man-made laws and constraints.

I will not repeat here details of lists of species, area, forest cover, human population numbers, climatological data, protected areas and other basic information, except where not yet easily available, for two reasons: much of this information varies in time and varies according to survey techniques used; second, much of this numerical information, often quoting varying sources is contained in recent publications. The most comprehensive, up to date summary of the mangroves of the world, listed country-wise, is to be found in the World Mangrove Atlas, published by ISME (Spalding et al. 1997). Published details of oceanic islands are not easily available in the literature; therefore a relatively great amount of space is given here to mangroves of small oceanic islands.

No two mangroves are alike. The only characteristic common to all of them, worldwide, is that the exclusive mangrove vegetation grows only in the intertidal zones, between the lowest and the highest levels of the tides at zigzag. World-wide there are only some 80–90 species that occur naturally in the intertidal zones and only there. Most of these species are trees and shrubs, which constitute the mangrove forest. It is the mangrove forest that creates the ecosystem mangrove. The forest is the dominant feature and the major primary producer of the ecosystem, where the substratum is alternately covered and uncovered by tidal waters. Other factors being suitable, mangroves occur around the world in warm brackish water, on coastal wetlands. If this specialised association is totally felled or otherwise destroyed, it will not easily regenerate itself. The restoration of mangrove ecosystems assisted by man is a costly, lengthy and labour-intensive process. Totally felled, abandoned mangrove areas become desertified. The sediment-rich muddy, low-lying, gentle slope intertidal zone is an ecologically demanding habitat where there are either mangroves or desert.

I will describe function and management of mangroves under the following rough classifications:

1. Arid coastal systems, including coastal lagoons (*kwar*, *hors*) irrigated or not by seasonal rains and seasonal rivers (*wadi*), or regular rivers.
2. Estuarine and deltaic mangrove ecosystems
3. Small deltas and coastal mangrove ecosystems
4. Oceanic islands
5. Systems dominated by socioeconomic factors – Management and Phenology

This very broad, imprecisely defined division covers practically all mangrove ecosystems of the Indo-West Pacific, either impacted or not by human activities. Two points should be raised here: First of all, there are dispersed over this vast area, localised groves of a few trees or even solitary trees that grow in isolation or semi-isolation; they are the result of seeds or propagules deposited by the vagaries of water movements. Isolated mangroves may be found on rocks and coral reefs or other odd places such as some mangrove patches in the Red Sea or in the humid estuary of the Nakama River of Iriomote Island of Japan. Second: Man as a species used the mangroves and their resources since the time when he first came in contact with them. Witness to this are shell mounds, remains of fireplaces, burial places and pottery found in mangroves all over the world. Sacred forests respected as areas of worship, concentration and meditation from where nothing can ever be removed are common throughout the world where man has ever lived, and mangrove forests are no exception. They are used for initiation rites, as burial places, for wilful preservation of biodiversity or sheer respect for Mother Nature and her gifts, including medicinal substances. For instance, *Excoecaria agallocha* has been used to combat leprosy throughout its range and was imported for this purpose by imperial Rome as early as the first century. The same use of this species of mangrove is common in South and South-east Asia and in the West-Pacific. Man has interfered broadly with the structure and dynamics of mangroves. However, in many areas, human population density was not large enough to substantially alter the structure and function of the forest, especially in vast areas or isolated islands like the Andaman-Nicobar Islands, or those of Malaysia, Indonesia, Papua New Guinea and Australia, where there may still be some virgin mangroves left. In a general way, the present geographic distribution of mangrove species is due to natural dispersion over geologic time, but on the other hand, man has also voluntarily transferred useful species from one place to another. Man may also have caused local disappearance of some species in some places due to over-exploitation and conversion of mangrove ecosystems to other forms of land use.

In trying to understand mangrove ecosystems and mangrove trees and shrubs in particular, it is important to look at the exclusive species of mangrove rooted plants as something unique. Throughout the world there are about half a million vascular plants, but only some 80 species belonging to about 18 different families are exclusively found in the intertidal zones world-wide. Two families are exclusively found in that habitat: they are the Rhizophoraceae and the Avicenniaceae, which are undoubtedly the best adapted. They have imprecisely defined species limits and hybridise readily. Some widespread species can survive in freshwater, such as *Sonneratia caseolaris*, Sonneratiaceae, which I found flowering and fruiting in a relict fresh water swamp near Lae in New Guinea, or *Bruguiera* spp. in glasshouses. There are some species of grasses, like for instance *Porteresia coarctata*, a form of wild rice, or *Spartina parviflora*, which are true mangrove species and colonise the intertidal zone readily and permanently.

3.2 The Mangrove Ecosystem: a Unique, Marginal and Fragile System

Throughout the world, there are about half a million species of vascular plants, but not more than about eighty species are specialists exclusively found in the tropical intertidal zone, not to be found anywhere else except in the mangroves. Only two families, the Rhizophoraceae and the Avicenniaceae are pantropical. They are the best adapted to, and exclusively found in the intertidal zone of the tropics, though some species extend into the sub-tropics. The pantropical species with the broadest distribution also in latitude is *Avicennia marina*, which as the other species of the two families and some of the genus *Sonneratia*, Sonneratiaceae, has also a great intra-specific variation. The species of both the Rhizophoraceae and the Avicenniaceae are best adapted to the intertidal environment, but the evolutionary response was different in the two families. For instance, adaptation to anoxic substrata was achieved by the development of aerenchyma in both the families, but in other mangrove species morphological and histological details vary as does also the salinity tolerance mechanism and foliar reaction to excessive sunlight. The Sonneratiaceae also is pantropical, but some of the species do not belong to the mangrove habitat. The Rhizophoraceae include four genera: *Rhizophora*, *Bruguiera*, *Ceriops*, and *Kandelia*. The genus *Rhizophora* includes 6 species and three hybrids; the genus *Bruguiera* has 6 species; the genus *Ceriops* two species and the genus *Kandelia* only one. The Avicenniaceae has only one genus with some 10 species with many varieties and hybrids. All the species of these two families and of the genus *Sonneratia* are rather ill defined; geographic varieties are numerous and hybrids occur frequently. Palynology shows that *Avicennia* and *Sonneratia* pollen make their appearance first in the Miocene in Borneo and *Avicennia* also in Nigeria; later *Rhizophora* appears in Nigeria, South America, Asia and Papua New Guinea. *Nypa* first appears in palaeontological records from the Palaeocene in Brazil and in Europe from the Eocene; *Ceriops* is also from the Eocene. These genera were fully evolved at that time as known at present. *Palaeobruguiera* is also recorded from the Eocene, but it was not yet fully evolved into recent forms of the genus *Bruguiera*.

The Pliocene route of dispersal of flora and fauna through the Tethys Sea, which separated the northern Laurasia continent, including present-day Europe and Asia, from the southern Gondwanaland, which included present-day Africa and India, was closed during the Miocene. There are no mangroves off the Mediterranean Sea at present, though *Avicennia* persisted for some time after the closure of the Tethys Sea. Eventually it disappeared from that sea, as other mangrove species before it. The disappearance of mangroves from Europe and the Mediterranean is probably related to the absence or very small amplitude of the tides in that sea. The palaeontological record shows a total absence of mangroves after the closure of the Tethys Sea and the emergence of a closed Mediterranean Sea.

The most obvious adaptations to the intertidal zone are well-known and repeatedly described, as is evident from the anatomy and histology of all their organs. The distribution of the aerenchyma in stilt roots, knee roots, pneumatophores, aerial roots, and all roots aboveground in general, in the leaf bases of *Nypa* fronds, also the large amount and often conspicuous size of the lenticels, present even on propagules and roots, and viviparity, which is not exclusive of Rhizophoraceae, but most developed in the species of that family, are some of the most conspicuous adaptations to the intertidal zone. Species of *Avicennia* are described as semi-viviparous; in fact their seeds have cotyledons and the beginning of a plumule or embryo contained in the strong fibrous shell of the seed, called *testa*. Actual germination, however, takes place only after the *testa* or shell of the seed rots and ruptures on the soil of the forest or on mud banks. In silvicultural practices, *Avicennia* seeds must be soaked in water for several days before they are capable of germinating. Viviparity is fully developed only in the Rhizophoraceae and may have been induced by the stressing conditions of adaptation to the intertidal zone. Viviparity is rare in wild species of plants, but common in species and varieties of plants subject to intensive cultivation, for instance papaya, apples and oranges. Result of stress induced by intensive cultivation could be compared to stress induced by the harsh, inhospitable environment of the intertidal zone and extreme environmental conditions of temperature, aridity, alternate flooding and exposure to air and above all, to stress induced by continuous fluctuation of all environmental parameters. Throughout the world, there are no two areas of the intertidal zone exactly similar to one another, accordingly, the mangrove forests may be comparable, but not exactly the same from one place to another. The immediate consequence is that there are usually not more than a few, or some eight or ten different species coexisting in the same area. Dominant species are not the same from place to place and zonation is peculiar to each site.

Some species of Rhizophoraceae and Avicenniaceae extend into the sub-tropics, northwards in Japan and southwards in Brazil, Australia and New Zealand. The pantropic species with the broadest distribution in latitude is *Avicennia marina*, which shows a considerable intra-specific variation. Though the species of the two families are undoubtedly the best adapted, the response to the newly conquered environment, the intertidal zone, varied in the evolutionary history of the species in terms of morphology and physiology. Some of the most typical adaptations are those of the Rhizophoraceae, while many of the details of the adaptations of the Avicenniaceae may be found also in mangrove species of other families, for instance pneumatophores, abundance and distribution of lenticels, excess salt exclusion or secretion, and the ability to coppice. All species of mangroves, however, tolerate anoxic, often acidic substrata and saline waters. Exclusive species of mangroves are capable of taking root, growing, flowering and reproducing in the intertidal zones world-wide, where other vascular plants would not be able to become established, or even germinate or survive if transplanted: only mangrove trees and shrubs are rooted in soils regularly bathed by tides. Many mangrove species survive in very low salinity water. Examples are known of isolated specimens growing and flowering in pure fresh water in glasshouses or relict marine lagoons with fresh water, but no flowering trees or shrubs other than mangroves grow in the muddy intertidal zone. For this reason, total felling of mangrove forests for whatever purpose spells desertification. Mangrove land turned into dry agriculture, paddy rice, sugar-cane plantations, or fish and shrimp ponds becomes degraded in only just a few years mainly because of acid soil formation, otherwise they may be kept running at great cost of energy inputs, labour and time.

It may also be noted that Soepadamo and Wong (1995) include in the Rhizophoraceae related pantropical genera that grow mostly in moist (primary and secondary) tropical forests. The inland species *Carallia brachiata* grows to timber size, is exploited for logs and is widely distributed from Madagascar eastwards to the Indonesian islands. It is found in lowlands and up to 1800 m, on poor organic soils in mixed dipterocarp forests, heath forests and on the margins of freshwater swamp forests.

Salinity of soils and waters is not the most important factor that limits the rooting and establishment of vascular plants. Salt marshes in the coastal zone and inland, other than the intertidal zone, have a number of species throughout the world, not only in the tropical zone, that have adapted to sea water with normal salinity up to 35 ppt and higher. However, mangroves do not become permanently established in saline water ponds inland, where there are no tides. Species of *Rhizophora* and *Avicennia* grow well only in the intertidal zone even in water salinity up to 42 or 45 ppt as measured in the Arabian Gulf (Kogo and Tsuruda 1996) or at places of the Indus River delta when the SW monsoon fails and there is scarcity of river water discharge (Qureshi 1999). On the other hand, along some open ocean beaches preferably with coarse sand or fine gravel, where tides have a minimal amplitude in the equatorial belt, it is common to find species of *Ficus*, *Erythrina*, *Thespesia*, *Hibiscus*, *Terminalia*, *Conocarpus*, *Calophyllum* or *Pongamia* with their roots and lower trunk bathed at high tide by full salinity sea water. Evidently there are factors other than salinity that exclude non-mangrove species from the intertidal zone. There is a special land vegetation in salinity and alkalinity-affected soils elsewhere than the intertidal zone, which only disappears when anaerobic conditions are extreme; salt marshes everywhere have a species association that never includes mangroves. There is no clear or single answer to the question of the mangrove's ability to colonise the inhospitable conditions of the intertidal zone. Research on the protective mechanisms developed by mangrove species against the hazards of the intertidal zone should continue. One example is the capability of mangroves to immobilise heavy metals present in the soil and in pore and surface waters (Lacerda 1998). Very little is known about the exchange processes going on in the rhizosphere of the mangroves' rooted plants; genetic research on these and other aspects is also still in its infancy (Parida et al. 1995; Parani et al. 1998), and only major tracts of the microbiology of mangrove soils and waters are known (Agate 1988). It is along these lines that scientific research should be focused on a better understanding of how it is that only a few species of vascular plants have been successful in overcoming and conquering formidable barriers to colonise the intertidal zone, and this only under warm environmental conditions. Adaptations to anoxic conditions of the soil are well-known. From the widespread occurrence of mangroves it would appear that it was a relatively easy evolutionary process to adapt to muddy anoxic, often acidic coastal soils. The greatest difficulty for rooted vascular plants to overcome was the presence of toxic chemicals in the soil. It seems likely that only very few species were able to adapt successfully. The exact mechanism of adaptation to toxic soils is not yet fully understood, and different species may have adopted different strategies. Lacerda (1998) described the immobilisation of heavy metals in the soils through the action of healthy mangrove forests, a feat which degraded forests are unable to perform. Evidently, for appropriate and efficient management of the mangrove ecosystem, the structure and function of the component parts should be well understood.

In soils in general, microorganisms and sulfides readily hydrolyse dead plant and animal tissues, and sulfates accumulate after microbial attack on sulfur containing amino acids. Sulfur is oxidised into sulfate in aerated environments, while under anaerobic conditions and waterlogged soils, like in the mangroves, sulfide accumulates. Sulfides are derived partly from sulfate reduction and partly from mineralisation of organic sulfur. Mangrove soils contain abundant biogenic pyrites, which under anoxic conditions caused by water logging, are not readily oxidised into sulfates and sulfuric acid. In the mangrove oxygen-depleted soil, there can be very high concentrations of sulfides, up to 150 mg l^{-1}, at the same time the population of sulfur oxidising microorganisms is only about 10^3–10^4 g^{-1}, but it may increase to several millions g^{-1}, when the soil is aerated. The most important micro-organisms involved in S and Fe cycles in mangrove soils are strict autotrophs and obligate aerobes, mainly *Thiobacillus thiooxidans* and *Thiobacillus ferrooxidans*, which are active at pH < 3 and below, while *T. denitrificans* is not an obligate aerobic microbe, which explains the different pathways of the N cycle in mangroves. The genus *Thiobacillus* also contains thermophilic species, which may be related to the occurrence of the mangrove species in warm environments. For more details on nutrient cycling in mangroves, see Agate (1988).

These are the reasons why clear felling of mangroves, exposure of mangrove soils to air, or blocking regular tidal inundation causes the formation of acid sulfate soils which are unfit for both agriculture and aquaculture. It is for this same reason that increasing amounts of calcium carbonate under any form have to be added as "soil additives" or "water neutralisers" to agriculture fields or fish and shrimp ponds made in former mangrove soils. In the intertidal zone, the surface layer of the soil is exposed to air during ebb tide; the *Thiobacilli* are active, as they normally are on land. During tidal flow, mangrove soil is underwater, *Thiobacilli* are inactive and sulfides accumulate in the mud. Only those plants that developed abundant lenticels and much aerenchyma tissue in various organs could actually evolve, develop in the intertidal zone and survive under regular and prolonged tidal flooding. The activity of digging crustaceans like crabs, the mud lobster *Thalassina anomala* and others play an important role in aerating sub-surface layers of the soil. The amount of soil actually brought to the surface by the activity of *Thalassina anomala* was recently quantified by Kawaguchi et al. (2000). This activity may be beneficial if carried out in areas regularly washed by tides or rain, which remove sulfides and sulfates and thus counterbalance the formation of acid sulfate soils.

There are other causes of man-made degradation of mangroves, besides the ill treatment of the soils. Such are tampering with the hydrological system or withdrawing the water supply by building dams for irrigation and power generation purposes, preventing tidal flush, over-exploitation of resources, and total felling of trees. However, these are frequently described in the literature and are better known.

Summing up, the mangrove trees and shrubs create the mangrove ecosystem, but the latter cannot be managed appropriately if the role played by its functional components is not understood. Man-made disasters and widespread desertification could be avoided if only developers would take the trouble to understand how the various parts of the system function. Other morphological, physiological and reproductive adaptations of mangrove plants to the intertidal zone will not be discussed here, as they are well-known and described in detail in the classic literature on mangroves, for example Tomlinson (1986). The conclusion may be drawn that only very few species of the thousands of vascular plant species could overcome the difficulties of becoming established in the intertidal zone. Probably there is no single determining factor that prevents the establishment of higher plants in the intertidal zone; it is rather a combination of factors, the most prohibitive being the peculiarities of the mangrove soils, rather than water salinity. Research along these lines is urgently needed, along with genetics research. In fact, there is not enough knowledge to distinguish phenotypic from genotypic-induced characteristics; species are ill defined, and there is great intra-generic and intra-specific variation, hybrids are common, but the extent of allopolyploids is not yet known.

3.3 Arid Coastal Systems, Including Coastal Lagoons

There are two main arid coastal zones in the Indian Ocean: the coastal zone bordering the northern Arabian Sea and much of the western and southern coasts of Australia. In the Arabian Sea, including the Gulf of Aden, the Red Sea, the Arabian Gulf, the Gulf of Oman and the coasts of Baluchistan and Sindh in Pakistan and of Gujarat in India, oceanographic conditions coupled with climate and air-sea interface phenomena create an especially arid climate with low precipitation. Seasonal upwelling of the Sub-Surface Indian Ocean waters in the northern part of the Arabian Sea contributes to the aridity of the bordering coastal zone. In other words, mangroves are restricted to the delta of the only great river, the Indus, to coastal lagoons irrigated by seasonal rivers, *wadi*, or periodic rains alternating with periods of drought, or are otherwise scattered as small groves or isolated groups of just a few trees. Upwelling conditions on parts of the Pakistani coast contribute to the high productivity of the coastal fisheries, especially where mangrove forests are well developed, as for instance at Sonmiani Bay (Miani Hor), Gawater Bay and the southern Omani coast. World-wide along dry coastal areas, the main source of available water for the mangroves is seawater diluted or not by fresh water from rains or rivers. The west coast of Australia has somewhat comparable conditions.

Historically, there is a good account of mangrove forests by contemporaries of Alexander the Great (356–323 B.C.), who described the mangroves of the Makran and Lasbela coasts, now Pakistan, the abundance of fish in the coastal waters and the absence of mangroves in the low lying areas of the then separate mouths of the Tigris and Euphrates rivers, which are now the 120-mile long Shatt-al-Arab at the northern end of the Arabian Gulf. Eratosthenes (245–204 B.C.), chief librarian at the then flourishing library of Alexandria in Egypt and member of its famous school of learned men, described in detail both the species and the forests of mangroves that bordered the entire coast of the Arabian peninsula. From his descriptions, species of the genera *Rhizophora* and *Avicennia*, which he compared respectively to laurel and the olive trees, are easily recognisable. There were no mangroves at the separate mouths of the Tigris and Euphrates rivers at that time, and there still are none at the mouth of the Shatt-al-Arab, because of the very large amount of silt carried by the two rivers and deposited on the prograding coast in large quantities. Seeds, propagules and seedlings would be quickly silted in and choked into oblivion if they took root there. The mangroves of the eastern coast of the Arabian peninsula were described as forests so high, that an archer ('and those people are not sluggish with their bow and arrows') would not have been able to shoot over them. The wood, it is reported, was used for construction of dwellings and boats and as firewood; probably at that time, as at present, camels and presumably goats would browse on *Avicennia* leaves. At present, most mangrove forests have disappeared from the Arabian Gulf, victims of excessive cutting and added insults, such as wars and crude oil pollution. Localised attempts at reforestation are progressing with moderate success.

Eastwards of the Straits of Bab-el Mandeb, along the Gulf of Aden, on the Somali, Yemeni, Omani and southern Iranian coasts there are numerous seasonal rivers, and the mangroves have historically been an important element of landscapes and source of wood and timber. In recent years, however, the decrease in fisheries was related to excessive felling of mangroves. At present, the Government and the Sultan Qaboos University of Oman join hands to reforest depleted areas and rehabilitate degraded ones. The Red Sea, unlike the Arabian Gulf whose greatest depth is less than 200 m, has a deep trench that runs its entire length and is over 2 000 m deep with troughs over 4 000 m deep. Unlike the Arabian Gulf, there are only few short perennial rivers in Erithrea that discharge in the Red Sea; otherwise there are only seasonal rivers, which altogether amount to only a small amount of river fresh water input into the sea. While salinity of surface waters is high in both seas, the Arabian Gulf waters carry a large amount of silt and nutrients from land runoff; they bear at present or had in the past relatively good stands of mangroves. In contrast, the Red Sea waters are poor in nutrients and silt, therefore Red Sea mangroves never developed much beyond the stage of isolated trees or small groves. The islands of the southern part of the Red Sea north of Bab-el-Mandeb have mangroves; Afonso de Albuquerque described those of the island of Kamaraan on the Arabian coast when he explored the Red Sea in the first quarter of the sixteenth century as sparse and scrubby "*mangues*" (mangroves). He attributed the lack of development of the trees to excessive browsing by camels and excessive cutting for firewood by the people of the area. On the other hand, coral reefs are abundant in the Red Sea, while there are no coral reefs in the Arabian Gulf where mangroves could grow, and in the past did grow into tall forests. In historical times, pharaohs of ancient Egypt dug canals for navigational purposes linking the Nile River with the Sea, the third and last one was dug during the Ptolemy dynasty. Each of the canals, completed or not, had a short life and each was eventually silted in by shifting wind-borne sands. What mangroves remain in the Red Sea, in the Gulf of Aden and the island of Socotra is surely much less than what was there five centuries or 2 000 years ago. Figure 3.1 shows the distribution and composition of mangroves of the Arabian Peninsula.

Though reduced, mangroves of desert coastal areas still play a significant role. Already Eratosthenes underlined the fact that the presence of such trees that grow in the sea is all the more remarkable because the land behind them is barren. In other words, the choice is: either mangroves or deserts. The contemporaries of Alexandre the Great related the rich fisheries at sea and the well-being of the dwellers of the Makran coast whom they called 'ichthyophagi' (fish eaters) to the presence of the mangroves in the coastal lagoons of the area. At present we know that seasonal upwelling of sub-surface oceanic waters contribute, together with the mangroves, to the fertility of the marine coastal waters.

Mangroves played an important role in the history of the Indian Ocean, especially in the Arabian Sea. They were used for construction of ships and housing. Planks, masts, ribs, rudders of the famous Omani ships used mangrove wood; somewhat modernised versions of the crafts used by the famous Yemenite and Omani mariners

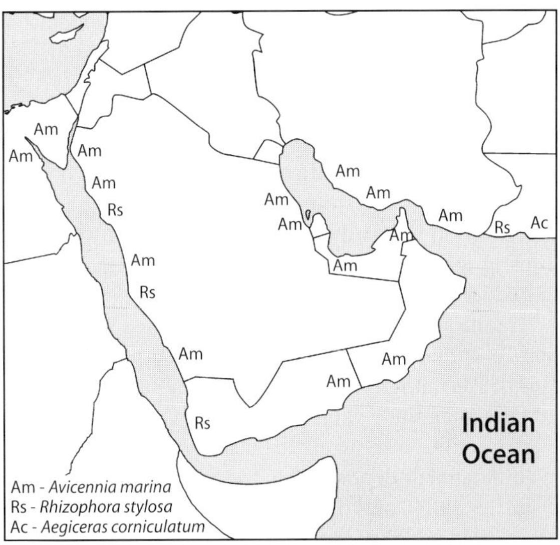

Fig. 3.1. Mangrove distribution along the Arabian Peninsula (adapted from Kogo and Tsuruda 1996)

and pilots still navigate the breadth of the Indian Ocean. For the past thousand years or so, Arab *dhows* and *felucas* sailed yearly from the southern Arabian coast to Zanzibar, from there directly to the south-western coast of India riding the SW monsoon. They then returned to south Arabia sailing with the NE monsoon. Much of the trade centred on mangrove timber.

The mangrove forests were not managed methodically, and excessive cutting caused impoverishment of much of the coastal area of what is now Tanzania and also caused degradation of the mangroves of the Arabian Sea. The mangroves of the coasts of Sindh and Baluchisthan, present-day Pakistan, and Gujarat in India have suffered from excessive cutting coupled with an arid climate. In the Karachi area, the mangroves suffered from natural shifting of the delta over the centuries and from reduced inflow of fresh water caused by barrages made across the River Indus upstream. The barrages were made to impound water to create reservoirs for power generation and agricultural irrigation. From 1955 to 1971, eight major dams were built reducing the river annual discharge from 101×10^9 m^3 to 12×10^9 m^3 (Qureshi 1996). Management of mangrove forests of Pakistan started with the Forest Act of 1957, which declared the two largest mangrove forests as "Protected Areas" and a Divisional Forest Officer was appointed to manage the mangroves. The "Working Plan of Coastal Forests", which was in fact the implementation of the management plan for mangroves, began in 1963 and was revised in 1985 (Ansari 1987). The management of the Karachi and Indus River delta areas was later expanded to also cover the coastal mangroves of the Sonmiani area and the entire Makran coast. Several national, bilateral and international projects have since worked in co-operation with Forest Department and IUCN/Pakistan. An estimate made some 15 years ago stated that there are about 10 000 fishermen along the Pakistani coastline who fish for prawns, shrimps, crabs, lobsters, and bivalve mollusks, both for domestic consumption and export. The catch comes from coastal and mangrove waters; directly or indirectly the bulk of the catch is mangrove dependent. In addition to this, the mangroves supply fuel wood for curing part of the catch, for instance shrimp; an average 0.03 m^3 of wood is needed to produce one kg of boiled shrimp.

The mangroves of the coastal lagoons of Baluchistan on the Makran coast receive fresh or nearly fresh water only from small, mainly seasonal rivers and not more than 10 to 20 cm of rainfall annually; nevertheless, these scattered groves, the largest forest with 20 m or more tall trees at Miani Hor, demonstrate that over many centuries it is possible to successfully have, use and manage the mangroves of arid coasts. Some species of mangroves seed themselves if the forest is not degraded and the main act of management is to avoid grazing by camels and buffaloes. Some degree of assisted regeneration is usually needed and practised, as for instance fencing, to protect from grazing. Details on the restoration and management of the mangroves of Pakistan are to be found in Qureshi (1996).

Traditional law and custom are observed also along the dry coastal area of Gujarat State in India, where an estimated 10 000 people live by selling mangrove wood alone (1987). The mangroves of this area are scrubby forests that never form a closed canopy, and are apparently valueless. However, in some areas firewood and tannin for curing leather come entirely from the mangroves. Tides are seven to nine metres tall, and there is abundant silt deposit. The customary law observed by all villagers is that only the branches and fronds that grow above high tide level can be lopped for fodder, and bark is removed without damaging the tree. Unfortunately, given human population pressure, even such wise laws are sometimes forgotten. A few years ago due to failure of the SW monsoon, fuel wood became so scarce that people had to cut many mangrove trees and even had to dig out the roots, though they were fully aware that there is no spontaneous regeneration in the area and that replanting is also a hazardous enterprise, because of the thick layer of soft muddy soil. Wherever the layer of mud is very thick, spontaneous regeneration rarely takes place, because the high rate of sedimentation chokes the seedlings before they have time to grow tall enough. Plantation of nursery-raised seedlings and saplings is possible, but equally difficult for the same reason. Planting seedlings on the convex shores of the creek or river meanders where sedimentation is very active is usually a wasted effort. Community level programmes were organised with some difficulty, because as a village elder said: "mangroves take too long, a few years to grow, and times have changed: "*Young people cannot wait, they go to the city for jobs for which they are not qualified, so we become poorer.*"

The other extensive dry coastal areas in the Asia-Pacific region are parts of the West and South coasts of Australia, though occasional drought conditions are likely to prevail on any particular part of the continent. There are few published studies of the reaction of mangroves to local droughts, and long-term monitoring of recovery after particularly dry spells appear to be absent. Bunt (1987) pointed out that 'many Australian mangroves respond to or depend as much upon inputs of fresh water by drainage from remote locations as upon rainfall.'

The relative importance of rainfall and river flooding for the welfare of mangroves is a subject in need of research. In fact, rainwater is usually almost pure water, while river water carries with it an array of particulate and dissolved organic and inorganic matter on which mangroves thrive, especially those of dry coastal areas. Nutrients brought to the mangroves by sea water are different from those brought by land runoff; the two types are marked by different C-isotopes and complete one another, a situation to which mangrove species are well adapted. This fact also contributes to the well-known high productivity of mangroves and coastal brackish waters in general. Mangrove plant species evolved from terrestrial plants, and it would appear that terrigenous substances are even now indispensable for the full development of mangrove vascular plant species. In fact, the mangroves of the Arabian Gulf receive abundant fresh water from the joint Tigris/Euphrates Rivers. It is interesting to compare the mangroves of the east and west coasts of the Arabian Peninsula. The seas on either side have high surface water salinity over 50 ppt, and sediment pore water may reach 100 ppt or more. Air temperature is equally high in the summer and low in the winter, down to 5 °C at Ras el Khafji, yet mangroves grow well in the Persian Gulf but do not grow beyond the stage of small groves of not too well developed *Rhizophora* and *Avicennia* in the Red Sea. One reason for the different response of the mangroves in the two seas could be the scarcity of river water input into the Red Sea, while the Arabian Gulf receives abundant, nutrient-rich waters from river discharge.

Australian mangroves are least developed on the West coast between latitude 20° S and 25° S. There are no mangroves along parts of the dry southern coast, though mangroves are present further south in the better-watered Melbourne and Adelaide areas. There is no river discharge on the dry Australian coasts; normal vegetative functions are restricted in dry areas by excessive solar radiation, soil aridity, and night and winter low temperatures. For instance, *Avicennia marina* tolerates temperatures down to 8 to 10 °C if exposure does not last more than two or three days, even down to 5 °C if only for one night, as measured in the Arabian Gulf and the Indus River delta area. Absence of river discharge appears to be a more serious handicap than low temperature or high salinity. Some species tolerate salinity well above normal sea water, for instance: *Avicennia marina*, *Rhizophora mucronata* trees and saplings taller than high tide level tolerate sea water salinity up to 47 ppt and interstitial water salinity up to 90 ppt, but perish due to infestation by barnacles when the latter entirely cover the leaf surface of short seedlings drowned by intrusion of sea water during droughts. This happened in the Indus River delta following three years of SW monsoon failure in the late 1980s.

Though estimates vary widely, it appears that the mangroves of the Makran coast at present cover an area less than 80 km^2. The importance of mangroves of dry coastal areas is that they grow where nothing else will grow. Along desert coastal areas the only forests with tall trees are mangroves; some old trees of *Rhizophora mucronata* at Miani Hor may be 30 m tall, while the supratidal vegetation is low, scrubby and sparse as usual in dry saline soils. Nevertheless there has been a steady decline in the vegetation of the Makran coast since 1960, probably due to increasing human population pressure. Reforestation of arid coastal areas is not an easy task. Almost 40 years ago, Ansari, then Chief Conservator of Forests of Sindh Province, initiated the work of rehabilitation of the degraded areas of the Indus delta. Originally, there were eight endemic species of mangroves in Pakistan: *Bruguiera gymnorrhiza*, *Ceriops tagal*, *C. decandra*, *Rhizophora apiculata*, *R. mucronata*, *Aegiceras corniculatum*, *Avicennia marina*, and *Sonneratia caseolaris*; regeneration experiments began by matching the right species to the right environment, especially soil pH, winds, rainfall, browsing by camels and buffaloes, and other factors. Summing up, it was found that there could be spontaneous regeneration of *R. mucronata* and of *A. marina*. However, success or failure in the management of the mangroves of the particularly demanding conditions along the northern margin of the Arabian Gulf depend primarily on the understanding and support received from the local people, the great majority of whom are fishermen. The greatest hazard is browsing by camels, particularly fond of seedlings and saplings, also excessive lopping of branches of *Avicennia*, winds and cyclonic storms, droughts and failure of river discharge. One of the main concerns with promoting natural regeneration or rehabilitation of dry coastal areas is protecting the area with barbed wire fences to exclude camels, buffaloes, goats or other herbivores from the area; growth of *Porteresia coarctata* and other grasses is to be favoured; it conditions the soil and prevents seeds and small propagules, like those of *Ceriops*, to be washed away by tides.

Small natural regrowth areas may result from seeds and propagules brought by the tides; if assisted with barbed wire fencing as protection against camel browsing and with old fishing nets or grasses to avoid tidal wash, patches develop well as small groves. One such patch that formed spontaneously near village Dam at Miani Hor was closely monitored since its beginning and is now several years old and tall, but participation of the local community was essential to save the seedlings and saplings from camel browsing. Qureshi (1999) emphasises the importance of social forestry for the success of management practices, especially important for arid coastal zones.

In conclusion, mangrove forests of arid coastal zones can grow and thrive if extraction does not exceed the turnover rate, or natural rate of production. Positive environmental factors are: upwelling conditions at sea, abundant equatorial and monsoon-induced precipitation, presence of permanent or seasonal rivers (*wadi*), sheltered areas such as coastal lagoons, and gentle slope of the continental or insular shelves.

About 50 km^2 have been planted during the past few years in the Indus delta alone. Details of management of the dry coasts of Pakistan can be found in Qureshi (1996), whereas Untawale (1986) described the arid coast mangroves of the State of Gujarat in India and Al-Muharrami (1994) and Fouda and Al-Muharrami (1996), those systems being dominated by socioeconomic factors.

3.4 Estuarine and Deltaic Mangrove Ecosystems

Estuaries and deltas have several characteristics in common; every single distributor of a delta functions as an estuary in itself. First and foremost, estuaries and deltas receive abundant river discharge from land runoff upstream and coastal seawater from tidal fluctuations. The brackish waters of estuaries and deltas world-wide are highly productive, especially in the tropics and subtropics. The reasons for the high productivity of estuarine waters, if they are not polluted by urban or industrial effluents, are: First is the mixing of waters of different origin, both from land and sea; these waters bring with them particulate and dissolved organic matter of different kinds, which as nutrients complement one another. Second, the brackish waters that come and go with the tides and eventually run out to the sea reside a long time, especially when and where the temperature is high, as is usual is in the tropics. Brackish waters can be compared to well-brewed nourishing broth where autotrophic and heterotrophic organisms grow and reproduce unhampered. The flora and fauna of tropical and sub-tropical brackish waters are the same world-wide, down to species, genus or family level. Environments being comparable, there are by and large parallel species or groups of species in the Indo-West Pacific brackish waters and in the Atlantic Ocean.

The major handicap for greater productivity of brackish waters is the amount of silt they carry. Suspended matter can be abundant to the point of reducing light penetration to a few centimetres or even less than one centimetre of the surface layer. Thus, the mangrove forest is by far the major primary producer of the mangrove ecosystem;

Table 3.1. Comparison of biomass aboveground and estimated carbon stock

Forest type	Biomass above ground ($t\ ha^{-1}$)	Estimated carbon stock ($t\ ha^{-1}$)[a]	References
Tropical rain forests	444.0	222.0	Tsutsumi (1990)
Temperate evergreen broadleaf forests	356.0	178.0	Tsutsumi (1990)
Temperate deciduous broadleaf forests	300.0	150.0	Tsutsumi (1990)
Sub-boreal forests	200.0	100.0	Tsutsumi (1990)
Crop fields	11.0	5.5	Tsutsumi (1990)
Freshwater swamps	150.0	75.0	Tsutsumi (1990)
Japan mangrove forests	102.7 – 203.3	51.4 – 101.7	Navasota (1979)
Indonesia mangrove forests	169.1 – 436.4	84.6 – 218.2	Komiyama et al. (1987)

[a] Composition of tree trunk; carbon, appr. 50%; hydrogen, appr. 6%; oxygen, appr. 44%; nitrogen, appr. 0.5%; ash, 0.3–1% (from Fujimoto, unpubl. Courtesy S. Baba).

Table 3.2. Rate of sedimentation and carbon accumulation in selected mangrove forests (data from Kiyoshi Fujimoto, unpublished, and S. Baba, personal communication)

Location	Duration of accumulation (^{14}C yr B.P.)	Depth of horizon (cm)	Rate of sedimentation (mm yr^{-1})	Carbon accumulation (t C ha^{-1} yr^{-1})	Remarks
Iriomote Island (Japan)	1 000	100	1.00	0.22	
Pohnpei (Micronesia)	2 620 – 1 800	129	1.57	0.55	Period of sea level sink
	1 800 – 1 380	83	1.93	0.93	Period of sea level rise
	1 380 ± 47	117.5	0.82	0.53	Sea level relatively stable
Southwest Thailand	710 ± 47	164.0	2.17	1.05	*Xylocarpus-Rhizophora* mixed forests
	1 310 ± 47	84.0	0.62	0.32	*Rhizophora* forests
	2 900 ± 46	147.5	0.50	0.25	*Rhizophora* forests

analyses of sediments show carbon content to be many times greater in the forest substratum than in the open water sediments. Tables 3.1 and 3.2 show carbon stock in the biomass of different mangrove forests and sediment and carbon accumulation rates, respectively.

Reduced light penetration in the open waters inhibits photosynthesis as well as number and variety of photosynthetic organisms; for this reason, open waters in the mangrove brackish waters often sustain a relatively large quantity of chemiosynthetic organisms and nanoplankton rather than phytoplankton. The minute organisms that constitute the nanoplankton are mainly unicellular green algae and blue-green Cyanobacteria and auto-heterotrophic flagellates capable of synthesising organic matter alternately by photosynthesis or by dark fixation of carbon, according to light intensity in the water. The production of organic matter through cycling and recycling of nutrients is an endless process in estuaries and deltaic waters, intensified by tide-induced convection cells, frequent in tidal dominated estuaries. Mangroves normally grow on the tidally flooded coastal plain and along river and creek banks as far inland as the reach of the tides. *Nypa fruticans* is the species that usually extends further inland in undisturbed mangroves.

Estuaries and deltas may be under a variety of environmental conditions: different tidal regimes and amplitude, different climates, and different physical and chemical conditions of the substratum; yet mangroves have adapted to all varieties of situations, within the limits of low temperature tolerance. Though species of rooted mangrove vascular plants vary from place to place, they are found only in the intertidal zone. Topographic and area distribution of species changes over time as a consequence of geologic, topographic, and climatic changes as well as mean sea level changes. But above all, the mangroves' species composition changes as a result of human interference. Mangrove ecosystems are often affected by forces that take place very far from the coastal plain, such as construction of dams, mining operations, snowmelt and precipitation on the higher reaches. The barrages on the Indus River have drastically changed the physiography of the Indus delta area; the mining operations for copper in the Fly River have caused considerable changes in the mangroves of the western reaches of the Gulf of Papua, (Toto Rau, personal comm.).

The tropical belt of the Indo-West Pacific Region has a countless number of small, medium size and large rivers; the greatest majority of them discharge into the sea after flooding mangrove swamps. In the northern part of the Bay of Bengal, all around the Malay peninsula, the Gulf of Thailand and all the way eastwards through the Java Sea and northward up to the southern China Sea, which roughly goes from the Equator to about latitude 6° N, the seas are shallow, less than the 200 m deep, while the lands behind them are mountainous. Rivers are plenty and water abundant mangroves are noticeable everywhere for their biodiversity, good development area wise and for the height of the trees. The main deltas and estuaries to be considered are: the Tigris-Euphrates estuary, the Indus delta, the Narmada River estuary, the Ganga-Brahmaputra joint delta, the Irrawaddy delta, the Mekong River delta, and the Fly River wide estuary in Papua New Guinea.

The Tigris and Euphrates Rivers had separate outlets at the time when Nearchus, in the fourth century B.C., coming by sea from the Gulf searched for the mouths of the two rivers. Nearchus and his pilots described the very large barren mud-banks that hid the actual mouth of the rivers and only at great pain they navigated upstream through the rivulets. No vegetation grew on the muddy banks. To this day 2324 years later, the coast continues to grow seawards in height and the estuaries of the two rivers have merged into a single one, the Schatt-al-Arab. Alluvium is deposited in such great quantities that seeds, propagules, and seedlings have no time to take root and grow before becoming irretrievably silted in. An ecological lesson: too much of anything, including favourable elements such as silt for mangrove spread, are an ecological hazard. The northern margin of the Arabian Gulf offers the contrast of a barren coast without mangroves but with abundant alluvial deposits bordering a rich agricultural hinterland. Developed mangrove forests grow further south along the western margin of the Gulf, in areas almost completely devoid of alluvium, seawards of a desert hinterland, where there is less silt deposition. Nutrient rich waters water both the environments, though at present they are polluted by crude oil. In the Arabian Gulf, and elsewhere, it is the nature of the waters, and the presence or absence of alluvium, which determine the presence or absence of mangrove forests. The growth and development of the trees and shrubs, which constitute the forest, are brought about by the quality of the alluvium that forms the sediments as well as the nutrient content of the brackish waters.

A somewhat comparable situation is that of the Narmada River estuary on the west coast of India at about latitude 30°21' N. The tides there have amplitude of a few metres, and alluvial deposits have built islands of soils compacted by mangroves and other natural processes. The largest of the islands is called Aliabet. The mangroves were browsed upon almost to extinction in recent times by herds of camels that annually swim to the island at the time of dropping their foals. Few shrubby mangroves remain, but a species of very nutritious wild rice, *Porteresia coarctata*, which normally grows in the intertidal zone seawards or landwards of the mangroves, has taken over much of the island. The camels and the tribals who own them continue to visit the island annually. The high layer of sediments and the high tides prevent natural regrowth of mangroves, and afforestation is difficult under these conditions.

The Indus River basin drains the water from the Karakoram and the western Himalaya ranges and discharges into the Arabian Sea after crossing the fertile agricultural lands of the Punjab. Descriptions of the delta have come to us from the fourth century B.C. and from the reports of Mohammad-bin-Qasim who arrived there over a thousand years ago. Trade between India and ancient Egypt, Sumer, Greece, and Rome was active since several centuries B.C.; the ports of Shah Bundar and Keti Bundar were recorded to be on the seafront in the first century. They later became nodal points for trade with the Arabs, especially from Yemen and Oman, who navigated also along the East coast of Africa as far south as Sofala at almost latitude 20° S on the Mozambique Channel. The mangroves provided poles, fuel wood, charcoal, timber, tannin, fodder and fish for the active trade throughout what was then called the "Erythrean Sea", which included the Red Sea, the whole Arabian Sea and most of the northern Indian Ocean. Details may be found, amongst others, in Wilfred H. Scoff's (1911) detailed and extremely interesting translation and notes of the "Periplus of the Erythrean Sea". In more recent times, the use of mangrove wood for boiling shrimp in the delta of the Indus River may be one of the causes of the large gaps in mangrove cover that are at present under restoration, at the cost of much time and patience. Both ports, Shah Bundar and Keti Bundar were important ports of call for seagoing and upriver navigation; remains of both can still be seen emerging from the mud. Other settlements have disappeared but would have existed in these areas. The mangroves had been felled to satisfy the needs of the settlements, but there was no spontaneous regeneration of importance in the cleared areas. Reforestation programmes have recently been met with success (Ansari 1987; Kogo and Tsuruda 1996; Qureshi 1990, 1996) though there still are extensive areas to be reforested.

In 1819 disaster struck, and damaging floods from the river destroyed the sun-dried brick-built cities, which melted away and were carried by the floods. The delta shifted its course many times in the area that goes from present-day Karachi to Siri Creek in the East at the border with India, which is the major part of the coastal area of Sindh Province. Later, when the colonial government built railways, mangroves were cut to fuel the locomotives; since *Rhizophora* wood has a higher calorific power than the wood of *Avicennia*, it was cut in preference to *Avicennia*. Still more recently, more mangroves were sacrificed for the construction of Port Qasim, which began in 1977. The result was that whatever mangroves remain in the former delta area of the River Indus are in most places forested with the single species *Avicennia marina*, and this too is often in a degraded condition, due to over-lopping of fronds used as fodder for cattle, goats and camels. Between 1955 and 1971, eight barrages were built at different places along the Indus River upstream; the annual discharge decreased from the earlier 101×10^9 m^3 to 12×10^9 m^3 from 1992 onwards. The average sediment load was originally about 400 million t yr^{-1}; it decreased to 35 million t in 1992 and thereafter (Qureshi 1996). The dams were built for power generation and irrigation of agricultural lands upstream; the Indus River was reduced to a trickle and the delta area saw the decline of agriculture, the death of mangroves, increased salinity, and intrusion of high-salinity oceanic water. The negative impact of these combined factors is particularly noticeable when the SW monsoon fails. Advancing seawater brings larvae of barnacles that would not survive in brackish waters, and when they develop into full-grown barnacles, they may totally cover the leaf surface of mangrove saplings, preventing photosynthesis, with consequent death of the young plants that are not yet tall enough to rise above high tide level. Recently replanted areas may thus be totally destroyed (personal observation).

Following destructive episodic events such as floods or failure of the monsoon locally and upstream, aggravated by anthropogenic impact locally, at present only three of the original eight species of mangroves present can be found in the delta. They are: *Avicennia marina*, *Ceriops tagal*, and *Aegiceras corniculatum*. The other species known from Pakistani coasts are: *Bruguiera gymnorrhiza*, *Ceriops decandra*, *Rhizophora apiculata*, *R. mucronata*, and *Sonneratia caseolaris*. Very negative socioeconomic consequences are serious because of the decline in fisheries production, urban migration of coastal dwellers, loss of traditional use, and management practices and skills. Destruction and degradation of mangroves result in decreased coastal protection and accretion, decreased fisheries production, and the silting of channels and canals that access the two main ports of the country: Karachi and Port Qasim; with the result that the expense of dredging to keep the channels clear are yearly on the increase.

The largest deltaic system in the world is formed by the combined discharge of two major rivers: the Ganga and the Brahmaputra. The basin of the river Ganga lies eastwards of the eastern-most affluent of the Indus River, the Sutlej. The Ganga and its tributaries drain the southern slopes of the Himalayas, while the Brahmaputra drains the northern slopes before turning south and entering India. The combined amount of water and silt carried by the two rivers had built over the millennia an immense low-lying plain with innumerable inter-connected waterways, channels, rivers, creeks, and canals, already described by Pliny the Elder (A.D. 23–79), who gives testimony of the trade of Greeks and Romans with the Gangarides, meaning the people of the Ganga River. There was a flourishing trade between the local population and Arab, Egyptian, Greek and Roman visitors. Over 20 ports and Roman *empori* are known from archaeological remains along the East coast of India alone. There are historical records of transport of goods from India and further east to the western Indian Ocean and to the Red Sea. The famous "Periplus of the Erythrean Sea" is only one of the documents, which specified goods and sometimes also the duties to be paid to the Romans and Arabs. The Greeks knew the SW monsoon wind already by the name of the mariner who first made it known to the Mediterranean pilots; it was called "*Hyppalos*". The name "Erythrean Sea", which means Red Sea, included all the Indian Ocean as known at that time, especially the Arabian Sea and the western Indian Ocean as far south as Sofala and perhaps further south. The latitudes of cities and ports are given, since latitude at that time could be determined within 10' to 20' of precision. The longitude between the cities that were on the coastline at that time, all at roughly the same latitude, was measured in *stadia* (roughly 200 m), which was the measure of the distance between two places, used at that time. The tropic of Cancer crosses the delta near present-day Dhaka, and the measure of latitude is particularly easy and accurate on the tropical line, especially with the instruments available at that time. Using these points of reference, it would be possible to estimate the amount and speed of accretion by sediment deposition, though tectonic, climatological and anthropogenic forces induce frequent changes in the geomorphology of the area.

The easternmost part of the delta opens into a wide-open bay that also receives the waters coming from the catchments area of the Chittagong Hills. The Sundarbans are the mangrove forests that occupy the seaward area under tidal influence. They are a highly dynamic system; because of the size of the area, the large amount of river water with great sediment load, the abundance of monsoon waters, the advance of tidal waters up to 40 km inland, the frequency of cyclones, the tidal surges and damaging winds, and tectonic movements. Biodiversity of plants and animals is very high and includes large mammals such as the Royal Bengal Tiger and other cats, otters, two species of dolphins, deer, wild boar and even rhinoceros and wild buffaloes in the upper reaches. Another peculiarity of the Sundarbans is the large resident human population, in addition to occasional or periodical visitors. In the eastern Sundarbans of Bangladesh alone there are now more than three million residents who live directly or indirectly from the forest.

The mangrove species composition and distribution in the Sundarbans is influenced by two factors: first, low salinity of Bay of Bengal Surface Waters, which is further diluted by river discharge and abundant monsoon rains. Second, the mean water level in the forest varies by one meter or more, according to the place and season; it is higher during the SW monsoon. Accordingly, the inundation regime in the inner reaches of the Sundarbans is seasonal rather than diurnal. Only those species which tolerate months of flooding alternating with months of exposure can become established, such are for instance *Excoecaria agallocha* and *Heritiera fomes*. When the land is underwater, sedimentation of silt is intense, and gradually the land rises above tidal level. Eventually it becomes dry land well washed by fresh water and suitable for agriculture. Third, the area is immense and differences from one place to another are significant and may be very great. The combination of seasonality of rains and high tides also contributes to abundant deposition of silt above mean high tide level and to a varying species composition. In the lowest reaches of the delta where tides are semi-diurnal, the islands are continuously eroded at the landward end while there is a continuous deposition of silt at the ocean end. In this way, the islands grow towards the ocean by accretion at the seaward end; they become 'walking islands' slowly moving towards the sea like giant amoebas. In fact, when seen from the air, they call to mind real *Amoeba limax* due to their slow displacement and their shape. This type of elongated island is also common in other major deltas of the world, but these islands are almost regular in size, shape and displacement in the delta of the Ganga-Brahmaputra system. All these and more peculiarities have created traditional laws and customs that are unique to Bengal; for instance, how can one compensate the fishermen, honey and fuel wood collectors or agriculturists whose houses and land are eroded away at the landward end? Who owns the accreted land? Who pays for its management? Community management of the mangrove ecosystem is vital for survival in the Sundarbans for the simple reason of its fluidity. More than anywhere else what a man, family or clan does will affect all others; were there no mutual understanding and consensus over major issues for the use and management of the system, it would have fallen to pieces long ago. Present-day threats to the longevity of the Sundarbans will be mentioned later.

Deposition of silt accumulates and increases the height of the river banks, what still remains as suspended matter when river water flows into the sea, is deposited and forms semi-fluid mud banks locally called *chars*. These can be consolidated by mangrove plantations using a technique well developed in Bangladesh to achieve land accretion. The indicator that the new islands are well consolidated is the rooting and spreading of the grass *Porteresia coarctata*, which is an edible variety of wild rice. It takes root spontaneously and develops grass beds on the margin of the mangrove plantation proper. As land accretion progresses seaward, the mangrove vegetation disappears in the centre of the islands that have become saucer shaped due to deposition of mud on the banks. As the mangroves disappear because the tidal flow no longer reaches the centre of the island, grasses take over and large herbivores like cattle and wild deer can graze peacefully until tigers swim to the island to hunt them for food. At the same time on a much larger scale, the Sundarbans as a whole turn into a fertile agricultural plain landwards, above the highest high tide level, while the delta as a whole continues to grow seaward. Spontaneously, land accretion goes on uninterrupted, especially in the eastern Sundarbans, for instance in the districts of Noakhali and Patuakhali. The greatest extension of the Sundarbans is about 200 km East to West; the tilting of the Indian tectonic plate favours shifting of gangetic waters towards the East, where there is greater deposition of less saline soils; monsoon rains further aid in washing the soils free of salts and sulfides, and this explains why, unlike former mangrove soils elsewhere, agriculture can be practised after the islands have been consolidated. It is common to see vegetables planted in between pneumatophores of *Heritiera* during low waters of the inter-monsoon period in the upper reaches of the Sundarbans or near to abandoned shrimp ponds. However, in the lower reaches of the delta, soils of newly felled mangrove forests can develop into acid sulfate soils, which here as elsewhere are not suitable for agriculture or shrimp farming.

The resident population of the entire Sundarbans including Bangladesh and India has in recent years grown to well over more than 3 million, and over-exploitation of resources is now on the increase. Human population competes also with wildlife for grasses, honey, fish and others, with losses at the wild animals' end. Along with increased salinity of the waters because of the diversion of freshwater upstream, the changing environment may be a cause of changed patterns of species composition. For several years there has been a heavy loss of *Heritiera fomes* due to a condition called "top dying", caused by as of yet poorly known factors associated with increased salinity that may impair the resistance of the trees to parasitic diseases. Human impact is particularly felt at the margins of the forest: in olden days the landward changes from mangroves to other species association was gradual, while it has now become abrupt where new areas are brought under agriculture or shrimp farming. Seawards, abrupt changes are brought about by sea walls and destructive intensive aquaculture. Both practices are to be condemned. Nature does not adapt easily to abrupt, drastic changes. Many islands where mangroves had been totally felled, for instance for shrimp farming, have been washed away to sea by cyclones or tidal bores, especially in the Chittagong Sundarbans. The disastrous cyclones that hit the East coast of India in the months of October/November 1999 have demonstrated in a tragic, practical manner, the efficiency of the protective role of the mangroves against the fury of nature.

Summing up, the present condition of the joint deltas of the Ganga-Brahmaputra system is in general not bad; management has been practised in one form or another over the past several centuries. The mangroves of India have been under some form of rational or non-rational use or protection for centuries, perhaps millennia. There are sacred areas, from which nothing can ever be removed. In the Sundarbans, tradition requires all men and women who enter the forest to seek the permission of the presiding deity to collect whatever they wish to collect, from honey and wax to animals for food or other. The Sundarbans have been managed by the Forest Departments of India and Bangladesh with success for the past 120 years, more or less. Nevertheless, problems arise from natural changes over time and from human impact, which has been particularly intensive in the nineteenth and twentieth century. Extensive to mild degradation has occurred in some areas of large estuarine and delta mangroves, induced either by natural episodic events such as cyclonic storms, tidal surges, climatic or hydrological changes in the river basin, or induced by man-made changes in the environment.

Whatever the causes, environmental changes have been growing in frequency and area destruction; increasingly rapid changes may have caused almost irreparable damages. On the other hand, countries have seen the wisdom of rational use and management of mangrove ecosystems. Reforestation, afforestation and re-afforestation practices have generally met with reasonable or full success, provided experimental methodology is combined with traditional wisdom and the communities of coastal dwellers participate in the process. On the other hand, destructive practices like the total felling of mangroves for whatever purpose should be proscribed by law. Afforestation or reforestation with appropriate species is known within site-specific environmental constraints; reforestation should be encouraged by all means, especially in those areas where without mangroves, there would only be deserts. The same applies to areas degraded by man or in accretion areas, which, wherever possible, should be put under cultivation or productive forestry practices. Where silt deposition is not too great, where river discharge is sufficient and the gradient of the coastal plain is gentle, mangroves will most probably compensate mean sea level rise and advance landwards at the same pace as the sea progresses. Finally, attention should be given and more studies done on the transition area between the mangrove forest and the supra-littoral vegetation, which survives exceptional and occasional high tidal inundation.

3.4.1
Mangroves of Bangladesh Sundarbans and Accretion Areas

Neaz Ahmad Siddiqi

Both natural forests and plantations represent the mangroves in Bangladesh. The natural forests include the Sundarbans and the Chokoria Sundarbans. The word Sundarbans means "beautiful forests". The plantations have been developed on new accretion areas along the coastline and offshore islands. The Sundarbans Forest is a highly productive ecosystem that immensely benefits the people living in the vicinity of the forest through the direct supply of various resources. A substantial proportion of the forest income is derived directly from the Sundarbans, which provide employment to about 300 000 people. Apart from the direct and indirect contribution of the Sundarbans to the forestry sector, they are noted for their protective role against cyclones and tidal surges. In fact, the protection offered by the forest against recurring natural calamities led the Forest Department to establish mangrove plantations in the coastal areas in order to minimise the loss of life and property of the coastal population. Figure 3.2 shows the coatline of Bangladesh and the location of major natural and planted mangrove forests. The natural mangroves have come into being through a process of evolution over a long time, and man-made plantations have been in the process of becoming established over the past three decades. The characteristics, nature of resources and management problems of the ecosystems vary considerably between the natural forests and the mangrove plantations. For clarity and convenience, the natural forests and plantations are discussed separately. Only plant resources are dealt with here.

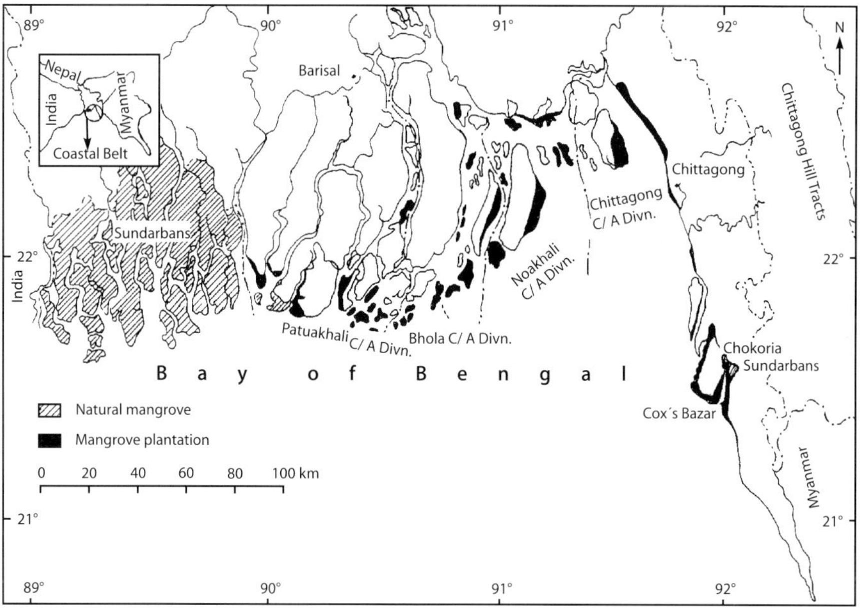

Fig. 3.2. Coastline of Bangladesh showing the location of natural and planted mangroves

3.4.1.1
Natural Mangrove Forests

The Sundarbans comprise an extremely flat deltaic swamp. It is the largest continuous mangrove forest in the world, spreading over an area of 10 000 km^2. Of the total area, 62% lies in Khulna Division of Bangladesh, while the remaining 38% is in India. The Sundarbans of Bangladesh lie between latitudes 21°31' N and 22°30' N and between longitudes 89° and 90° E (Fig. 3.2). Rivers and canals criss-cross the entire forest area. The salinity of the area varies greatly; it is higher towards the west and south. Unlike many mangroves of the world, the Sundarbans are rich in floral and faunal diversity. The forest has about 70 species, including high economically important trees, shrubs and palms (Table 3.3). Of the available plants, approximately 55% of the species are exclusive mangroves, while the remaining 45% are obligate mangroves. *Heritiera fomes* and *Excoecaria agallocha* are the dominant species. The height of the forest generally varies from 15 metres in the east to 5 metres in the west. In regards to fauna, the animals are rich and diversified. The aquatic environment also supports plenty of commercial fish, shrimps, crabs, gastropods etc. The insect fauna is also varied.

Mangrove Resources. Timber wood is an important source of forest revenue and contributes over 80% of the income generated by the Sundarbans. A number of industries including Khulna Newsprint Mill, Khulna Hardboard Mills, match factories, saw mills and boat building industries are greatly dependent on the Sundarbans for the supply of raw material. Based on an old survey, it appears that the yield of the forest is

Table 3.3. List of economically important plants of the Sundarbans and their uses (Das and Siddiqi 1985)

Family	Scientific name	Type of plant	Main uses
Avicenniaceae	*Avicennia officinalis*	Tree	Fuel wood, anchor, logs
Combretaceae	*Lumnitzera racemosa*	Small tree	Fuel wood, posts
Euphorbiaceae	*Excoecaria agallocha*	Tree	Matchsticks and boxes. Raw materials for news-print and other papers
Leguminosae	*Cynometra ramiflora*	Small tree	Fuel wood, charcoal
Malvaceae	*Xylocarpus granatum*	Tree	Furniture
	X. mekongensis	Tree	Furniture, bridges and house construction
Palmae	*Phoenix paludosa*	Thorny palm	Post, rafters for huts
	Nypa fruticans	Recumbent palm with underground stem	Thatching for houses
Rhizophoraceae	*Bruguiera* spp.	Tree	Furniture, bridges and house construction
	Ceriops decandra	Shrub or small tree	Fuel wood, houses, posts, charcoal
Sonneratiaceae	*Sonneratia apetala*	Tree	Packing boxes, paneling fuel wood
Sterculiaceae	*Heritiera fomes*	Tree	House and boat building, electric poles. Hard boards

$1.12\ m^3\,ha^{-1}\,yr^{-1}$ (Davidson 1984). This is almost 6/7 times lower as compared with the high production of mangroves of Malaysia and Thailand. However, it may be noted that when the price of the wood resources is taken in consideration, the income is not low. Highly valuable timber yielding species like *Heritiera fomes*, *Xylocarpus mekongensis* and *Bruguiera sexangula* occur in the Sundarbans in good proportion.

The Sundarbans are an important source of fuel wood for the local population as well as for distant markets. Species used as fuel wood mainly comprise *Ceriops decandra*, *Cynometra ramiflora*, *Amoora cucullata*, and *Hibiscus tiliaceus*, which thrive in the under story of the forest. Fuel wood is also obtained from the branches of taller trees like *Heritiera fomes*, *Avicennia officinalis*, *Sonneratia apetala*, and others. Even the bark of *Excoecaria agallocha* is used as fuel wood. *C. decandra*, which is abundant in the western part of the forest, is a good quality fuel wood with high calorific value.

Leaves of *Nypa fruticans* are used extensively as thatching material for the construction of huts in the coastal area. They are harvested once yearly. *Phoenix paludosa* is the other species of the family Palmae; its stems are used as house posts, jetties and rafters. There are three species of grasses; since they are annual plants, harvesting is done every year to avoid loss. The bark of *Ceriops decandra* is rich in tannin. The locally extracted tannin from the bark is used for dyeing fishing nets.

Honey and wax are harvested yearly; they are made by the honey bee (*Apis dorsata*). Approximately 185 000 kg of honey and 44 400 kg of wax are harvested annually (Hussain and Ahmed 1994).

In the Sundarbans, there are 49 species of mammals, 282 of birds, 63 of reptiles and 10 of amphibians (Rashid et al. 1994). Hendrichs (1975) mentioned 350 individual tigers (*Panthera tigris*), 80 000 spotted deer (*Axis axis*), 20 000 wild boars (*Sus scrofa*), 40 000 Rhesus macaques (*Macaca mulatta*), and 20 000 Smooth Indian Otters (*Lutra perspicillata*). The Sundarbans ecosystems support aquatic mammals, both dolphins and porpoises. The Gangetic Dolphin (*Platanista gangetica*), which occurs in some of the larger waterways, is quite commonly seen. The Irrawaddy Dolphin (*Orcaella brevirostris*), Little Porpoise (*Neomeris phocaenoides*), Malay Dolphin (*Stenella malayana*), and Pilot Whale (*Globicephala macrorhyncha*) are not common. Estuarine crocodiles (*Crocodylus porosus*) in the rivers are still quite common, although their population size has drastically declined in recent years. The beauty of the forests and their diversified floral and faunal composition are ideal for the development of ecotourism, but required facilities have not yet been developed in the Sundarbans.

Management. The Sundarbans have been managed over a period of 120 years. Alternation in the felling cycle, inclusion of more species for management and adjustment of exploitable diameter have been made during this period. The last working plan came into being for the period from 1960 to 1980. The forest was intended to be managed on a sustained yield basis under a selection-cum-improvement felling system with a 20-year felling cycle (Choudhury 1968). However, there has been a considerable decline in the production of forest resources.

In the absence of exact data on the increment, it is difficult to determine the allowable annual cut. A new working plan is yet to be prepared. The forest is still operated under the previous plan. Several interim prescriptions have been put forward and a temporary moratorium on timber harvest with some exceptions was imposed recently (Chowdhury and Ahmed 1994). The Sundarbans, being a mangrove forest, are a relatively unstable ecosystem. Changes of various aspects of the abiotic and the biotic environments are constantly occurring. Many changes are unpredictable and beyond the control of the foresters. Apart from this, human induced stresses on forest resources in a poor and densely populated country like Bangladesh are unavoidable. Under the prevailing socioeconomic conditions of the country, sustainable management of different resources of the Sundarbans ecosystem is not an easy task. It is quite difficult to strictly follow the management prescriptions. Ultimately, the result is the decline of forest resources. Some important problems affecting sustainable management of the Sundarbans are discussed below.

Geomorphological changes. Erosion and compensatory accretion are occurring continuously along the bank of large rivers, but rivers of the Sundarbans are more stable than the main stream of the Ganges or the Brahmaputra further east. The erosion and accretion balance in the Sundarbans has been estimated to be 145.30 km^2 for the period 1960–1984 (Jabbar et al. 1992). When there is erosion along the sea face, new islands are also emerging in the waterways. Erosion along the riverbanks causes the disappearance of matured and valuable stands, in other words, loss of forest cover. On newly formed accretions, it takes time to develop forest crops, particularly those with commercial value, which appear at later stages of succession (Das and Siddiqi 1985). Even large-scale engineering works cannot prevent this unstable situation.

Deposition of silt is another factor that affects vegetation. On the western part of the forests, silt deposition is low and the forest floor is compacted and does not support vigorous tree growth. Again, too much deposition of silt in the northeastern part of the forest poses a threat to the existence and continuity of mangrove vegetation. Deposition of silt causes a rise of the forest floor and due to irregular flow of tidal water, mangrove regeneration does not take place (Siddiqi 1986). These areas are usually covered with some non-mangrove and commercially unimportant vegetation.

Salinity increase. Generally, members of the family Rhizophoraceae are the climax species of the mangrove forests. However, in the Sundarbans, *H. fomes* is the climax species; it has preference for low-level salinity. Due to the increase in salinity, *H. fomes* is no longer common in the Indian Sundarbans. The forest does not support healthy stands of this species, and regeneration of *H. fomes* is also poor. In Bangladesh Sundarbans, the water salinity varies between 5 ppt in the east and 30 ppt in the west. Healthy stands of *H. fomes* are well distributed in areas with low and moderate levels of salinity. Fresh water flow has been diverted upstream in the Indian Bengal, where a dam was built at Farakka to impound Ganges river water. It is reported that this results in the increase in salinity level of the waters of the forest (Shafi 1982a). The commercially important species of the Sundarbans have preference for different levels of salinity for survival, distribution, reproduction and optimal growth. Naturally, any change in the level of salinity will affect the standing vegetation, the productivity of the forest and the entire ecosystem.

Ecological succession. *Heritiera fomes* is the climax species of the forest and alone contributes to about 65% of the standing volume of merchantable timber. Due to siltation, increasing salinity or other environmental factors, the species is under stress. Almost 17% of the stems of *H. fomes* are affected. The future of this species is uncertain.

Next to *H. fomes*, *E. agallocha* is the most abundant species, which is used as raw material for newsprint, match sticks and boxes. Long-term records of data indicate that *E. agallocha* alone constitutes 54% of the seedling recruitment (3 months old) in the permanent sample plots (Siddiqi 1994). Latif et al. (1992) reported that the rate of ingrowths for *H. fomes* was 15 trees per hectare per year, while the natural death and removals were 27 trees per hectare per year. The corresponding values for *E. agallocha* were 22 and 8, respectively. The data suggest that the proportion of *E. agallocha* was increasing and *H. fomes* was decreasing. There is a need to study in detail whether a process of retrogression has started, which may cause the replacement of a highly valuable species like *H. fomes* by a comparatively less valuable species.

Mangrove regeneration. Plenty of seedlings of different species appear every year on the forest floor. Siddiqi (1994) reported 27 750 new recruits (3 months old) per hectare per year. However, recruitment density varied considerably in different parts of the forest. Besides, most of the seedlings soon disappear from the forest floor. Even 5% of the recruited seedlings did not last up to a period of 33 months. Failure of the seedlings to survive and establish themselves is a problem.

Chaffey et al. (1985) recorded 29 000 seedlings (height < 1.3 m) for the Sundarbans based on data collected from the temporary sample plots. According to them, the proportion of seedlings, saplings and poles was 69.8, 25.3 and 4.9% respectively (Table 3.4). It appears that regeneration was satisfactory for the greater part of the forest to ensure a rich stocking for the future, other factors remaining favourable.

Table 3.4. Regeneration status (%) (seedling, sapling, and pole) in the Sundarbans (after Chaffey et al. (1985) (average number of seedlings, samplings and poles was 29 090 ha^{-1})

Regeneration		*Heritiera fomes*	*Excoecaria agallocha*	Others species	Combined species
Seedlings	(Ht < 1.3 m)	71	85	42	70
Saplings	(Ht > 1.3 m, dbh < 2.5 m)	26	10	50	25
Poles	(Ht > 1.3 m, dbh 2.5–4.9 cm)	3	5	8	5

Table 3.5. Decrease in vegetation density based on canopy closure as apparent from different inventories

Year	Crown closure (ha)	
	<70%	>70%
1959 (Forestal 1960)	81 829 (22%)	295 748 (78%)
1983 (ODA 1958)	140 337 (35%)	255 177 (65%)
1996 (FRMP 1997)	409 945 (99.8%)	660 (0.2%)

The fact however, remains that canopy closure of the forest has decreased between 1959 and 1983 and that 65% of the forest was classified as having a canopy closer to 70% or more in 1983. There was a further alarming situation in 1996 as shown in Table 3.5 (Palis 1998).

Timber production. The yield of the Sundarbans is low in terms of wood production. There are sizeable areas which are less or poorly productive areas in different parts of the Sundarbans. Moreover, stocking and regeneration are inadequate in some parts of the forests. Bringing the vacant areas under vegetation cover through artificial regeneration by planting might improve the stocking and consequently the productivity of the forest. *E. agallocha* and *C. decandra* are found to be suitable for planting in less productive but tidal inundated areas (Siddiqi et al. 1993a). Nipa (*Nypa fruticans*) plantations can be established on vacant canals or riverbanks over the greater part of the forest (Siddiqi 1995). The raised lands that do not support mangrove vegetation can be planted with non-mangrove species like *Samanea saman*, *Albizzia procera*, *Lagerstromia speciosa*, and *Acacia nilotica* (Siddiqi 1998).

Merchantable growing stock of *H. fomes* and *E. agallocha* was depleted by 40% and 45%, respectively between 1959 and 1983. Depletion of standing volume is the result of over-cutting, which may be due to overestimates of past increment. The standing volume of *S. apetala* and *A. officinalis* remains unchanged. *Bruguiera sexangula* has decreased in stocking but *Xylocarpus mekongensis* has increased due to a moratorium imposed on the felling of this species (Chaffey et al. 1985).

Since 1988, there has been a ban on harvesting of timber with special exception of top-dying *H. fomes* trees for poles, sow-logs and hardboard, and of *E. agallocha* for pulpwood to be used in the Khulna Newsprint Mill. The results as reflected during a survey carried out in 1996 do not appear to be encouraging. For *H. fomes* trees 10 cm and above in dbh (diameter at breast height), the reduction in number of trees be-

Table 3.6.
Comparison of number of trees per hectare over time for *Heritiera fomes* and *Excoecaria agallocha* as found in different inventories (after Canonizado and Hossain 1998)

Species	Inventory/year	Trees >10 cm dbh	Trees >15 cm
H. fomes	Forestal (1959)	511	211
	ODA (1983)	296	125
	FRMP (1996)	124	106
E. agallocha	Forestal (1959)	345	61
	ODA (1983)	224	34
	FRMP (1996)	41	20

tween 1959 and 1996 has been reported to be 76%. For *E. agallocha* trees 10 cm and above in dbh, the reduction in number of trees is 88% for the same period (Table 3.6). Timber smuggling appears to be a major cause of the drain. Induced by timber shortages, timber smuggling during the moratorium was probably pushed to a level higher compared to earlier times. There was also a lot of damage inflicted by top dying of *H. fomes*. About 41% of the utilisable volume has been lost to top dying (Larsen 1994). Besides, significant losses are most likely to be from erosion of river embankments. With regard to *E. agallocha*, the current status is worse than *H. fomes*. The prescribed allowable annual cut exceeds the natural increment in the *E. agallocha* production areas by 50%. Fishermen also use the species as floating outriggers for their boats and the fishermen's use is 10% of the increment. Thus, the harvest is 60% higher than that of the accepted increment (Canonizado and Hossain 1998). To arrest depletion of *H. fomes* and *E. agallocha*, illicit removal of trees and fixation of a risk free allowable annual cut is to be made.

Ceriops decandra is mainly used for fuel wood and house posts. There are prescribed rules to harvest this plant. The stumps are often cut too low, which causes damage to the plant. Sometimes, all the stems in a clump are harvested; this may kill the clump. Cutting is concentrated near the waterways: this leads to over-exploitation near the water and under-exploitation in distant places. During harvesting of *Cynometra ramiflora*, virtually no prescription is followed, and this affects the sustainability of the resource.

Nypa fruticans is a valuable resource of the Sundarbans. As per prescription, the unopened centre leaf and leaf next to it in each clump must be retained at the time of harvesting. It needs to be checked whether this rule causes over-cutting of the plant. Besides, the cutters do not go to areas that are distant from the waterways. Some clumps appear to be over cut while others are under-exploited, which ultimately affects sustainable yield of this resource. A decline in stocking of *Phoenix paludosa* has been reported in recent years.

Heritiera fomes is one of the most important tree species of Bangladesh. The incidence of top dying, a disorder causing death from top downwards, is spreading throughout the forest. Approximately 17% of the stems of *H. fomes* are affected by top dying; the exact cause for top dying is unknown. Studies do not indicate insect infestation (Choudhury and Baksha 1983) or pathogenic attack (Rahman et al. 1983), which could be the direct cause of top dying. Trees could be exposed to stressing conditions due to a single factor or a combination of factors including salinity, reduction of critical levels of inundation, loss of canopy due to cyclone storms, water logged condition in the soil, and excessive sediment deposition on the pneumatophores (Karim 1994).

3.4.1.2
Chokoria Sundarbans

Besides the Sundarbans proper, the other natural mangrove of Bangladesh is the Chokoria Sundarbans. The original area of the forests was 18 200 ha; they lie in the delta of Matamuhury River between latitude 21°36' N to 21°45' N and longitude 91°58' E to 92°05' E. Initially the number of species was 53 belonging to 42 genera. *Ceriops decandra* and *Avicennia officinalis* were the main association of the forests. *Heritiera fomes*, *Excoecaria agallocha*, *Sonneratia apetala*, *Bruguiera gymnorrhiza*, *Aegialitis rotundifolia*, and *Phoenix paludosa* were also common (Cowan 1921). This was a moderately dense forest that underwent tremendous human pressure for exploitation of fuel wood, fishing, grazing of cattle etc. However, overnight conversion of a greater proportion of the forest for shrimp farming was the main cause for the virtually complete destruction of this natural mangrove (Siddiqi et al. 1994). The soil has become acidic, with pH less than 4.5. These soils are rich in the minerals pyrite and jarosite and have a high content of sulfur, and become acid sulfate soils (Richards and Hassan 1988). Investigations showed that the area was suitable for rehabilitation with mangroves only with the co-operation of the local people and shrimp farm owners (Shadidullah and Siddiqi 1994).

In fact, the Chokoria Sundarbans are not suitable for aquaculture. Besides, ponds were not made in a proper manner that allows regular inundation and drainage of the tidal water. During the rainy season, salinity decreases drastically while in the dry season it increases too much due to evaporation. Salinity is not at the optimum level for growth of shrimps, and rapid changes affect the survival and productivity of shrimps. Ironically, the area was transferred to shrimp farming on the grounds that the salinity was too high to allow for mangrove vegetation; in fact it was under the influence of vested interests. Shrimp farming in the coastal zone is undertaken in an area of over 0.14 million ha (Ali 1998). The average productivity of shrimp in the western and eastern coastal area is 344 kg ha^{-1} and 328 kg ha^{-1} respectively (Shawpan Chandra Paul, pers. com.). Against this, the maximum production recorded at Chokoria Sundarbans was 180 kg ha^{-1} (Choudhury et al. 1990). Furthermore, shrimp farming greatly reduces employment opportunities for the local people. The Chokoria Sundarbans had a great positive socioeconomic role locally in addition to offered protection to the people against frequent cyclones and tidal surges. The possibility of reforesting the Chokoria Sundarbans' original area with mangroves should seriously be brought under consideration.

3.4.1.3
Plantations in Accretion Areas

The coastal areas of Bangladesh are prone to cyclones and tidal surges. To mitigate the loss of lives and property against these natural calamities, mangrove afforestation was initiated in 1966 along the barren shoreline and offshore islands of the country. An area of about 1.3 million hectares has been planted by the Forest Department, although a considerable proportion of this area is now without tree cover due to plantation failure. Subsequently, the role of plantations in land stabilisation, employment generation, socioeconomic development in remote areas and wood production also came under consideration.

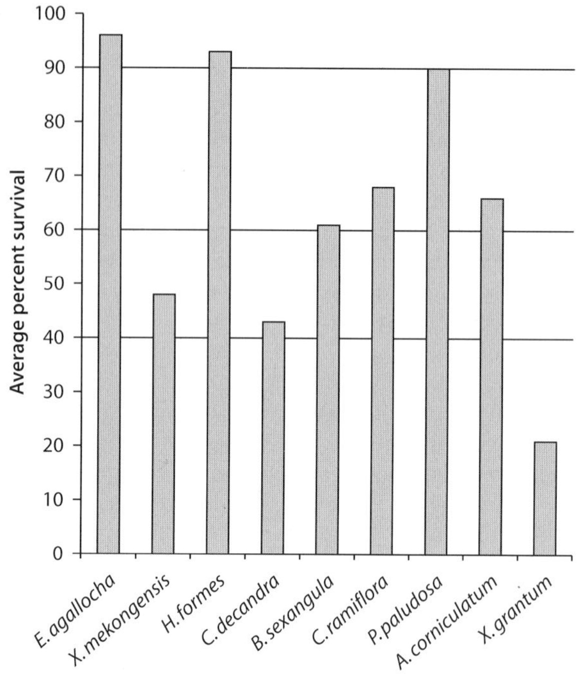

Fig. 3.3. Survival rates of major mangrove species used in the massive afforestation programme in Bangladesh

No systematic studies were initially carried out for the selection of species for the massive planting programme. Virtually all the commercially important mangrove species, viz., *Sonneratia apetala*, *Avicennia officinalis*, *A. marina*, *A. alba*, *Amoora cucullata*, *Bruguiera sexangula*, *Excoecaria agallocha*, *Xylocarpus mekongensis*, *Heritiera fomes*, *Ceriops decandra*, and *Nypa fruticans* were planted on new accretion areas (Siddiqi and Khan 1990). Figure 3.3 presents survival rates of major mangrove species used in the massive planting programme of Bangladesh. Of these, *S. apetala* proved to be the most successful. *A. officinalis* showed good performance only in the eastern coastline (Drigo et al. 1987). At present, *S. apetala* constitutes 94.4% and *A. officinalis* 4.8% of the successful mangrove plantations (Ahmad 1997). Planting seedlings raises plantations of *S. apetala*, while dibbling seeds for *A. officinalis* do this.

The coastal areas of Bangladesh lie between latitude 21°–23° N and longitude 89°–93° E. The shoreline is about 700 km long. Soil texture ranges from silty loam to silty clay loam, pH varies between 7.5 and 8.2 (Drigo et al. 1987). Soil salinity reaches the peak in April–May and declines with the monsoon. Salinity is higher in the southern and eastern part (3–10 mS cm^{-1}) and lower in the west (1.0–4.0 mS cm^{-1}). The salinity of the surface water is influenced by the large amount of water that flows through different rivers into the Bay of Bengal. The total volume of sediments carried by the Ganges and Brahmaputra rivers into the Bay of Bengal is about 2.5 billion t yr^{-1}. A small portion of this volume contributes to spontaneous delta building (Nishat and Haque 1984). The coastal afforestation activities are managed by four administrative forest divisions, namely, Chittagong, Noakhali, Bhola, and Patuakhali Coastal Afforestation Divisions. Figure 3.4 summarises the mangrove afforestation effort in Bangladesh.

Fig. 3.4.
Mangrove afforestation effort in Bangladesh from 1965 to 1995 (adapted from Siddiqi and Khan 1996)

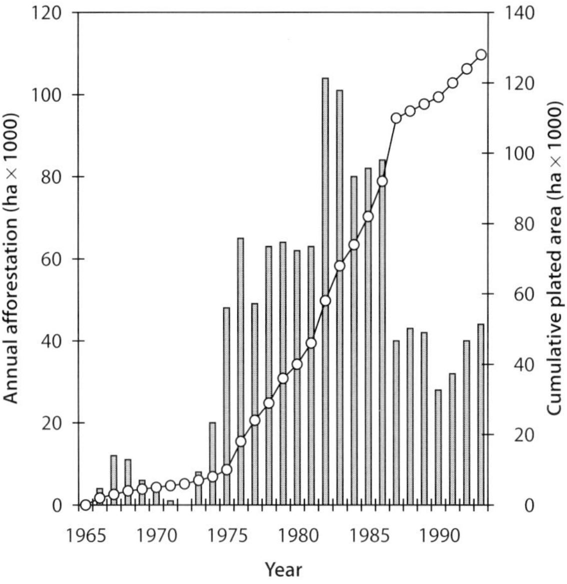

Site selection for plantations. Proper selection of the site is most vital for ensuring successful plantations. Geomorphological changes in the coastal areas are rapid and unpredictable, making it difficult to correctly identify ideal sites. McConchie (1990a) reported four types of changes, viz., rapid accretion, sand smothering, sediment winnowing, and erosion that adversely affect the mangrove plantations in Bangladesh. To reduce the loss of damage to plantations, "A guideline for assessing land stability conditions at potential coastal afforestation sites in Bangladesh" was developed by McConchie (1990b). Dalmacio et al. (1991) pointed out some practical considerations in site selection. These include:

- Avoid mud-free tidal sand flats. Mud free sands can be readily redistributed by wind or wave action, which causes burial or uprooting of the seedlings.
- Presence of tidal flats with 50% grass or algae is a good indicator of land stability, as algae and grasses bind and protect the sediments.
- Presence of well-developed channels indicates that the site may not be sufficiently stable and suitable for planting.
- Presence of ripples usually indicates instability of the site.
- Presence of mud-cracks on dried sediments implies that the sediment is cohesive, stable, and has sufficient clay content.
- Burrows in the sediments commonly made by crustaceans indicate that the sediment mass is reasonably stable.

The afforestation programme is carried out in a very unstable environment. Thus, there will always be a risk of some plantation loss during the time it takes for the trees to reach maturity. For initial success, the experience of the field staff is very important. Planting is undertaken following the appearance of grasses (*Porteresia coarctata* and *Myriostachya wightiana*) on the newly accreted land. Besides, while selecting the site (Nuruzzaman 1962), consideration should be given so that planted seedlings get at least 4–6 h daily of exposure to sunlight for photosynthesis.

Silvicultural practices. Different species have different silvicultural requirements. In the coastal areas, level of salinity, depth and duration of tidal inundation, maturity of soil and intensity of light are the most important factors for determining the performance of the species. At the beginning of the afforestation programme, 12 species were tried (Choudhury 1971; Alim 1974). Growth of *Excoecaria agallocha* and *Sonneratia caseolaris* was good (Siddiqi and Khan 1990) but they survived only in limited areas. *Avicennia officinalis* showed good survival and growth performance on the eastern shoreline, and there it constitutes 22% of the plantations. Plantations of *A. officinalis* were developed by sowing seeds. Possibly greater success could be achieved using seedlings. *Sonneratia apetala* is the most successful species. It alone constitutes 94.4% of the total mangrove plantations. In addition to mangroves, some non-mangrove species have also been planted on slightly raised lands (Siddiqi et al. 1993a). Proportions of various species in different areas is shown on Table 3.7.

Both *S. apetala* and *A. officinalis* are pioneer species in the ecological succession of the natural mangroves of Bangladesh (Das and Siddiqi 1985). These species grow well on new accretion areas with regular inundation. They are strongly light demanding. This might be the reason why these species marked better performance. Alternate species are not suitable for afforestation. This resulted in the creation of mono specific plantations either with *S. apetala* or *A. officinalis*.

Nursery Technique for *Sonneratia apetala* – Ripe fruits are usually collected in late August by climbing trees. A fruit of *S. apetala* contains 25–125 seeds, and one kilogram contains 140–150 fruits. After collection, the fruits are allowed to rot. Seeds are collected by rubbing and washing the rotten mesocarp in water. One kilogram of green fruits produces about 8 000 seeds (Siddiqi et al. 1993b).

Nurseries containing several beds each with size 1.2 × 1.2 m are prepared on the shoreline. Embankment is provided around the nursery to regulate water level. For one bed, 0.5 kg of seeds is sufficient. One bed yields about 3 000 useable seedlings,

Table 3.7. Species distribution of mangrove and non-mangrove plantations in the coastal divisions of Bangladesh

Species	% of species in plantations of four coastal divisions				% of total mangrove plantations
	Chittagong	Noakhali	Bhola	Patuakhali	
Sonneratia apetala	74.6	98.8	98.0	95.0	94.4
Avicennia officinalis	21.4	0.1	–	–	4.8
Excoecaria agallocha	–	–	1.6	0.4	0.4
Bruguiera sexangula	1.1	–	–	0.8	0.4
Non-mangrove	2.9	1.1	0.4	3.8	0
Total	100	100	100	100	100

which are adequate to raise a plantation of over 0.4 ha. The viability of seeds declines quite fast (Siddiqi and Islam 1988). A higher level of salinity hinders germination (Siddiqi et al. 1989), thus sowing on the nursery beds should be done during the monsoon when salinity is quite low all along the coast.

Site preparation and planting. New alluvial lands are flat and are found along tidal rivers, estuaries and the seacoasts. There is little site preparatory work in coastal afforestation activities. After the land rises above water, grasses (*P. coarctata* and *M. wightiana*) make their appearance. Planting is undertaken before the grass gets thick. In case of dense grass cover, cutting grass at ground level is done along the line of planting. The plantation area is divided into several sections, and planting spots are fixed using ropes to ensure uniform spacing. No pit needs to be excavated before planting. The roots of seeds are simply pushed into the soft soil. When soils are not soft, holes are made with the help of a pointed stick and roots are inserted. Soils around the roots are pressed well so that the seedlings are not washed away by wave action.

Planting is carried out during the monsoon (July to September). The monsoon season has been found to be a physiologically optimal time for planting (Islam et al. 1990); however, many newly emerging sites often cannot be planted during the monsoon when the weather is rough and wave action is strong. In such areas, planting during the winter (December to February) may be undertaken when weather is calm and the planting sites are approachable without risk. But such sites are unstable and subject to frequent wind and wave action, making it difficult to develop successful plantations. However, Shafi (1982b) noted that planting in winter is also successful as seedlings are least affected by natural calamities.

The seedlings are planted when they are about 0.6 m in height. Initially these were uprooted forcefully by hand from the nursery beds. This damages the root systems and it is believed to have been one of the causes of initial mortality after out-planting (Siddiqi and Das 1988). Now the uprooted seedlings are packed and taken to the planting sites by boat. Neap tides are chosen for planting. This reduces the chance of tidewater stagnation for a long time and enables proper and efficient planting work for a longer time. Vacancy filling and replanting are frequently carried out to establish well-stocked plantations. Weeding is not required, as initial growth of planted seedlings is quite fast.

Initial plant spacing. Fixation of initial plant spacing is an important silvicultural decision in the establishment of plantation. It affects rising of forest costs and wood production. As success in establishing mangrove plantations is often unpredictable, a closer spacing was chosen. Spacing was adjusted from time to time during the last 30 years of plantation history. At the beginning, spacing of 2.4×2.4 m was done. A greater portion of plantations was later raised at a spacing of 1.2×1.2 m. Presently, a spacing of 1.5×1.5 m is practised. Siddiqi (1987) suggested a wider spacing (1.7×1.7 m) as more appropriate.

Thinning and climber cutting. Initial growth of *S. apetala* is fast. Crowns spread sufficiently and compete for light in a well-stocked plantation. Siddiqi (1988) found high tree mortality if the stand remained unthinned. In some plantations, first thinning was arbitrarily undertaken after nine years by the Forest Department. Long-term collection of data and their analysis showed that thinning could be avoided by adjustment of initial spacing increasing to 1.75×1.75 m (Islam et al. 1991). It was found that mortality was more age-related than density dependent and that thinning would affect the survival marginally. At low densities, a small enhancement of diameter increment was apparent for 5–6 year old plantations. *S. apetala* is a good fuel wood but not popularised (Siddiqi 1993), thus the Forest Department could not market the thinned products. Thinning is no longer practised.

On some *S. apetala* plantations, the growth of climbers is profuse. They need to be removed for full release of trees. A climber cutting operation was thus introduced. The effect of this operation was not studied. At that time, severe damage to the wood of trees by insects was reported. The need for climber cutting became questionable from an economic point of view. Subsequently, the climber cutting operation was discontinued.

Growth, yield and rotation. The growth of *S. apetala* varies greatly from place to place. Productivity tends to increase from east to west of the coastal belt (Dirge et al. 1987). On a good site the stand may attain an average height of 20 m with a diameter of 20 cm at breast height in 20 years. Productivity may be up to 10 $m^3 ha^{-1} yr^{-1}$ (Latif and Del Castillo 1990). In determining the rotation of the plantations, a number of factors have been considered. A rotational age of 12 years for site indices of 15 m and 18 m dominate height at a reference age of 12 years, and 15 years for site indices of 6.9 and 12 m dominant height at the same reference age has been prescribed (Latif and Del Castillo 1991). A protective role of the plantation, devising a method for the maintenance of tree cover, wood production and alternative use of the reclaimed land, etc., need to be taken into account for a decision on the method of harvesting and fixation of rotation. However, a final decision on the harvest of the crop has yet not been taken.

3.4.1.4
Sustainable Management of Plantations

The plantation area is extended every year by planting newly accreted lands. A huge volume of wood is available in the plantations; however, S. *apetala* does not produce good quality timber. Wood is usually used for panelling, making packing boxes, parts of boats and general construction work (Das and Siddiqi 1985). Research findings indicate that this species holds a bright prospect as raw material for making hardboards (Khan and Shafi 1992). The protective role of mangrove plantations is well appreciated in Bangladesh. Initially it was planned to transfer the reclaimed land for agriculture after harvesting the trees; this however may affect the stability of the fragile ecosystem. A decision is yet to be taken on the exploitation of the forest. In the absence of a decision, it is not possible to develop a silvicultural system for sustainable management of the newly established forest. Meanwhile, the plantations encounter many problems (Saenger and Siddiqi 1993; Siddiqi and Khan 1996) that are in need of attention for the continuity and efficiency of the programme.

Insect infestation. Insect infestation in *S. apetala* plantations is common all along the coast. The overall infestation was estimated to be 52%. Raising mixed plantations instead of monoculture has been suggested (Islam et al. 1988). The pest was identified as a bee hole borer, *Zeuzera conferta* Walker (Cossidae, Lepidoptera).

The larva of the pest enters the wood through the bark and makes profuse tunnelling inside the wood of the standing tree. The bole of an apparently healthy tree may harbour hundreds of active larvae. The tunnelled and weakened stem or branch often falls down with the increase in wind velocity (Baksha et al. 1990; Baksha 1996). Integrating the use of pheromones and nematodes together with improvement of silvicultural management and encouragement of natural enemies has been suggested to suppress the pest population to a tolerable level (Baksha 1996). Similar infestation by caterpillars at recurring intervals of about seven years are known to occur in *Avicennia* spp. of natural mixed forests in Brazil (M Vannucci, pers. com.).

Raising mixed plantations. Monoculture either with *S. apetala* or *A. officinalis* is characteristic of the coastal plantations. Both are pioneer species in ecological succession, but other species are not found suitable for initial plantation. Scope exists only for establishing mixed plantations in combination with *S. apetala* and *A. officinalis* for silvicultural control of insect pests. Research is underway in this direction. It seems that both of the species can grow all along the shoreline. However, *A. officinalis* can withstand a higher level of salinity, while *S. apetala* has preference for low salinity. A more important factor that affects the simultaneous raising of a mixed plantation is the depth of tidal inundation. *S. apetala* prefers a higher level of inundation. So, seedlings of *S. apetala* are planted first. On the following year, seedlings of *A. officinalis* are planted between the *S. apetala* seedlings, when depth of inundation is lower due to deposition of silt in one year. Experimentally, this method of raising a mixed plantation appears promising, but it requires further investigation.

Natural regeneration. Artificial regeneration is expensive and it is inconvenient to develop multi-species forests by this method. It can be avoided if natural regeneration is adequate for the development of a second rotation crop. In natural mangroves of the Sundarbans, new regeneration of *S. apetala* is hardly found under old stands of the species. During a particular period of forest growth, inundation level and soil undergo changes that prevent further establishment of *S. apetala* seedlings (Troup 1921). Latif et al. (1992) reported no recruitment of *S. apetala* in the established stands of the mangroves of the Sundarbans.

By and large, seedling regeneration of *S. apetala* is found in many mangrove plantations and adjacent areas, but the new seedlings fail to establish themselves. The seedlings usually disappear before growing into saplings and the pole stage. Siddiqi et al. (1994) reported that there was no alternative to the present practice of artificial regeneration for raising mangrove plantations or developing second rotation crops.

Second rotation crop. In order to maintain a continuous forest cover in the coastal areas and to enhance the production of the coastal forest while *S. apetala* is maturing, under planting is highly desirable. A successful under planting will also result in the creation of mixed and multi-storied forests with more valuable species. Ten commercially important species were planted under the canopy of thinned *S. apetala* plantations in different places along the shoreline. *Excoecaria agallocha*, *Xylocarpus mekongensis* and *Heritiera fomes* appeared promising. For under stories, small trees or shrub species, viz., *Ceriops decandra*, *Cynometra ramiflora* and *Phoenix paludosa* can be planted (Siddiqi et al. 1992). However, for the eastern shoreline, the prospect of

Table 3.8. Performance of mangrove species in the under planting trials one year after planting in the established *Sonneratia apetala* plantations (Siddiqi et al. 1992; Siddiqi and Shahjalal 1997)

Species	Survival rate (%) after one year			Annual height increment (cm)		
	West coast		East coast	West coast		East coast
	Patuakhali	Chittagong	Bhola	Patuakhali	Chittagong	Bhola
H. fomes	73	97	64	36	09	17
X. mekongensis	73	57	37	16	–	–
E. agallocha	99	100	48	97	73	03
C. decandra	71	97	61	10	06	12
B. sexangula	95	93	44	15	09	06
A. corniculatum	97	97	86	70	24	14
P. paludosa	96	100	40	32	57	10
C. ramiflora	48	93	–	48	37	–
L. racemosa	81	–	–	81	–	–
X. granatum	64	97	–	64	86	–

under planting is uncertain (Siddiqi and Shahjalal 1997). None of the seven mangrove species in the under planting trial shows success (Table 3.8). On the western coast, successful species attained a height of 6–7 m in six years, while in the east, the planted seedlings failed to establish themselves.

Raising nipa palm. Nipa (*Nypa fruticans*) is a highly useful mangrove palm, particularly for the rural people who use its leaves for thatching houses. Nursery and planting techniques for nipa have been developed (Siddiqi et al. 1991; Siddiqi 1995). In the Sundarbans natural forests, plantations of *N. fruticans* can be established on vacant canals or riverbanks. But the shoreline is geomorphologically much more dynamic, and it is risky to establish nipa plantations. Site selection appears to be the most important and difficult task for success of the plantations. A relatively stable site, minimum wave action and low rate siltation are required. Desirable characteristics of a suitable planting site are:

- Colonisation by grasses and sedges indicates that the site is relatively stable to support nipa plants
- Since growth of nipa is slow, accretion or erosion should be as minimum as possible
- Site must be stable against current and wave action. A bank of narrow canals is usually less disturbed by wave action
- Nipa does not tolerate much inundation. No inundation during the dry season is necessary, and 10–12 days inundation per month for a period of 4–6 hours per day in the wet season is preferable
- Site should be flat with gentle slope. A soil with lower salinity (4 mS cm^{-1}) is favourable for the survival and growth of nipa plants

Growth of grasses is vigorous in sites suitable for nipa. So regular weeding is needed up to three years following planting. Seedlings are generally planted at 2×2 m spacing. Pruning of dead or damage fronds is made. The successful raising of nipa will be a big achievement towards sustainability and heterogeneity of the forests and a source at regular earning.

3.4.1.5
Concluding Remarks

The Sundarbans are managed on a sustained yield basis. The decline in stocking of various wood resources is a matter of great concern. Besides, the protective role of the forest and the value for its aquatic resource and rich wildlife fauna cannot be overemphasised. Problem oriented short-term and long-term research should be undertaken to improve the condition of the forest. Available research findings that are helpful to improve stocking and productivity of the forest need to be utilised.

With regard to afforestation, Bangladesh is a leading country. The country now possesses the largest mangrove plantation in the world, comprising mainly *Sonneratia apetala* and *Avicennia officinalis*. The mono-specific plantations of *S. apetala* are facing some serious problems. The silvicultural operations and plantation management must be shaped to ensure sustainability of the programme. Research studies to this end need to be further intensified. Moreover, available research findings should be extended and utilised at the field level for successfully raising plantations and for their sustainable management.

3.4.2
South and Southeast Asia River Deltas

Other major deltaic mangrove areas in the Indo-West Pacific are the Irrawaddy and Mekong River deltas. The Irrawaddy delta is heavily impacted as a source of fuel wood and charcoal production. All around the Gulf of Martaban including Moulmein, coastal areas are heavily over-exploited. Results of reforestation attempts through mangrove plantations have not been followed long enough to assess successes and failures. The Ayeyarwaddy (Irrawaddy) River delta has a population density of 442 persons km^2, which is three times that of the whole country, or a total of about five million for the delta division (Ohn 1999). This figure includes 'economic refugees' who find more work opportunities in the delta than elsewhere in Myanmar. The delta supplies firewood and charcoal to local residents and to Yangon. Quoted as important for this purpose is *Heritiera fomes*, while *Ceriops tagal* is used for house posts, poles, timber. After 60 years of over-exploitation, in 1984 the percent area under cover of mangroves was only a little over 29% of the original. At present, conversion to paddy-rice cultivation and over-cutting for fuel wood are the main causes of an alarming rate of deforestation; in 1991 there were only 14% of closed canopy forests and 10% were degraded mangroves, while over 64% were non-forest lands. The study by Ohn (1999) concludes that land policy use of forestry, agriculture and fisheries uses has to be applied urgently since the people of the delta depend on this ecosystem. To avoid breakdown of the system and ensure survival of the population, immediate rehabilitation of denuded mangrove areas should be methodically undertaken. The mangroves of the Irrawaddy

River Delta were already heavily impacted and degraded some twenty years ago (personal survey, unpublished).

In the past, the Mekong River delta was covered by extensive well developed mangroves. Changes in the hydrological pattern, construction of sea walls, reclamation of land for agriculture (mainly paddy-rice cultivation), and extensive silviculture of *Hevea braziliensis* for rubber and wax have drastically changed the pattern of distribution and development of mangroves. Floral and faunal abundance and species diversity of the mangrove ecosystem and its productivity are at present unknown. In Vietnam, especially in the Mekong delta, minor forest products were held as important resources for the daily use of coastal dwellers, a loss to be reckoned with (Hong and San 1993). One of the important resources was *Nypa fruticans* used for a variety of purposes, from fishing net floats made from the lower part of the fronds, to food, vinegar and alcohol made from the seeds, to roofing and house partitions made from the shingle. This useful element of the ecosystem is also all but lost. From the '80s onwards, efforts are underway to rehabilitate degraded areas. Different methods were attempted with varying degrees of success and failure; at present the method that appears to be more successful is the tambak and tumpang-sari approach, on which we will return later.

During the Vietnam war (1962–1972), the mangroves of the Mekong River and elsewhere were destroyed by defoliants and by napalm. In the delta alone, over 1 000 km^2 of forest were destroyed Since soon after the war, efforts have been underway to restore the mangroves. These, however, have had to compete with utilisation of the land for rice cultivation and shrimp ponds, although it was well-known that mangrove soils exposed to air become acid sulphate soils because of the oxidation to sulfuric acid of the abundant biogenic pyrites and other sulfides present. Attempts at recuperation of the damaged areas is underway by various methods, inclusive by turning extensive areas into tambaks, which are a traditional system of silvi-aquaculture originally from Sumatra. However, results of these long-term experiments are not yet publicised. Also, other methods were attempted, such as reforestation of embankments with selected species of mangroves that tolerate relatively long periods without tidal inundation; results vary. All the methods attempted are labour intensive and usually relatively expensive, yet the resources from the mangrove ecosystem are essential as food and materials for a growing population. As elsewhere, the main resources from the Mekong and Irrawaddy river deltas are: poles, timber, thatching materials, fuel wood, charcoal, fish, mollusks, crustaceans, honey and wax, dyes and tannin. Engineering works in the Mekong River delta had gone on for decades, even before the Vietnam War, mainly for the purpose of gaining land for agriculture. The whole area is now heavily impacted by man, and rehabilitation is difficult and time-consuming.

3.4.3
Mangroves of the Irrawaddy and Mekong River Deltas

M. Aizpuru · F. Blasco

The mangroves of Myanmar are practically unknown to the international scientific community for several reasons; field surveys have been extremely scarce during the last decades. Today the only available recent data are those that can be extracted from high-resolution satellite data. We have been able (Blasco and Aizpuru 1999a,b) to obtain a complete wall-to-wall coverage with satellite products (mainly SPOT data) from the northern border with Bangladesh near Cox Bazaar to the southern end of the country up to Ranong in Thailand.

The coastline of Myanmar (±2 500 km long) can be arbitrarily subdivided into three main components; the Arakan Coast, the Irrawaddy River Delta and the Tenasserim Coast.

The Arakan Coast is where the mangroves are presently being depleted at an accelerated pace. For the first time we are able to locate, map and quantify the magnitude of deforestation processes in this part of Myanmar especially between Sittwe and Taungup where *Heritiera fomes* stands constitute presumably the main timber resource. According to satellite data, at least 1 700 km^2 of these mangroves have been clear felled recently in this part of Myanmar. Further north, up to the border with Bangladesh, all mangroves have been virtually eradicated and converted to agricultural land. The total area extent of mangroves along the Arakan Coast is about 700 km^2.

The Irrawaddy River Delta has one of the largest deltas in Asia with an average flow of 12 000 m^3 s^{-1} and an area of about 20 000 km^2. This is also an area where the original mangroves are disappearing rapidly. Recent satellite surveys compared to earlier data clearly show their unabated decline due to demographic pressure. Figure 3.5 is to our knowledge the most recent. Cartographic inventory for the mangroves of this delta show that irrigated crops and aquaculture ponds are expanding especially on the eastern corner of the delta (Kyonkadun area). According to our maps, the present extension of the mangroves of the Irrawaddy does not exceed 2 400 km^2. From a floristic point of view, these remaining mangroves seem to be almost identical to the mangroves of the Sundarbans (Ganga-Brhamaputra deltas).

The Tenasserim Coast, southern coast of Myanmar, is much more diversified in terms of geomorphology and mangrove habitats, from the Sittang estuary to Victoria Point (border with Thailand). This is a very complex physical situation where it is extremely difficult to produce an accurate mapping of mangroves due to the innumerable amount of islands known as the Mergui Archipelago. From an ecological point of view, such important data as interstitial water salinity or soil textures or redox potential are totally missing. The assumption is that high annual rainfall (>2 000 mm from

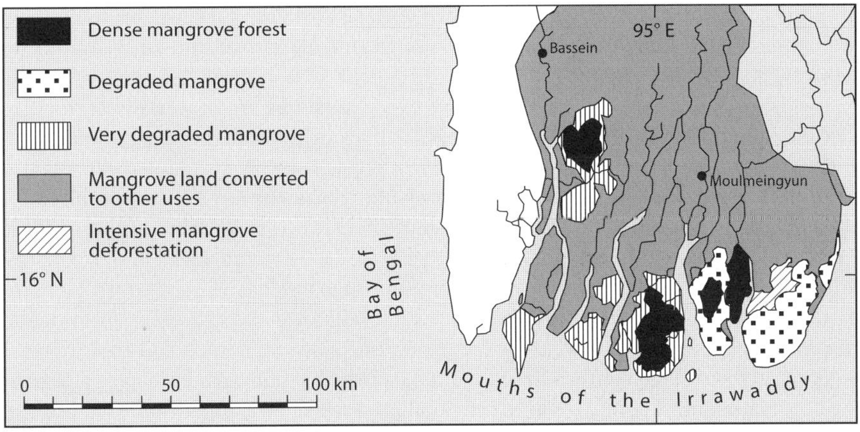

Fig. 3.5. The mangroves of the Irrawaddy River delta

April to October) and a high rate of groundwater flow may reduce the average salinities, increase primary productivity and accelerate regeneration processes, even during the pronounced dry season (January to April).

Our screen analysis of satellite data leads to a couple of important remarks: a) most of these mangroves seem to have been underestimated by previous surveys and statistics, and b) at least 85% of these mangroves are degraded or very degraded. No mangrove restoration seems to be carried out in this coastal stretch. Altogether, the mangroves of Tenasserim coast cover 3 900 km^2.

3.4.3.1
The Mekong River Delta

With a deltaic area of 90 000 km^2 and an average flow of 14 000 $m^3 s^{-1}$, the Mekong Delta is one of the largest river mouths in the world. In natural mangroves, differences in vegetation structure and floristic composition indicate that ecological conditions (including soil structure, nutrient availability, hydrological regime etc.) and succession patterns vary from one zone to another. In monospecific artificial mangrove stands, which are now the most conspicuous mangroves in South Vietnam, these ecological differences are by and large erased, although they are locally evident wherever the size of even-aged trees change or in localities were the paucity of seedlings and saplings is conspicuous. On the other hand, it is impossible to analyse today the consequences of this extreme floristic simplification in the mangroves of the Ca Mau Peninsula on the total biodiversity including the diversity of benthic invertebrates.

Our map (Fig. 3.6) may seem over-simplified. However, the case of the Mekong delta is very peculiar, as most degraded mangrove stands are intermingled and can't be individually mapped. For the sake of clarity, transitional or intermediate mangrove stages have not been shown on the map. They were included in a single class under "very degraded mangroves".

For many years, especially during the wars (1950–1970) and during the post-war period, the mangroves of this delta have been massively destroyed. The situation has drastically changed since 1983 with the beginning of an extensive reforestation programme with *Rhizophora apiculata* Bl. Our latest satellite survey gives the following figures: Dense mangrove: 25 km^2 – Very degraded mangrove: 800 km^2 – Mangrove reforestation areas: 1 300 km^2.

It is remarkable that the original mangrove ecosystems are practically extinct today. Vietnam is now one of the rare countries in the world where the area of mangroves is expanding.

In conclusion, the present status of the mangroves of the large deltas of the Indo-West Pacific is not too good. There has been extensive to mild degradation in some parts of all of them, induced either by natural episodic events such as cyclonic storms, tidal surges, climatic or hydrological changes in the river basin, or induced by man-made changes in the environment and by wars. Whatever the causes, environmental changes have been on the increase in frequency and size of the impacted area, increasingly rapid changes may have caused almost irreparable damages in many places. On the other hand, most countries in the region have seen the wisdom of rational use and management of mangrove ecosystems. Reforestation, afforestation and re-afforestation practices have by and large met with reasonable success, provided sufficient time

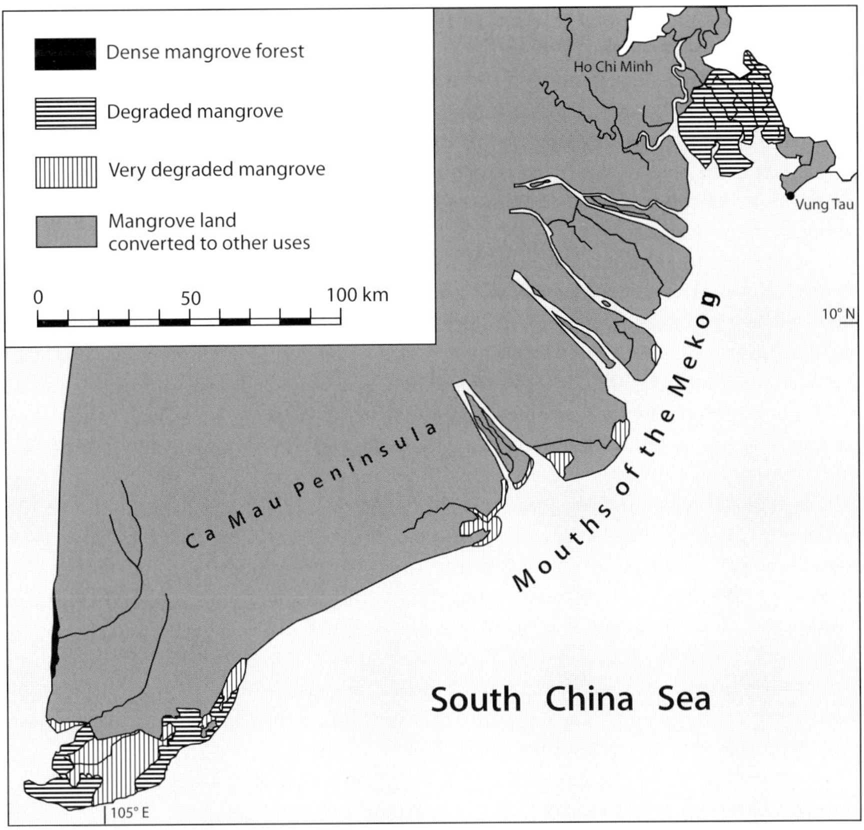

Fig. 3.6. The mangroves of the Mekong River delta

is allowed to assess the successes and failures and provided experimental methodology is combined with traditional wisdom and co-operation of the littoral communities. On the other hand, laws should proscribe destructive felling of mangroves. Where silt deposition is not too great, where river discharge is sufficient and the gradient of the coastal plain is gentle, mangroves will most probably advance landwards and compensate areas lost to mean sea level rise.

Attention should be given and more studies done on the transition zone between the mangrove forest and the supra-littoral vegetation, especially those species that survive exceptionally high tidal floods.

3.5 Small Deltas and Coastal Mangrove Ecosystems

A considerable amount of mangroves occur along the extremely long coastal zone of the Indo-West Pacific, where sufficiently sheltered areas favour the setting of seeds and propagules The most important features for the development of mangroves in open ocean coastal areas are the gradient of the shelf and the pattern of coastal currents around islands and along the continents, the slope of the coastal plain, fresh water, and sediment input. Where coral reefs or scattered islands function as natural breakwaters, or behind man-made breakwaters where the gradient is gentle, mangrove seedlings can take root, and the mangroves usually can establish themselves even where freshwater from land run-off is scarce, but rainfall is abundant and temperature uniformly high. For this reason, coastal lagoons or otherwise protected areas may bear flourishing mangrove stands in monsoon areas even in the absence of perennial rivers.

Coastal mangrove systems occur in the equatorial and tropical belts, north and south of the Equator. I use here the term: "tropical belt" to designate the area comprised between the Tropics of Cancer and Capricorn, which includes the equatorial belt that has oceanographic and climatic characteristics of its own. Local environmental factors such as winds, rainfall, temperature, soils, and hydrology are superimposed on the general global climate and determine the features of each site as well as the presence or absence of individual species. At low latitudes, there may or may not be well marked periods of drought alternating with periods of winds and heavy precipitation as is typical of monsoon systems. As a rule, the equatorial belt has abundant rains, distributed along the year. From this point of view it roughly comprises the coasts and islands from latitude 10° N to 10° S in the Indo-West Pacific, which has the longest coastline covered with mangroves. Indonesia is the country with the largest area of mangroves, all of them on or around the Equator and divided over about 17 000 islands; the total area of mangroves is over 43 000 km^2. Though there are periods of higher rainfall and seasonal winds, the coastal climate is best defined as tropical humid, where rainfall is never less than 60 mm per month, or where a short dry season is compensated by heavy rainfall during the rest of the year.

The coastal zone of the equatorial and much of the tropical belts under whatever pattern of rainfall are characterised by abundant precipitation, lack of well marked or prolonged dry seasons and uniformly high temperature. All these are factors that favour fast and continuous growth, speciation and self-regeneration of damaged areas of the mangrove forest. Biodiversity is high, and many species can be found coexisting. The mangroves of the equatorial belt enjoy favourable conditions from all points of view, but among the most important is that they receive nutrient-rich waters from the large number of small perennial rivers or coastal lagoons as well as pure water from rainfall that leaches away salts and prevents the formation of acid sulfate soils. The mangroves of the extensive coastal zone of the tropical Indo-West Pacific share many general characteristics, but differ in site-specific details related to biodiversity, structure and function. The trees of Sabah and Sarawak on the island of Borneo are being described in detail and distribution and abundance of species can offer clues to environmental characteristics (Soepadamo and Wong 1995). For instance *Dolichandrone spatacea* (Bignoniaceae) in Sabah colonises banks of tidal rivers and estuaries, but in Sarawak the same species grows only on the inland limits of the mangroves on clayey soils. Also, there are three species of *Sonneratia* in Borneo: *S. alba*, *S. caseolaris* and *S. ovata*, but the latter was so far recorded only in Sarawak and not in Sabah. *S. ovata* is widely distributed elsewhere in the region and the hybrid *S. alba* × *S. ovata* occurs in Brunei where also the hybrid *S. alba* × *S. caseolaris* is found; both hybrids are also found in Papua/New Guinea. Speciation, hybridisation and intraspecific variations are the rule, suggesting incomplete genotypic isolation, which is readily apparent in the species-rich exuberant equatorial belt. Zonation of mangrove species is also as varied as the places where it occurs.

Along the east coast of India, there are few relatively small deltas that were or still are heavily forested with mangroves. Though all of them are at present undergoing some degree of stress due to human impact for habitation and resources and are frequently hit by cyclones, some are relatively well preserved, because they were earlier kept untouched for centuries as hunting grounds for royal families, such as those of the delta of the Mahanadi. River. The mangrove deltas of the Godavari, Krishna and Cauvery rivers are now adversely affected by over-exploitation. The mangroves of the delta of the Cauvery, in the south, are under pressure from man and nature: failure of the NE monsoon has been frequent in the recent past; there is diversion of water upstream for agricultural purposes and siltation downstream: all are matters of concern. Many of the distributaries of the deltas and estuaries on the East coast of India are drowned estuaries because of the slow tectonic tilting of the Indian Plate and abundant discharge of river water. Species diversity and species composition is different on the east coast of India as compared to the west coast. The Bay of Bengal Surface Water (BBSW) is entirely different from the Arabian Sea Surface and Sub-Surface Waters (ASSW and ASSSW). BBSW has lower salinity, there are no upwelling areas and BBSW are richer in terms of nutrients; the zooplankton composition also is different (Pereira and Vannucci 1991). On the eastern coast, the NE monsoon dominates but is weaker than the SW monsoon that dominates the seasonal climate fluctuations on the West coast; depressions and tropical cyclonic storms are much more frequent on the East coast and may be immensely destructive, as happened in October–November 1999, when cyclones hit the east coast of peninsular India. The structure and dynamics of the mangrove ecosystems of the East and of the West coasts of India are entirely different.

The word monsoon is from the Arabic *mausim*, means "season" and "time of year". Not all coastal mangroves are in the equatorial belt or under monsoon regimes, but they are more extensive, frequent and well developed in those areas where seasonal rainfall is well marked and where the forest substratum becomes well drenched and washed by rainwater. Some tracts of the Atlantic Ocean coasts of Brazil and Africa are under monsoon regimes, where dry or very dry periods alternate regularly with rainy periods. There are not many coastal mangrove areas affected by regular winds in the northern hemisphere or in the equatorial belt, but they are important, because mangroves could be absent or reduced were it not for the regular seasonal rains brought by the seasonal winds.

The distinctive characteristic of monsoon dominated coastal areas, is that the winds and rains they bring are seasonal, usually alternating with dry periods. South and SE Asia have the most characteristic monsoon systems. The SW monsoon blows regularly from May to October in the North-Indian Ocean, while the NE monsoon blows from October to different times in various places. The SW monsoon winds carry with them large amounts of humidity that precipitate as rain, which are stronger along the coasts than inland. The NE monsoon carries less humidity; it is usually not as strong and regular as the SW monsoon. Monsoon winds were well-known to mariners of ancient civilisations, especially from ancient Sumer, Greece, Rome and Arabia. The Greek name for the SW monsoon was *Hyppalos*, from the pilot who first described it. Over the past few centuries the SW monsoon has been fairly regular, and in the State of Kerala, India, even the date of the "bursting of the monsoon" was known to be sometime during the last week of the month of May, except when it was early, or more often late by a few days. Rarely the SW monsoon failed or was weak. Over recent centuries, the monsoon trough has shifted eastwards over South Asia, rendering the Thar Desert of Pakistan and the Rajasthan desert of India drier. Dated ^{14}C water underlies the Rajasthan desert, where palynological studies reveal broad-leaf forests to have existed up to the Upper Quaternary, a sure sign of abundant rainfall. The shifting of the monsoon eastwards may be one of the causes of the decrease of precipitation over the Indus river basin, and of the dryness of the coasts of the Gulf of Kutchch eastward up to the Gulf of Khambat, as well as the present dryness of the Saurashtra coast of the Kathiawar peninsula in western India. In monsoon irrigated coastal areas, both the groundwater level and the surface water level at the mouth of the distributaries may be higher by one metre during the SW monsoon. Rains and high waters may advance as far inland as 40 km or more, where the slope of the coastal plain is gentle, bringing seeds and propagules of mangroves with them. In some of these places, there are extensive areas where the *Nypa* palm is the dominant species.

Typically, in coastal areas where tides have reduced amplitude or are sheltered from seasonal oscillations of water level, there is a well-defined boundary between mangroves and the upper littoral flora. On the western coast of Sri Lanka, where tidal amplitude is only about 50 cm or less, one can stand with one foot in the mangrove and another on the hard soil of the supra-tidal. On the other hand, the higher levels of extensive mangroves with seasonal fluctuations of inundation level have an ill-defined boundary between them and the supra-tidal vegetation. The supratidal vegetation is typically composed of some mangrove species particularly tolerant of a few months of exposure and several species of non-mangrove vegetation, which is tolerant of occasional flooding, by diluted tidal waters. Mangrove species more frequently found in the upper reaches are *Heritiera fomes*, *H. littoralis* and *Excoecaria agallocha*, while supra-littoral common trees are *Barringtonia*, *Albizzia* and the ubiquitous wild rice *Porteresia coarctata* and *Myriostachya wightiana* and other grasses. This situation is often found in coastal monsoon dominated areas, or in abandoned shrimp farms in the Chittagong area of Bangladesh. In these degraded areas there are scattered scrubby *Excoecaria agallocha* surviving in what water is left of the ponds, along with weeds and grasses of the supra-tidal, while landwards in the area that is flooded during the monsoon, the villagers plant different species of vegetables during the inter-monsoon, in between pneumatophores of *Heritiera fomes*. Mangroves of monsoon-dominated areas have a greater relative abundance of species and individuals of low salinity-loving species, such as *Sonneratia*, *Nypa* and grasses. Along desert coastal areas, like the Makran coast in Pakistan, the vegetation immediately above the mangrove is a pure association of halophytic species.

Mangrove soils are known to accumulate large quantities of sulfides, especially biogenic pyrite, which produces sulfates later oxidised to sulfuric acid after exposure to air. For this reason, clear felling of mangroves for conversion of the land to agriculture or intensive aquaculture is usually unsuccessful; the commonest conversion is for the use of the land for rice cultivation or shrimp farming, which invariably fails after the third or fourth year. This has been the sad recent experience of some countries, which have followed the ill-advised hasty conversion of former mangroves into paddy rice cultivation or sugar cane plantations. For the same reason, shrimp and fishponds become unproductive after two or very few years of intensive production. Soil additives are an expensive way to correct soil acidity, and it is not more than a temporary rather than definite corrective measure. However, in monsoon areas where periodical rains wash away the accumulated acids, it is possible to maintain dry land agriculture or paddy rice cultivation. Elsewhere for the maintenance of neutral pH and productivity of previous mangroves soils, it is necessary to maintain the alternation of periods of inundation and exposure to air. If tidal flooding is not maintained, extensive practices for leaching away the unwanted salts and acids from the soil are necessary, and may take much time and money. The former mangrove islands of the Sundarbans, especially the eastern Sundarbans and the accretion areas, receive abundant fresh water both from river discharge and monsoon rains. After they are raised above tidal inundation by sediment deposition, fresh waters leach away both salts and sulfates. The same situation prevails in the Kerala backwaters where copious rainfall ensures relatively high soil pH. This peculiarly fortunate situation whereby former mangrove areas may be converted to agriculture are exceptions to the rule and prevail only because of special local features such as very copious rains.

Aquaculture in the equatorial belt and in monsoon dominated areas is an ancient practice, especially in South and SE Asia, where connection with the sea is preserved and the land is regularly inundated by tides and washed by the monsoon rains. In South Asia, mangrove forests have been used for human habitation, food and materials for thousands of years; capture and captive fisheries are practised in all forms, from keeping fish, mollusks and crustaceans alive in under-water baskets to sophisticated systems of traditional aquaculture. We will briefly mention three such systems: the silvi-aquaculture *tambaks* and *tumpang-sari* of Indonesia now practised over extensive areas in many part of the world, the silvi-agri-aquaculture *pokkali* and *bheri* systems of India, canal fishing of south India (Selvam et al. 1999) similar in many places of South and South-East Asia.

The pokkali system of fish and prawn cultivation was already in use when the Portuguese first landed on the west coast of south India in the last years of the fifteenth century. The word *pokkali* is the name given to a variety of rice that over centuries has been selected to tolerate water salinity as high as 2–3 ppt or more. The *pokkali* system of aqui-agri-silviculture is typical of the Cochin backwaters which occupy over 250 km^2, and extend from latitude 9°20' N and latitude 10°20' N, and lie between extensive consolidated sand-bars now also under cultivation, that run parallel to the coastline, and the Western Ghats to the East. A chain of lagoons that are the main communication and transportation link of the entire coastal zone takes much of the coastal plain. Short rivers bring runoff water from the hills to the backwaters. Tidal range is about one meter and mean water level difference may be even greater between the monsoon and inter-monsoon seasons. The system functions as a vast estuary under monsoon regime. The rainy season is the SW monsoon that sets in late in the month of May and extends through September. During the monsoon period, paddy-rice is cultivated in the fields; in the inter-monsoon period, from October to May, the fields are used to rear shrimp and fish. During the inter-monsoon there is intrusion of oxygen rich seawater that brings with it shrimp larvae spawned at 30–40 m depths over the continental shelf, in the Arabian Sea. Intruding seawater invades the estuary and can be identified up to 15–20 km upstream. Over centuries, the area has been managed by man into a mosaic of plots about 2–7 ha each contoured by low earthen bunds or dikes about one meter above high tide level and used as footpaths by men and animals. The bunds are cut at appropriate places and set with small sluice gates to allow flooding by incoming tides and partial drainage at ebb tide. When open, the gates allow for the entrance of incoming waters; they are provided with filters that prevent the escape of shrimp larvae and small juvenile fish during ebb tide. The floor of each pond is dug down to some depth such as the whole system remains at sea level and waters do not need to be pumped in, nor are the completely drained away during ebb or when the little sluice gates are open.

Special salinity-tolerant varieties of rice have been selected over the centuries, and the most common is the *pokkali* variety. In addition to these seasonal ponds, there are more permanent ones such as the *chemmin-kettu* that are permanently flooded over a period of two or three seasons to ensure full growth of bigger fish species. Throughout the system of interconnecting lagoons along the southwestern coast of India, small islands are left intact, with the mangrove forest growing untouched. Some of these are preserved as "sacred forests" (Vannucci 1999). Coconut palm trees are planted all over, where there is enough space in the elevated terraces among the plots. This type of silvi-agri-aquaculture is very efficient in that direct benefits and goods provided by the mangrove forest are preserved to some extent; in addition rice, coconut, plantains, bananas, pepper and other spices, as well as vegetables are produced in the area, and fish is taken from the open waters. Though the production of fish from the ponds in terms of kg /ha/yr is smaller than what is obtained from intensive aquaculture elsewhere, the production has continued over centuries, while the productivity of intensive aquaculture is well-known to be unsustainable. The production of the Cochin backwaters, shown on the Table 3.9, is that of an average year before the construction of the oil terminal at Cochin (Vannucci 1987).

The lagoon areas of the southwestern coasts of India harboured a very high human population density, up to 1550 inhab. km^{-2} about ten years ago; it is higher at present. Man lived in the area since unrecorded times thanks to the availability of abundant food. Gradually, man changed his environment by cutting mangroves for firewood and other utilities, and he introduced to the area other species of plants, amongst the most important was the coconut (Table 3.10), *Cocos nucifera*, which has given its name to the State of Kerala. He also introduced the mango *Mangifera indica*, the breadfruit, *Artocarpus altilis*, and others from SE Asia. Other species, like the sweet potato, *Ipomea batatas*, the cassava, *Manihot esculenta* and *Amaranthus* spp. were introduced from the Americas and *Ricinus communis* from Africa in relatively recent

Table 3.9. Total production estimate of Cochin backwaters

Shrimp from backwaters	Average 865 kg ha^{-1} $season^{-1}$
Shrimp by traditional methods	A total of 1500–2000 t
Artisanal capture shrimp from the sea	3600 t (1983)
Commercial capture shrimp from the sea	25000 t mechanized vessels
Mollusks	About 10000 t, mostly clams
Paddy rice	500–1200 kg ha^{-1} $season^{-1}$

Table 3.10. Coconut production in the Ernakulam District, India (*source:* India Coconut Development Board)

Area under production of coconuts for the year 1982–1983	62 916 ha
Production of nuts (million)	344
Productivity nuts ha^{-1}	5 468
Average yield per palm $year^{-1}$	36
Yield of dry fiber per 1 000 nuts	81 kg
Average wt of 1 000 nuts	1 000 kg
Wt of water of 1 000 nuts	100 kg
Wt of copra of 1 000 nuts	150 kg
Wt of oil/dry wt of copra	68–70%
Moisture content of copra	6%

times; some of these plants now grow wild. Though the earliest records of bananas are from India and are dated 2 500 B.P., their cultivation is considered to be much older. Earlier still, fruits were harvested from the wild, perhaps tens of thousands of years before they were cultivated.

Paddy rice has been cultivated since prehistory, and locally adapted varieties were developed. Inside the lagoon, in between the plots, there are small terraces elevated above highest tide levels where permanent or semi-permanent dwellings are built for the local residents and workmen. Human society in Kerala is highly structured, for instance the *pokkali* fields are worked by fishermen during the monsoon, while the owner of the fields cultivates paddy rice in the intermonsoon; such a complexly managed ecosystem would not function as systematically, I would almost say automatically, if the human population were not equally regulated in tune with the environment. The abundant dissolved and particulate organic matter produced in the backwaters, including the rice stems which are left to rot in the fields are a food for the growing shrimp and benthic fauna, including benthic fish, and contribute to enrich the adjacent coastal marine waters which sustain rich fishing grounds. At the end of the season the sluice gates of each field are opened, and the remaining immature fully metamorphosed small Penaeid shrimps return to the sea with ebbing tides. They will complete their growth at about 30–40 metres depth at sea on the continental shelf, and when mature will spawn the next year's class of larvae.

Though the yearly production is less than what is obtained by intensive rice cultivation or shrimp farming, the important fact is that it is ecologically sound, since the area was not degraded over centuries of traditional management and it has remained productive for as long. Also, from the point of view of mangrove swamp conservation, the practice is interesting because much of the mangroves were left in place, species diversity is probably not greatly affected, the mangrove islands are a source of seeds, propagules and seedlings, they produce nutrients, they are a nursery area for fish, capture fishery, and in general, they maintain the ecological equilibrium of the area. The entire system was methodically developed and works like clockwork according to the vagaries of the monsoon. Comparable systems also exist in other parts of India: there are over 1 500 ha of *bheries* in West Bengal, over 2 300 ha of *gawani* in Karnataka, and 1 800 ha of *kazan* in Goa. The lessons to be learned here are that whenever conversion of mangroves is carried out, some wholesome rules are to be observed: first, parts of the mangrove ecosystem should be left untouched; second, changes should be gradual; third, polders should not be built below mean tidal inundation level; fourth, abrupt and major changes in the hydrological and topographical systems should be avoided. Large-scale engineering works bring about significant changes over a short period of time, but natural forces cannot adapt easily to abrupt changes, which usually have negative effects on the environment. A Latin proverb says: *Natura non-facit saltus* (Nature does not make jumps).

The study of the Kerala backwaters is particularly interesting because it is an example of a strongly impacted but not degraded mangrove area about which there is historical record. Through the year, surface water salinity ranges from almost fresh water during the monsoon, to 34 ppt at the mouth of the estuary in the intermonsoon and following months (Vannucci 1987). The fields are inundated during the SW monsoon, and when the monsoon rains are over, water drains away from the fields that are exposed to air for a brief period. During the intermonsoon, seawater advances in the estuary, bringing in shrimp larvae that were spawned over the shelf at sea; sluice gates are opened to admit advancing waters in the *pokkali* fields. Thus in a one year cycle, in each field or plot, the paddy is cultivated when the water level is high and rain water washes salts away, then shrimp and fish are reared in the same fields, when the water level is low and water salinity higher. The alkalinity of seawater neutralises the acids that may have been formed. Tidal forces inundate and drain the fields, without the need to elevate water, recruitment of fish and shrimp is done from the wild stock, a reasonable amount of young shrimp are allowed to return to the wild stock at sea, and the main work consists of managing the soil adequately.

The Kerala backwaters are an example that shows how mangrove areas in monsoon-dominated areas can be profitably managed for a sustainable production of a variety of resources. The example also focuses on the role played by rainwater to leach soil sulfates. The role of river waters is different, as they are carriers of suspended and dissolved organic and inorganic matter. Mangrove ecosystems of areas where precipitation is scarce would not lend themselves to such a type of management. Finally, human ingenuity is shown here at its best; it makes use for its own purposes of the largest possible number of ecological niches available in the system and contributes to creating more rather than reducing the number of plant and animal species and homogenising the non-living environment to the extent of creating a monotonous system with a reduced number of ecological niches. It enriches the system by increasing the complexity of the system's structure and dynamics. Exposure of soils to air in the *pokkali* or in the *bheri* fields is never long, and the reduced soil acidity, which may be formed, is compensated by the low pH of incoming tidal waters. The net result is that these empirically well-managed fields, a checkerboard of mangroves and cultivated plots, have become some of the highest sustainable productive areas on the globe.

The Southwest coast of India is not the only area of the Indo-West Pacific where monsoons are the determining climatic factor. By and large the entire west coast of peninsular India northward up to the estuary of the River Narmada sustains mangroves that are much better developed than would otherwise be, were it not for the SW monsoon. Along this coast roughly from Mumbai (Bombay) southwards, the long-shore oceanic current has a peculiar behaviour in that its direction is reversed seasonally under the influence of the monsoons, it flows northwards during the SW monsoon and the trend is reversed with the current flowing southwards during the NE monsoon. The most noticeable results are the occurrence of strong seasonal erosion at some places and the seasonal formation of mud banks at other points off the coast. Especially in the State of Kerala where the temporary mud-banks are locally called '*chagara*', marine *currents* cause the upwelling of deeper waters that bring large amounts of fish to the surface, which allows for exceptional catches. The condition, however, does not last for more than just a very few weeks, at the time of the reversal of the currents.

On the other hand, the southern part of the East coast of India is under the influence of the NE monsoon, which produces less precipitation, is more irregular and fails more often As a result, the mangroves are not as developed as they could be with more regular and abundant rains and river discharge: if damaged, they take longer to recover, as is the case of the Pichavaram mangroves at the mouths of the Cauvery River that meet the Bay of Bengal south of Madras. Much of the waters of the river have been diverted upstream for other purposes at a time when growing population numbers have caused gross over-exploitation of the mangroves. Further north along the east coast of India, mangroves are very well developed because of the relatively large river deltas of the Godavari, Krishna and Mahanadi rivers, lower salinity of the Bay of Bengal Surface Waters, the gentle slope of the continental shelf, and because much of the forests had been preserved during centuries as hunting grounds of Royal families. The East coast, however, receives more frequent and much stronger cyclones than the west coast. Witness the two immensely devastating cyclones, which followed one another in quick succession in the months of October and November 1999. Even better growth and extension of mangroves is found in the northern part of the Bay of Bengal, in the Sundarbans, where the SW monsoon is active and where there is a great amount of river water discharge. Unfortunately, the northern part of the Bay of Bengal also falls victim to several cyclones every year.

Around the coasts of India, mangrove species composition varies in relation to tidal amplitude, salinity of surface and pore water, soil composition and a combination of climatic and socioeconomic factors. Accordingly from west to east and around the Bay of Bengal along Myanmar, Thailand and Malaysia, coastlines there contain a great variety of mangrove ecosystem types (Untawale 1986). Most probably the mangroves of the Indian sub-continent are those that have been longer inhabited by man. Mangrove ecosystems of India are varied. Only palynological studies may reveal what they were originally; at present they arc only remnants of centuries, perhaps millennia of interaction between man and nature. For instance, the once abundant *Nypa fruticans* is now reduced to scattered stands of not so well-developed fronds. Captive fisheries of many kinds vary from low salinity aquaculture farms to high salinity salt farms for rearing *Artemia salina*, used mainly for feeding shrimp larvae reared in hatcheries. Captive fisheries in mangrove areas are an ancient tradition throughout S and SE Asia, and socioeconomic implications of the management of mangroves have very ancient roots.

Coastal mangroves in Sri Lanka occur mainly in well-protected bays or lagoons and small river estuaries. In SE Asia, the open ocean is in fact a major inland sea bordered by the main Indonesian islands. Year-round high temperature, abundant rainfall and relatively low salinity of marine waters are factors that contribute to the geographic spread of the species and to the luxuriant development of the mangroves, especially where there are extensive coastal plains. On the other hand, volcanic, seismic and other telluric activities also prove to be a hazard to mangroves. In this context, it may be recalled that one of the first plant species that colonised the new island formed after the eruption of the Krakatoa, in between Sumatra and Java Islands, was the mangrove. The exuberance and rapid growth of mangroves in the equatorial belt of the Indo-West Pacific is a factor that facilitates management and enhances productivity.

Along the east coast of peninsular India, on the coast from southern Myanmar, Thailand and Malaysia eastwards to the Mekong River delta and throughout SE Asia in general, there are coastal lagoons and a multitude of medium-sized, long and short rivers, the discharge of which fluctuates significantly with the season. The major rivers are the Chao Phraya, which traverses Bangkok and is at present highly polluted, the Irrawaddy, which is highly impacted by firewood extraction and the Mekong, whose delta has been widely damaged by man for conversion of mangroves to rice cultivation and was also destroyed by man during wars. In a general way, coastal areas at low latitudes are maintained and well watered by monsoons, thus even in the absence of large estuaries and deltas, the littoral zone is or was extensively covered with mangroves. Species diversity is high, and individual development of trees and of the forest is very good. The tallest mangrove trees in the world grow in the equatorial belt; in Ecuador, South America, they reach over 60 m and in SE Asia may reach 50 m or more. The Ranong area on the Andaman Sea coast of Thailand is one of the best studied during the past 10–15 years. The South and North coasts of the Island of Sumatra are a good example of the key role that rains play for the development of the mangrove forest; the southern coast is dry, since there are only few short rivers, and the SW monsoon is weak. The northern coast has abundant equatorial rains, the monsoons are active and there are numerous rivers all along the coast; this, which in a way has been their disgrace, since mangroves have been extensively cut for other purposes, mainly aquaculture. Coastal mangroves of the Philippine Islands are exposed to the open ocean but grow in well-protected coastal lagoons and inland seas of the many islands.

In Southeast Asia, coastal mangroves occur throughout the equatorial belt north and south of the Equator. At low latitudes there are usually no well-marked periods of rain as under monsoon climates, where patterns of wind and rainfall are well marked alternating regularly with periods of drought. All the islands of Indonesia as well as Malaysia, Papua/New Guinea and the Island of Mindanao (Philippines) lie in the equatorial belt. Indonesia has about 17 000 islands; with a total area of about 43 000 km^2. Species diversity is the highest world-wide. There are periods of higher rainfall and seasonal winds, and some areas have very short periods of drought, up to two months in the southern coast of Sumatra. In general, the coastal climate is best defined as tropical humid where rainfall is never below 60 mm per month or where a short dry season is compensated by heavy rainfall the rest of the year. The mangroves of the equatorial belt, whether coastal, estuarine or deltaic, whether they have a monsoon pattern of rainfall or otherwise, favour fast and continuous growth and self-regeneration of the forest, frequent co-existence of many species, great intraspecific variability, and formation of hybrids.

Both in peninsular Malaysia and in Sabah and Sarawak, the coastal area has a typical equatorial climate. The SW and the NE monsoon periods are marked by winds that may be strong enough to cause some damage to the mangroves; however, the force of these winds is not comparable to that of cyclonic storms. An Anoxic condition of soils prevails everywhere, but due to abundant rainfall, pH may be relatively high, up to 5.5 or higher, especially where the substratum is calcareous and of coral origin.

The mangrove forests of Malaysia, apart those of Sabah and Sarawak on the Island of Borneo, are best developed on the West coast facing the Straits of Malacca, which is more sheltered than the East coast. The best examples of well-managed mangrove forests are those at Matang in peninsular Malaysia, facing the Straits. After a series of trials, management techniques were improved by making them appropriate to the place. Improvements continue as the ecosystem changes with time. It should be noted that the system that proved to be successful for the management of Matang mangrove forests would not necessarily be successful elsewhere. For instance, the management system used in the Sundarbans of Bengal is totally different; each place needs a type of management appropriate to the particular ecological conditions of the place and of the local natural species.

In SE Asia including Malaysia, Thailand, Cambodia, Vietnam, parts of the Philippine Islands and most of the Indonesian Islands and the southern coast of Papua/NewGuinea, the geographic position, in addition to climatic factors such as year-round high temperature, abundant precipitation and relatively low salinity of marine waters contribute to the widespread and luxuriant development of mangroves, especially where wide coastal plains extend to the intertidal area. On the other hand, volcanic, seismic and other telluric activities prove to be just as much a hazard to mangroves as to other formations. The exuberance of mangroves in the equatorial belt of the Indo-Pacific is a factor that favours good management of mangrove ecosystems when it is made appropriate to the site. Another characteristic of the equatorial belt mangroves that must be considered when management is planned is that under such conditions, there may be no clear-cut line dividing the above-mangrove vegetation from the mangrove vegetation proper. The lack of defined separation between the two biocoenoses is due to the relatively high pH of the equatorial mangrove soil, up to pH 6, to the relatively low salinity of the pore waters and to the presence of the upper-littoral flora tolerant of some degrees of pore water salinity. The presence of a transition zone between mangrove and upper-mangrove in the equatorial and sometimes also in the tropical zone is the major difficulty to be overcome by the scientists studying mangroves with remote sensing techniques. Typical of the equatorial and tropical coastal areas are the *pokkali*, *bheri* and other silvi-aquaculture systems of the northern part of the East coast of India; the *tambak* of Java, a system now extended to other parts of the Indonesian archipelago and brought to Angola and Mozambique by the Portuguese, are other examples of sea-level shrimp and fish farming that do not cause degradation of the mangroves.

The *tambak-sari* (Fig. 3.7) are ponds about one hectare in size, usually with an island in the centre where mangrove trees are planted. The permanently flooded area occupies about two thirds of the total area and is excavated down to a depth of about one meter or less below ebb-tide level, all around the central mangrove island which is inundated during flood. Fish and shrimp congregate in the peripheral ditch and are easily fished out. A specially dug main feeding canal brings tidal waters to the ponds; small sluice gates are opened during flood to allow shrimp and fish fingerlings to enter the pond, while fine meshed filters prevent their return to the feeding canal and the sea. The fishermen operate the small sluice gates manually at the correct time of

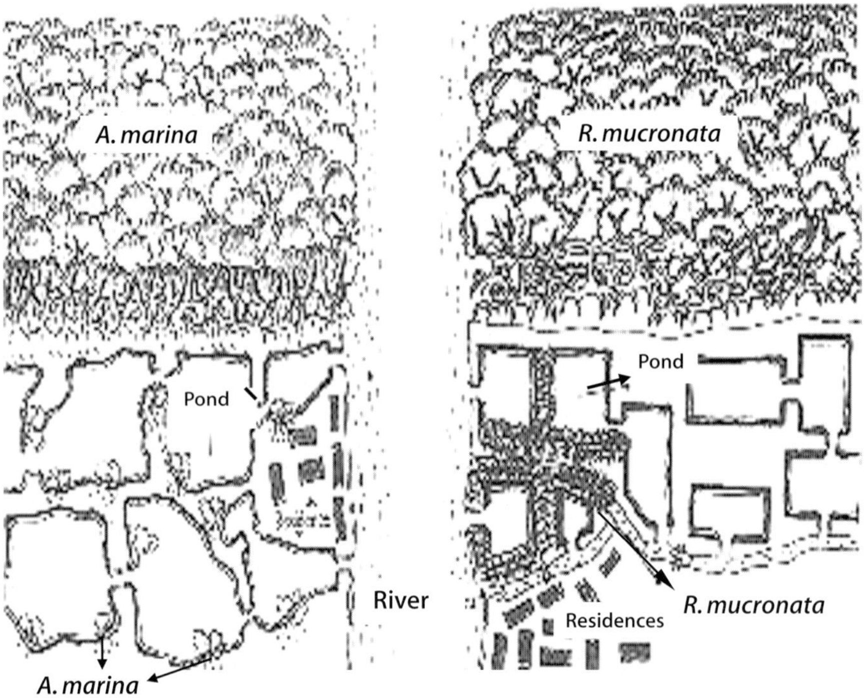

Fig. 3.7. Two examples of designs of Tambak-Area of mangrove plantation to *tambak* (1:1). Option 1: preferable for *Avicennia* spp.; Option 2: preferable for *Rhizophora* spp. (modified after Inoue et al. 1999)

the tidal cycle. Such ponds do not require artificial aeration, soil additives, fertilisers or added feed. Every few years, depending on the place, some sort of overhauling for maintenance of the soil and of the mangrove island is also practised. In 1980 in Indonesia, the total tambak area was 188.601 ha of which 18% were embankments. In 1991, the total tambak area was 290.983 ha of which about 15% were embankments.

The important lessons to be drawn from these examples are: first, any changes to be made in an ecosystem must be gradual to avoid degradation and often irreparable damage; second it is possible to change one ecosystem into another more suitable to human needs, provided indigenous or introduced species and manipulations work for and not against natural forces such as tides, winds, currents or species present; and third, mangroves should be partially preserved. After much testing and long periods of acclimatisation, new species can be introduced and varieties can be developed locally. In the Kerala backwaters, for instance, coconut, pepper vine, other spices and vegetables were introduced from abroad and had to be acclimatised to the area, while varieties of salinity-resistant rice had to be produced locally by artificial selection. Recently, species of fishes new to the area were introduced experimentally, as for instance *Chanos chanos*, the milkfish.

In conclusion, coastal mangrove ecosystems are best developed in the equatorial belt, where abundant rains wash the soil clear of excessive seawater salts and locally produced sulfides, sulfates and sulfuric acid. Mangrove coastal forests may extend inland as far as tidal intrusion is felt along rivers and creeks, where the coastal plain covers a vast area and the slope is gentle. The Equatorial belt of the Indo-West Pacific, where there is an extensive shallow coastal shelf and abundant runoff from land, favours sediment deposition and progradation of the land. For instance, Johore beach on the Straits of Malacca coast is an example of a protected open ocean coast where sea currents bring seeds and propagules that take root. The mangrove coast develops as best it can with alternate periods of partial destruction by long-shore marine currents and storms followed by regrowth. Experiments in assisted regeneration by protecting the area from winds and surf action with a fence parallel to the shore made vastly with discarded tires have proved successful but costly and limited in extent. Abundant organic matter in the sand of the beach sustains the densest population of mudskippers that I have ever seen. Many more examples could be quoted of highly dynamic and rich coastal mangrove areas of the Indo-West Pacific; for instance, coastal mangroves of Sri Lanka are exposed to the open ocean and grow best in well-protected areas such as lagoons or small river estuaries. In SE Asia, the open ocean is in fact a major inland sea from the Straits of Malacca lying eastwards in the equatorial belt. The geographic situation coupled with climatic factors contribute to the widespread luxuriant development of mangroves, especially where the coastal plains extend to the intertidal area and rivers and creeks are numerous. Extensive stands of *Nypa fruticans* may be found far inland. *Nypa fruticans* is everywhere the species that advances farther upstream. In conclusion, coastal mangroves are best developed in the equatorial belt where there are abundant rains that wash the soil clear of noxious chemicals and salts and where land runoff favours sediment deposition and progradation of the land.

The Matang mangrove forest (latitude 4°45' N to longitude 100°35' E) on the West Coast of peninsular Malaysia facing the Straits has a forest cover of 40 711 ha, 85% of which is managed as productive forest while the remaining 15% is 'unproductive', this includes *Avicennia* and dry land forest. The productive forest is primarily composed of two species of *Rhizophora*, with considerable loss of the original biodiversity. The type of management of the forest is clear felling in strips combined with natural regeneration and replanting. The main uses of the mangroves in the imposing Matang forest are for charcoal, firewood and poles.

The highly successful silvi-aquaculture systems originally from Sumatra and East Java have recently spread to other islands of Indonesia, to Vietnam and parts of Africa. The traditional *tambak* or tanks and *tambak-sari* are arranged in a row of plots, each two and up to seven ha in area, with a peripheral ditch dug down to about 1 m depth and occupying about 2/3 of the total area of the plot The centre of the *tambak-sari* plot is planted with mangroves usually of the genus *Rhizophora* and occupies about one third of the total area of the tank. The tanks are separated from each other by bunds raised to about lm above high tide. A feeding canal brings tidal waters to the tanks, each with a small sluice gate manipulated by hand. It is opened at flood tide to admit seawater, the shrimp larvae and fish fingerlings that come with it. Similarly to the *pokkali plots* of the Cochin backwaters in Kerala, the *tambaks* are at sea level; no pumping of seawater, aeration systems or additional feed or soil correctives are needed. Collection of the fish is easily done with push-nets or drag-nets. The waters are fertilised by the small individual mangrove forest in the centre of the tank or on the banks. On a rotational basis, some tanks are left unused for management of the soil, general overhauling and replanting of the island or other services as needed. The *pokkali*, *bheri tambaks*, *tambak-sari*, and other traditional systems produce less per unit surface than artificial ponds for intensive aquaculture, but they continue to be productive for centuries, while fish or shrimp ponds under intensive aquaculture are well-known to become unproductive after a few years, at which stage they are abandoned. An evident ecological hazard was created by intensive culture.

The coastal zone everywhere is a highly dynamic system, especially so in the tropical belt world-wide.

3.6 Oceanic Islands

The present sub-chapter is focused on the least known mangrove ecosystems of the Indo-West Pacific region. Continental margins, large deltas and extensive coastal mangroves are dealt with in the scientific literature in greater detail than mangroves of oceanic islands. At present, though there is much published information on the mangroves of the coasts of large landmasses, next to nothing has been published on oceanic island mangroves, though the islanders themselves have centuries-old experience in the uses to which they can be put. Because of scarcity of general knowledge on small island mangroves and because the relative importance of the coastal zone is greater the smaller the island, and furthermore in view of the threat of mean sea-level rise, the mangrove ecosystem's response to fluctuations in sea level and tidal fluctuations is ever more under scrutiny (Woodroffe 1999). The response of mangroves as an ecosystem to sea level fluctuations is not easily predictable; there is no simple answer to the question, and a widely applicable and standard predictive model is unlikely to be produced. It is not yet clear whether in past millennia mangroves responded adequately to changes in sea level, whether they are already affected or what will happen in the future at different places under varying scenarios. Like everything else about mangroves, each single site has peculiarities of its own, and the reaction of mangroves as individual species and as an ecosystem is linked to local factors such as: species composition, topography, land runoff, precipitation, winds, amount and nature of suspended matter in waters and air (micro drops of wind-born marine water), tidal regime, oceanic currents, and slope and extent of the shelf. What is certain, however, is that low-lying islands may become the first casualty of sea level rise, especially atolls. For these reasons much space is given here to this topic. From the point of view of the mangroves, the oceanic islands of the Indo-West Pacific may be classed as high rise islands formed mainly by volcanic activity or originated by growth and uplift of fossil coral reefs, and low lying islands and atolls. Very few oceanic islands of the Indo-Pacific have extensive mangrove forests, one exception being the Andaman-Nicobar Islands in the Bay of Bengal, around latitude 7–14°. The oceanic islands of the Western Indian Ocean, Madagascar, the Comoros, and Mauritius are treated separately under the Africa-chapter. Only the equatorial Seychelles, Socotra, the Lakshadweep and the Maldives from the Indian Ocean will be mentioned here, followed by mention of the oceanic islands of the Pacific. Oceanic islands are very numerous, indeed and are scattered over the entire Indian Ocean, SE Asia and western Pacific Ocean. In recent years, Untawale (1986, 1987), Jagtap and Untawale (1999) have studied the mangroves of the islands of the Indian Ocean, and Chinnaraj (1993) contributed the only paper we are aware of on the manglicolous fungi of the Maldives.

3.6.1
Socotra

The main island and a few smaller ones lie around latitude 12° N. off the southern coast of the Arabian Peninsula. Socotra was well-known as a convenient stop-over or landfall for trading, especially of "dragon's blood" or cinnabar, a product of *Dracaena* spp., and transfer of goods; it was also a hiding place for mariners who sailed the breadth of the Indian Ocean since ancient times. The mangroves of this island, other small islands of the Gulf of Aden as those of the Red Sea islands, and coasts were used for building shelters, houses and as a source of firewood, for emergency repairs of the crafts, and for medicine. What mangroves left now on Socotra are mainly on the southern coast of the island and certainly are not representative of what the original forest must have been; ports and harbours were on the northern coast, where little if any mangroves are left. Palynological studies should provide interesting information; there could well have been dense mangrove forests some centuries ago.

3.6.2
Seychelles

They are a group of some low lying and some high rise islands at about latitude 14° S. High rise oceanic islands usually have a steep coastal slope on land and a precipitous drop at sea; there is not much space for mangroves. The Aldabra atoll bears the largest amount of mangroves, with nine species (Untawale 1986). The mean annual temperature is 25–27 °C, and precipitation is 2 375 mm annually. Mangroves were common in 1964 when we visited the islands during the UNESCO-organised International Indian Ocean Expedition. Travelers, explorers, collectors and scientists had visited most islands and fauna and flora had been described. In more recent years, supply ships and fishing crafts, mainly long liners for tuna fish, visited the main island famous for its giant turtles and several plant and animal curiosities. Since then, the islands have become a tourist paradise, and what the condition of the mangroves was before and during the "age of discoveries" is difficult to say. Spalding et al. (1997) gives 29 km^2 as the area of mangroves in Seychelles.

3.6.3
The Republic of Maldives

These islands have played a significant though quiet role in the history of the Indian Ocean. Located on the main maritime trade route, ships often made landfall at the Maldives because of storms, unfavourable winds or other reasons. The Portuguese ships often visited them on their return travel from the "*spice islands*", the Moluccas, back to the harbours of Cochin and Cannanore on the SW coast of India, present-day State of Kerala. In India, the goods would be transferred to the larger vessels sailing back to Portugal. The Dutch who came later followed the same pattern: sail to India by the SW monsoon from May to October; the following year the NE monsoon was available for the return journey. In between the two major journeys from and back to Europe, smaller craft and Arab dhows went to the spice islands and returned loaded with goods that were usually unloaded at Cochin, at latitude 10° N, longitude 76°15' E. On the way, anchors were often lowered at Male if weather was rough or pirates menacing. The culture of the Maldivians, their unique language, called *dhivehi* and script, reflect the influence of Arabs, Portuguese, Singhalese and Keralites, which were the people with whom they traded. Very little is available in the scientific literature on the mangroves of this island nation exclusively formed by atolls; it is taken here as an example of the role that mangroves play in equatorial small low-lying islands where there is reduced vegetation cover. One of the most recent accounts of the mangroves of the Republic of Maldives, which is particularly interesting from the point of view of mangroves of equatorial atolls, is the paper mentioned above recently published by ISME (Jagtap and Untawale 1999).

3.6.3.1
The Mangroves of the Maldives Islands
M. Shafiu

Introduction. The Republic of Maldives consists of 1190 coral islands, which form a chain 820 km long and 130 km wide, set in an area of 90 000 km^2 of the Indian Ocean, including the exclusive economic zone. Geologically, the chain forms part of the Lakshadweep-Chagos submarine ridge that extends to the Central Indian Ocean from the Southwest coast of India (Fig. 3.8).

The islands form 26 natural atolls administratively grouped into 19 units. Most of the islands are small, few with a land area in excess of 1 km^{-2}. They are low-lying with an average elevation of 1.6 m above sea level. The islands extend 765 km north to south; therefore, there is a fair degree of variation in rainfall between northern, central and southern atolls, which warrants an arbitrary division into three ecological zones. Table 3.11 shows mangrove species present and physical parameters of major atolls of the Maldives. Annual precipitation varies from 1600 mm to 2200 mm. The average monthly temperature is 25–31 °C, which is fairly constant throughout the year. The soil is relatively young and consists of coral sand to loamy sand with a layer of dark soil varying in depth from 20 to 40 cm. There is a central depression in every island where the water table is high and the organic content of the soil is higher than on the fringe.

Farming system. The agricultural sector consists of small-scale production crops: coconut, fuel-wood and fruits as well as poultry and other small livestock. Its contribution to the GDP has fallen steadily from 19.7% to 7.7% from 1982 to 1995, probably due to relative expansion of tourism, trade and fisheries. However, in real terms, agricultural production has increased from 1988 to 1994, due to improved agricultural techniques, good quality seed, usage of fertilisers, and pest management. The farming system includes a mixed home gardening and bush fallow system of shifting agriculture. Most plants needed for domestic consumption are grown in mixed home-gardens, which is a regular practice.

Mangrove-related ecosystems. The benthic fauna at sea is the usual coral reef fauna at low latitudes. However, depletion of the fauna is reported as associated with intensive erosion in some places. In the mangrove forest in general, the principal photosynthesisers are the trees; however, in these islands where the mangroves are reduced and often do not form a closed canopy, other angiosperms, phytoplankton and

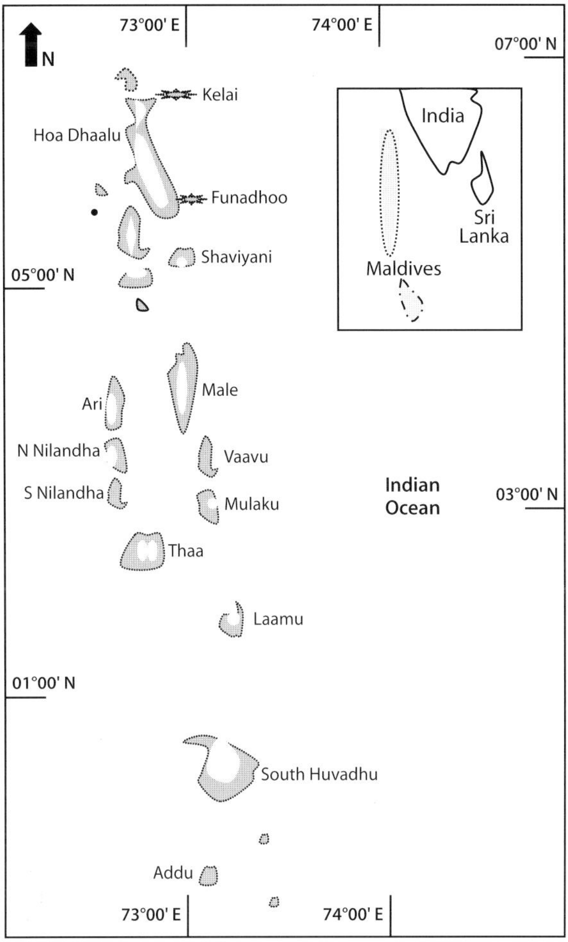

Fig. 3.8.
Map of Maldive Islands

marine algae, including the zooxanthellae commensals of the corals, may contribute significantly to the primary production of the islands. Detritivorous species of fishes and crabs in particular may play a role in the production of the mangrove stands. The mangroves only rarely inundated by high tides, are: *Lumnitzera racemosa* and *Ceriops tagal*. The mangroves grow in depressions, and adjoining to these the supratidal community is composed of the usual Pemphis *acidula*, *Scaevola taccada*, *Thespesia populnea*, *Hibiscus tiliaceus*, *Suriana maritima*, *Pandanus* sp. (screw palm) and the coconut. Numbers vary from island to island but must be counted for the purpose of ecosystem management. The sandy soil is of coral origin having very little or no organic matter. The mangrove coastal community has a strategic location to check the impact of storms or high waves.

Gaafaru Island. Gaafaru is a small island on the north of Kaafu atoll; it has a population of about 800. Length and width are approximately 240–360 m. It has two very small

Table 3.11. Mangrove distribution in Maldive Islands. Average daily temperature and seasonal range 26 to 30 °C. Tidal amplitude ranging from 0.7 to 0.9 m at spring tides. Annual precipitation range between 1500 to 1950 mm (further details in Jagtap and Untawale 1999)

Island (atoll name)	Mangrove area	Population	Species present
Kelai (Haa Alif)	200 ha	1 604	*Sonneratia caseolaris* *Bruguiera cylindrica* *B. gymnorhiza* *Lumnitzera racemosa* *Rhizophora mucronata* *Avicennia marina* *Derris heterophylla*
Landhoo (Noonu)	120 ha (20 ha)	–	*Sonneratia caseolaris* *Bruguiera cylindrica* *B. gymnorhiza* *Rhizophora mucronata* *Lumnitzera racemosa* *Excoecaria agallocha* *Acrostichum aureum*
Gaafaru (Kaafu)	9 ha (113 ha)	800	*Bruguiera cylindrica* *Rhizophora mucronata*
Kaashidu (islands and atoll)	(3 patches)	1 000	*Sonneratia caseolaris* *Bruguiera cylindrica* *Rhizophora mucronata* *Pemphis acidula*
Farukolhu, Funadhoo (Shaviyani)	89 ha (7.2 ha)	–	*Ceriops tagal* *Bruguiera cylindrica* *R. apiculata* (?) *Pemphis acidula*

patches of mangroves, about 1800 and 700 m^2 each. Species present are *Bruguiera cylindrica* and *Rhizophora mucronata* growing on hard sandy soil. According to local people, these patches are inundated by seawater only during spring tides of the months of May and June. However, they receive floodwater from the island during heavy rains, since both these spots are located in small depressions. These mangroves could become established because of deposition of silt and clay brought by rain run-off. The stands are very dense, with about 6–7 plants m^{-1}; the patches are estimated to be 12–15 years old; though there are gradations of all ages, there seems to be no regeneration, probably due to the dryness of the substratum, although the plants appear to be healthy. The stand is too thick, and thinning is needed to induce an increase of girth of the trunks. The tallest pole is over 3.6 m high. Conservation is needed to preserve biodiversity; the stand is just landwards from the coral sandy beach. It is suggested to plant seedlings of *Bruguiera cylindrica* on the fringe of these patches during the rainy season to explore the possibility of expanding the patches. If successful, the practice could be extended to other islands to increase natural protection. It is also proposed to dig a very narrow canal to connect the sea to the stand to promote more frequent flushing by seawater at spring tides. Such a canal has to be excavated only after a proper level survey is done to avoid erosion or loss of silt and clay from the island to the ocean.

Kaashidu Island. For administrative purposes, this is included in Kaafu atoll. It is a sizeable island with over 1 000 inhabitants. There are three patches of mangroves on the island. Mangrove species present are: *Sonneratia caseolaris*, *B. cylindrica* and *R. mucronata*, all lying in depressions. *Sonneratia* grows towards the centre, while *Bruguiera* and *Rhizophora* grow towards the shore. In the mixed stands, *Bruguiera* is dominant, and there are small stands of pure *Bruguiera*; they all grow in association with tall screw-pines (*Pandanus* sp.) about 9 m high, *Scaevola taccada* and ferns. The narrow belt between the fringe of the depression and the deposit of coral rock on the beach is mostly occupied by *Pemphis acidula*. The floor of the mangrove patch is moist and full of thick litter, and soil depth is about 15 cm. The tallest *Sonneratia* observed is about 11 m tall with 1.5–1.8 m dbh. Natural regeneration is poor and poles are few. There are abundant stumpy pneumatophores of *Sonneratia*. The size of the largest patch is estimated to be somewhat less than half a hectare. *Bruguiera* is the most common species, regeneration is in patches, young plants over 2 m high with dbh greater than 7.5 cm grow back to back, and there are about 25 plants m^{-2}. The largest *Bruguiera* observed was about 6.5 m tall and 0.75 m dbh. Stocking is too dense and plants are too congested to grow, so branches are found dying. The area is a depression lower than the average height of the island. Tides do not reach the area, but during spring tides of the monsoon season, seawater percolates into the depression from below. During the rainy season, runoff water from the island accumulates into the depression bringing silt, clay and humus thus the area remains moist, soft and spongy, giving increased water holding capacity to the site. The water table is about one meter below ground level as a thin lens of fresh water floating over the saline water. A local islander had planted *B. cylindrica* in the 1950s by direct planting of propagules, primarily to forestall any probable famine situation, since the propagules can be used for human consumption after boiling. Later, five other islanders also planted *B. cylindrica* on very small patches; these plots are now abandoned because the propagules had been planted too close to each other, just 10–15 cm apart. Local people use *Bruguiera* and other mangrove species for boat building, house construction, boat plying poles, fencing stakes, firewood, and others. Bark is used for medicines and for dyeing fishing lines and nets. Fruits of *S. caseolaris* are refreshing and fetch a good price, up one rufiyaa each at Malé.

The fringe of the depressions harbouring mangroves has been brought under farming. Screw pine (*Pandanus* sp.) and other vegetation is being cleared for the purpose. The soil is very shallow with a depth of about 15 cm at most. After a few years the farmer is compelled to abandon the farm, because all the topsoil is finished. These marginal lands should either be put under tree cover or be used for silvi-pasture. In and around these depressions an estimated area of about 2–2.8 ha appears to be suitable for mangrove afforestation.

Small-scale afforestation at the rate of about half of a few hectares per year may be undertaken. Species such as *S. caseolaris* and *B. cylindrica* may be tried on these sites. Results of such operations need to be closely monitored for an on-the-spot improvement of the techniques. The congested stands of *B. cylindrica* urgently need thinning; community participation should be incorporated with these activities. Details can only be defined after a study of the site and a complete inventory of the growing stock has been made. Rules for cutting *Pemphis acidula* growing on the seashore must also be framed; the present procedure of harvesting only the younger branches of less than 2.5 cm diameter is likely to be injurious to the growth and permanence of this species.

Farukolhu Funadhoo. This is an island of Shaviyani atoll, where the atoll Chief is stationed. The island is approximately 2 200 × 423 m long and wide. It encloses a shallow lagoon about 1 200 × 60 m long and wide, situated on the eastern side of the island (Fig. 3.9). At ebb tide, the average depth is barely over 20 cm; the lagoon is open to the sea at the north end; the land on its east side is about 7 m wide. The narrow belt of mangroves grows only on the eastern side of the lagoon; the dominant species is *C. tagal*; there are also three trees of *B. cylindrica* and few *Rhizophora* (probably *R. apiculata*) scattered and mixed with *C. tagal*. A younger population of *Ceriops* is near the water line of the lagoon, whereas the older ones are further away, which suggests that this species is slowly advancing towards the lagoon. The mangroves look vigorous and healthy, but regeneration is very poor. The crown is closed at about 0.5; they grow in a belt of about 3–4 rows. The adjoining community to the east is composed mainly of *Pemphis*, which extends up to the shoreline. The soil is soft, sandy, of coral origin; there is almost no silt or clay. Local islanders report that the lagoon has 40 species of fishes, 10 species of mollusks, 12 species of crabs and 4 species of small shrimps. One species of crab is reported to grow up to half a meter in diameter. The island Chief told us that these crabs eat the young seedlings of the mangroves. At present, the exploitation of mangroves is restricted, very little or no extraction is done, and only occasional cutting takes place. Attempts could be done to afforest the lagoon front by assisted regeneration with Ceriops and *Rhizophora* spp. In between the older trees of *Ceriops* that grow away from the lagoon front, some *B. cylindrica* may be planted on a trial basis. Close monitoring of results is required to develop a site-specific technique. The *Pemphis acidula* population, abundant to the east of the mangrove patch is an important feature, but most of these bushes have grown old. At present, branches about 3.5 cm diameter are cut to be used as wooden nails in boat construction, but it is not a judicious practice, and most bushes are over-exploited. Present practices must be regulated to achieve maximum benefit on a sustained yield basis. Destruction of the *P. acidula* stands will impair the adjoining mangroves.

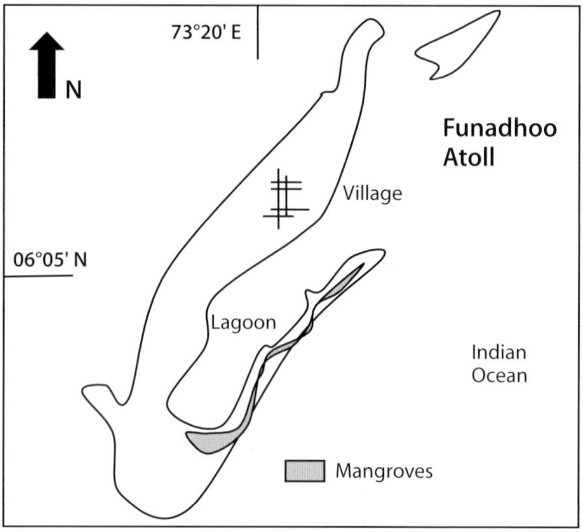

Fig. 3.9.
Mangrove distribution at Farukolhu Funadhoo Island

Kelai Island Haa Halif Atoll. This atoll has 16 inhabited and 24 uninhabited islands. The following 6 islands of this atoll are reported to have mangroves: Kelai, Filladhoo, Muraadhoo, Thakandhoo, Baarah and Madulu. Kelai has an area of about 200 ha, it has three patches of mangroves, the one at the north-eastern side is the largest; Fig. 3.10 shows their approximate location; all the mangroves grow in the depressions which receive rain run-off from the island; the rains bring silt and clay, which accumulate in the depressions in which the mangroves grow. These ground depressions also receive marine water from underground porous connections with the sea during the high tides in the months of June and July.

The species of this island are: *S. caseolaris*, *B. cylindrica*, *L. racemosa*, *B. gymnorrhiza*, and *R. mucronata*. Zonation is distinctly visible with *Sonneratia* on the fringe of the depressions towards inland. The bottom of the depressions with more silt and clay are occupied by *B. cylindrica*; the less clayey area has *L. racemosa*. Few *B. gymnorrhiza* and *R. mucronata* grow mixed with *B. cylindrica*, especially in the two southern patches. The soil under *Sonneratia* is almost dry, with lots of humus but is not soft. Almost all the *Sonneratia* trees show signs of top dying, and a few are already dead. The soil under *L. racemosa* is comparatively sandier and less wet than under *B. cylindrica*; the soil is about 0.3 m deep. Mangroves are mostly scrubby, and the average height of *Lumnitzera* is about 3.5 m; on an average there are 4 scrubby bushes on a 9–10 m^2 area; they look healthy but are too thickly branched, and the floor is covered with dead twigs and branches. Stakes for fences are collected from this area; regeneration is poor and scanty. The soil under *B. cylindrica* is soft down to 0.6 m, and wet with a large quantity of humus and litter. This species occupies the largest area under mangroves on this island. *B. gymnorrhiza* and *R. mucronata* trees are few and scattered in the *B. cylindrica* area. A rough estimate indicates that *B. cylindrica* covers about 16 ha; it is very dense, and the height of the trees is about 5 m (Fig. 3.11).

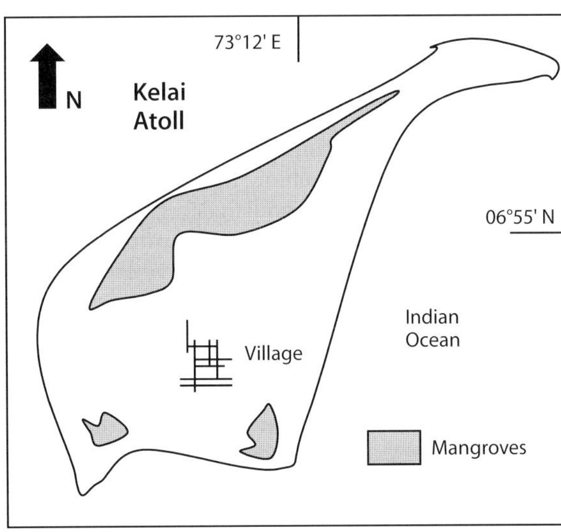

Fig. 3.10.
Sketch showing the approximate location of mangroves in Kelai Island

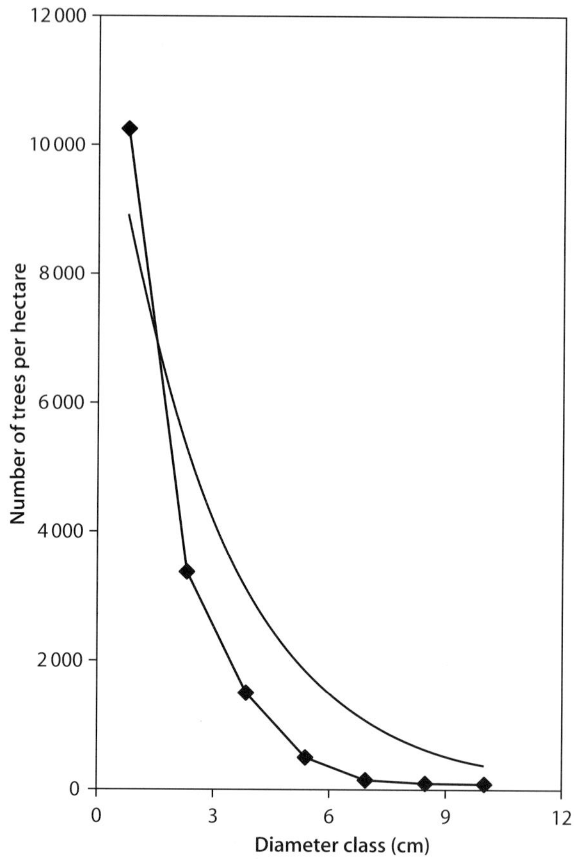

Fig. 3.11. Diameter class distribution of *B. cylindrica* stand at Kelai Island

Landhoo Island Noonu Atoll. The following islands bear mangroves: Kendhikhuladhoo, Dhonerikandoodhoo and Landhoo. The latter has the largest mangrove vegetation; the species present are: *S. caseolaris* (Kulhlhava), *B. cylindrica* (Kan'doo), *B. gymnorrhiza* (Bodu Kan'doo), *R. mucronata* (Ran'doo), *L. racemosa* (Burevi), and *Excoecaria agallocha* (Thella). The area is roughly 120 ha. Mangrove patches are in depressions; the one in the north is longer, and the southern patch is wider. Both patches together occupy an area of about 20 ha; the patch in the north has all six species, but the one at the south has only *S. caseolaris* and *B. cylindrica*. *S. caseolaris* is found on the inland side of the depressions. *B. cylindrica* and *B. gymnorhiza* occupy the wet area in the centre of the depressions. *L. racemosa* occupies the more sandy area towards the sea. *Excoecaria agallocha* is found in between *Lumnitzera* and *Bruguiera* stands. *R. mucronata* is found mixed with *Bruguiera*. The uses of these species are – *Sonneratia caseolaris*: Timber for boat and house construction; fruits are edible. Branches are used as firewood, and the pneumatophores as bottle-corks. *Bruguiera cylindrica*: Timber for boat-building. Poles with 7–9 cm diameter are used as boat plying poles. Small pieces are used for docking the boats for repairs. The propagules are boiled and consumed as food. Bark is sometimes used in traditional medicines and for dyeing fishing lines. *Lumnitzera racemosa* is used as house posts and in the roof structure. The wood is mostly used for fencing stakes. Small wood is used as firewood. Butt logs may be used in boat building. *Excoecaria agallocha*: Poles are used as fencing stakes. It is rarely used in roof structures as purlin. It is used as firewood, but demand is low. *Rhizophora mucronata*: The uses are similar to those of *B. gymnorhiza*, but it is not frequently available.

The depressions have stagnant water in the centre and have no apparent direct link with the sea, but they receive seawater from underground connections during high tides, down to a depth of about half a meter. During the monsoon they receive rain run off and become flooded. Apparently they have no fishes, but there are three species of crabs; one of them is up to 30 cm diameter. There is one species of mollusk in the marsh. The *S. caseolaris* stand is sizeable; the average dbh of older trees is about one meter and the height 16 m. Some trees appear to be hollow. There are few young trees on the southern patch that are reported to have been planted by the islanders around 1986; some of these bear fruits. The soil is dark, deep and soft down to about half a meter; it contains much humus. The pneumatophores are stumpy and are up to 0.5 m tall. The fruits can be sold; as a norm anyone of the community can collect them and sell them. No natural regeneration of *Sonneratia* is visible. It is reported that during the wet season natural regeneration shoots up, but the seedlings die as the dry season sets in. The stand of *B. cylindrica* is almost uniform. The average dbh is about 15 cm. The height is about 6 m. The crown closer is 0.4. Natural regeneration is scanty; the floor is wet and contains much humus. The stand of *B. gymnorrhiza* has an average height of about 7.5 m; the tallest tree measured is 12.3 m tall with dbh of 12.7 cm. No natural regeneration was noticed. The soil is comparatively drier but contains enough humus. The stand of *L. racemosa* has an average height of 4.5 m with an average dbh of about 5 cm. Most are scrubby, but a few here and there have grown to tree-shape. The floor is covered with branches and twigs, indicating that extraction is taking place. The estimated crown closer is 0.8. The floor is soft and humid, but no natural regeneration was noticed. The stand of *E. agallocha* grows as scrub; the average height is about 5.4 m, the average dbh is about 5 cm. The crown closer is about 0.8 and there is no natural regeneration. *R. mucronata* trees are few and occur mixed with *Bruguiera*.

Traditional uses. Different species of mangroves have different uses by the local islanders. *B. cylindrica* wood is mostly used in boat building or for boat plying poles. The wood of *Rhizophora*, *Sonneratia* and *Ceriops* is used for house construction, especially in the roof structure. *Lumnitzera* is mostly used as fencing stakes; the larger ones are sometimes used in boat building and for making roof frames. The use of *Sonneratia* is for making planks and for house construction. The fruits, especially the propagules of *B. cylindrica* and the fruits of *Sonneratia caseolaris* are used for human consumption, and they can be collected by any member of the community; they are popular and fetch a price of one rufiyaa each at Male, but islanders sell them at the rate of 5–10 to the rufiyaa. The bark of *Bruguiera* is occasionally used in traditional medicine; those of *C. tagal* and those of *Bruguiera* are sometimes used for tanning fishing lines. Large scale use of mangroves has decreased since the felling and extraction of naturally grown plants was prohibited, except when permission is granted by the government. The law of the land permits the grower of a tree to be its owner if he has planted it. Many islanders have planted mangroves, especially *S. caseolaris* and *B. cy-*

lindrica in the inhabited islands. The communities live close to the mangrove patches. In some cases, as in Landhoo island, the islanders planted *S. caseolaris* in their backyard. Some of these trees have grown to over 15 m in height and reached a dbh of 1 m.

Mangroves play a very important role in the protection of human habitations. In some of the inhabited islands the mangrove patch is the only thick vegetation cover; it serves as a natural protective barrier to the inhabitants and the island itself. Severe tropical storms, cyclones or typhoons are not too common, but the country experiences rough and stormy weather in the months of April and May and again in October–November. Maniku (1990) reported on the recorded occurrence of cyclones: October 1733; 7 June 1752; 7 May 1812; 9 October 1819; 29 December 1819; 8 December 1821; 1898; 25 December 1923; 9 January 1955; 23 June 1987. There were also "freak storms" between 1971 and 1988 that affected 97 islands, with different storms hitting different islands. There are no detailed records of the protective role of mangroves, but surprisingly enough the long list of the storm-affected islands between 1812 and 1988 does not include the names of Kalai and Landhoo Islands, which harbour a good mangrove vegetation. Mangroves also enhance silt deposition in the depressions and reduce wash-off, a fact of vital importance in any small island; accumulation of humus in the depressions may reach over half a metre in thickness, especially under *Sonneratia* and *Bruguiera*, whose roots retain the silt carried from land by rain run-off; silt might otherwise get lost to the sea. The mangroves of the Maldive Islands are not big enough to harbour large animals; among mammals, bats are important as pollinators of *Sonneratia*, and large frugivorous bats are common in the not-too distant Comoro Islands; rats and some birds are present; no detailed reports of the fauna are available, and population and species diversity vary from island to island. On the other hand, the marine fauna is rich in species diversity and number of individuals. Though during some seasons there is water in the depressions, they are reported to be devoid of fish; however, there are fishes in the lagoons bordered by mangroves; these include milkfish, tilapia, mullet and flag-tail fish; shrimps are also present. The exact relationship between mangroves and fishes in the Maldives is yet to be studied, as well as the influence of the mangroves on species diversity and size of the fish population.

Biodiversity. The Biodiversity of mangroves of the Maldives has not been studied from the genetic point of view. The "Catalogue of Plants" published by the Ministry of Fisheries and Agriculture, October 1992, includes 486 species of plants found on the islands. Most species are exotic and were introduced at some time in the recent past. Some of the few indigenous plants are: *Pandanus* sp., *Pemphis*, *Suriana*, *Hibiscus*, and the reported 13 species of mangroves; on the Island of Landhoo alone there are six species. Biodiversity of mangroves should be maintained, and the genetic, physiological and morphological adaptations of the species should be studied. They evidently have the genotypic potential to adapt to neutral or high pH soils, low tidal range, seasonal rather than daily alternation of fresh and sea water inundation, and alternation of dry and humid conditions. As elsewhere, the Rhizophoraceae are the best pre-adapted to vegetate under very stressing conditions; however, the scarcity of Avicenniaceae and presence of a single species, *Avicennia marina* with a restricted distribution in the Maldives is unusual, because among all mangroves this is the species spread over a wider range world-wide and tolerant of the greatest environmental variations. This may suggest that the species composition of mangroves, as of other plants of the

Maldive Islands is neither natural nor fortuitous, but is mainly the result of human colonisation. In fact, all the species present serve a specific useful purpose.

At present, the maldivian mangroves have no commercial use and are also of no tourist interest. Tourists visit the islands mainly for diving and for water sports, which do not include mangroves, but they could eventually become an added attraction for ecotourism. Commercial uses have at present been restricted by law; imported timber is now used for boat building, also because it is cheaper and easier to procure. The size and the extent of the mangroves of the Maldives are not big enough to sustain a regular commercial supply; their basic value and the reason for which they must be maintained and nurtured into further development is entirely ecological and for internal consumption and traditional uses. From the scientific point of view, mangroves of atolls require detailed studies; these could lead to a better understanding of the ecological role they play in the world at large, of the peculiarities of the genome and details of the genetic processes that facilitate their adaptation to various sorts of unusual stressing environments. The adequate management of the mangroves of the Maldive Islands is a serious matter to be considered.

3.6.4
Mangroves from Other West Pacific Islands

Lakshadweep Islands. Also known as the Laccadives, these form a chain of low lying islands, have volcanic origin and are relatively young and still emerging (Ahmad 1970; Jagtap and Untawale 1999); they continue southwards into the Maldive atolls and the Chagos Islands. Like the Maldives, they are low lying; many are inhabited and have played an active role in the history of communications in the Indian Ocean and adjoining seas. The main difference between the two archipelagos is that the Lakshadweep belong to India and therefore are not as isolated as the Maldives, and they receive most support and supplies from mainland India. The Lakshadweep archipelago is formed by two main groups of islands, one to the north and the other to the south of the 10° N parallel; the two groups are separated by the "9 Degree Channel", which was a well-known route between the west coast of India and the Arabic peninsula. Mangroves are limited to Minicoy Island in the south, which is at a distance of only about 60 nautical miles from the northern-most atoll of the Maldives, the Tiladummati. The mangroves of Minicoy are dominated by *Bruguiera cylindrica*, mixed to stray plants of *Ceriops tagal* and *Avicennia marina*. These, except the last mentioned, are the most common and useful species in the Maldives. Though propagules and seeds could theoretically have been carried by the SW monsoon ocean currents to Minicoy, the best guess is that these species were purposefully transferred by man centuries ago, from one group of islands to the other; in what direction is anybody's guess at this stage. People from South India and from Sri Lanka were surely involved in the process. Linguistic affinities point to a close relationship between the Maldives and Sri Lanka. Jagtap and Untawale (1999) compared the mangrove formations of both groups of islands and found them to be similar.

Andaman-Nicobar Islands. These islands form an archipelago of about 570 islands and islets that extend over a length of 700 km, roughly from about latitude 6 to 14° N and between longitude 92 to 94° E in the Andaman Sea of the Bay of Bengal. Recently,

Dagar et al. (1991) published a complete and updated report on the mangrove ecosystems of the islands. The book includes detailed analytical studies of the geography, climate and hydrology of the island's environment, including soils, waters, topography, flora, and fauna as well as ethnobotany. Earlier, pedologic and edaphic studies had been too few and scattered, and many generalisations were built on an all too small number of actual observations. Special emphasis is on factors that determine the presence, abundance and distribution of mangrove species and on the adaptations of the mangrove ecosystem to the various islands of the archipelago. The yield in volume, weight and kind of the various resources produced by the mangroves is given in detail. For instance, Balachandran (1988) gives the yield in number of boles of different diameter classes in the Andamans. The number of boles of North, Middle and South Andamans is respectively 1431, 734 and 1140. The yield in terms of $m^3 ha^{-1}$ is given respectively as 115.1, 120.8 and 58.9 $m^3 ha^{-1}$. The main timber producing species are *Rhizophora* spp., *Heritiera littoralis* and *Bruguiera* spp., in decreasing order of number of boles and volume of wood produced (Balachandran 1988). There are at present 34 mangrove species belonging to 17 genera and 13 families.

An important contribution of the book is the attention given to the many mangrove-associated species, also called "non-true mangroves". These species are often by-passed by field workers and authors alike. They are neglected on the assumption that they are not important, because they are not exclusively found in the mangrove ecosystem itself. While the timber of these species is in many cases negligible in terms of quantity and quality, their worth as producers of minor forest products and their ecological or ethnic significance is often considerable. The ecologically mangrove associates, preferably called "above-tidal species" should be regarded as species not yet fully adapted to the intertidal zone. In other words, many of the above-tidal or supra-tidal or "littoral" species are those that, in evolutionary terms, have not yet been able to make the final jump over the ecological barrier that separates the exclusive mangrove species from those that may be close to becoming mangrove tidal species but are not yet fully adapted to the intertidal conditions. The "above tidal" or "supra-tidal" terms include the zone inundated only by spring tides; they occupy the flat strip of land along the coast with a width of a few to several metres, according to the local slope of the coastal plain. There are some stray examples of salinity adapted inland species, as for instance species of *Prosopis*, which are at present invading the intertidal zone (pers. com. by K. Bandyopadhyay and V. Selvam) also in the area of the Vellar River estuary in South Tamil Nadu, India. The conquest of the intertidal zone would have been favoured by the climatic conditions of the equatorial belt world-wide.

The mangroves of the Andaman and Nicobar Islands have remained virtually untouched up to the twentieth century and are therefore an ideal place for the study of structure and dynamics of insular mangrove ecosystems. The foundations of these studies have been laid by Dagar et al. (1991). At the time of writing the book, aquaculture practices had already begun in the Andamans. Areas cleared for fish and/or shrimp ponds were correctly placed behind the mangroves.

The main products of the Andaman/Nicobar mangroves are: timber for construction purposes, fuel wood from a variety of species, coal making from *Cerbera manghas*, *Avicennia* spp. and others. Several mangroves and supra-tidal species are used for making boats (Dagar 1986). Additional uses of mangroves are for tannin and dye extraction, thatching materials, fodder, cordage, ropes, and mats, including those taken from such species as *Pandanus* spp. or *Hibiscus tiliaceus*. Harpoons for hunting the dugong and for making elephant gear are usually made from *Hibiscus* wood. Fish poison, besides *Derris* spp. is extracted from *Dodonaea viscosa*, *Barringtonia* spp., *Aegiceras corniculatum*, and *Excoecaria agallocha*. Different species are used as food, beverages, for making wrapping materials, plates, and objects of arts and crafts. Bows are made from stripped wood of *Manilkara littoralis*. Medicines are made from many different species and used for curing various ailments, some of them not yet included in the classical treatises of *Indian Materia Medica*, such as those of Chopra et al. (1956) and Nadkarni (1976). Mostly they refer to "minor forest products". Bandyopadhyay (1986a,b) valued one hectare of mangroves in 1985 at US$11 364 as compared to agricultural production in the same area, which was of the order of a maximum of US$455 ha^{-1} yr^{-1}. The book concludes with a valuable, and to this day valid chapter on management of mangrove forests, especially those of oceanic islands.

The Andaman and Nicobar Islands were relatively untouched until recently except for a Penal Colony established during colonial rule at the capital city of Port Blair in the Great Andaman. The Nicobar Islands' mangroves are mostly in a pristine state and poorly known. In olden days, mangroves were mainly cut for poles and piles. Species diversity is high, and the tallest trees may be up to 25 m tall. The archipelago has an average precipitation of over 3 100 mm yr^{-1}, an average humidity of almost 80% and mean minimum temperature of about 23 °C. Tides are semidiurnal with amplitude up to 2.25 m at places. The main islands are the raised peaks of a submarine chain of mountains, but numerous coral islands also dot the archipelago. There are no perennial major rivers, but creeks and abundant rains and runoff from land provide sufficient freshwater and organic matter to enrich the substratum. There is a marked low precipitation season from December to March. Mangroves occur almost along the entire periphery of the main islands, and the substratum on which they grow is accordingly varied. *Avicennia* and *Sonneratia* have trees up to 20 m tall and are more common on coral reef flats, while *Rhizophora apiculata* and *R. mucronata* are more abundant on the muddy seaward area with maximum inundation. The landward end where there is maximum exposure time is colonised by *Heritiera littoralis*, *Excoecaria agallocha*, *Acrostichum* sp., *Nypa fruticans*, and *Phoenix paludosa*. Water salinity, including tidal water, is usually below or well below normal seawater salinity (35). Twenty-six or twenty-seven species of mangroves are known from the islands; *Avicennia officinalis*, *Bruguiera gymnorrhiza*, and *Rhizophora apiculata* are among the tallest and most common species. The estimates of the area covered by mangroves vary from nearly 45 000 ha to about 115 000 ha. There are numerous mangrove associates. The great biodiversity of mangroves in the Andaman/Nicobar Islands as compared to the coastal zone of the Bay of Bengal as it is at present, considering their distribution and relative abundance, is indicative of the original biodiversity of the mangroves of the Indian Ocean.

Twenty-three non-exclusive mangrove species are represented in the Andaman/Nicobar Islands. There are 34 true mangroves; they belong to 17 genera and 13 families. The list of species present is Acanthaceae: *Acanthus ebracteatus*, *A. ilicifolius*, and *A. volubilis*. Arecaceae: *Phoenix paludosa* and *Nypa fruticans*. Avicenniaceae: *Avicennia marina* and *A. officinalis*. Cesalpiniaceae: *Cynometra ramiflora* and *C. iripa*. Combretaceae: *Lumnitzera littorea* and *L. racemosa*. Euphorbiaceae: *Excoecaria agallocha*. Meliaceae: *Xylocarpus moluccensis*, *X. ranatum* and *X. mekongensis*. Myrsinaceae:

Aegiceras corniculatum. Plumbaginaceae: *Aegialitis rotundifolia*. Rhizophoraceae: *Bruguiera cylindrica*, *B. gymnorrhiza*, *B. parviflora*, *B. sexangula*, *Rhizophora apiculata*, *R. lamarckii*, *R. mucronata*, *R. stylosa*, *Ceriops tagal*, *C. decandra*, and *Kandelia candel*. Rubiaceae: *Scyphiphora hydrophyllacea*. Sonneratiaceae: *Sonneratia caseolaris*, *S. alba*, *S. apetala*, and *S. griffithi*. Sterculiaceae: *Heritiera littoralis*. The most common species are *R. apiculata*, *R. mucronata*, *B. gymnorrhiza*, *H. littoralis*, *N. fruticans*, *S. caseolaris*, and *E. agallocha*. This extensive list of species, which is probably the minimum number of species of the original mangroves of the Indian Ocean, suggests that throughout the Indian Ocean, the mangrove flora is at present only a relict of a much greater variety of species. It seems evident that there would have been a very rich mangrove biodiversity in the Tethys Sea. A short list of the closest mangrove associated species points at the same conclusion.

The Andaman-Nicobar Islands are most likely a refuge of the flora and fauna of the eastern Tethys Sea. The Botanical and the Zoological Surveys of India, located in Calcutta, have been busy completing the inventory of species of these islands, which are still to a great extent in pristine conditions. Until now, the impact of human population density has not exceeded the natural regeneration capacity of the mangrove ecosystem and probably of other ecosystems as well. However, growing industrialisation and urbanisation pose a threat to the future.

Ryukyu Islands. Among the islands of the West Pacific, we include here the Ryukyu Islands of southern Japan and the mangroves of New Zealand. The archipelago of the Ryukyu extends from the Tropic of Capricorn to latitude 31°22' N, which includes in this count also a small stand of *Kandelia candel* at Kiire, near Kagoshima on the southern Island of Kyushu in mainland Japan. Eleven species were recorded from the Ryukyu Islands; their northern limit of distribution in the archipelago is as follows: *Kandelia candel* to Satsuma Peninsula, Kagoshima Prefecture; *Bruguiera gymnorrhiza* to Amami-oshima Island (Kagoshima Prefecture); *Rhizophora stylosa* (Okinawa Island); *Avicennia marina* (Miyako Island, Okinawa Prefecture); *Sonneratia alba* (Ishigaki and Iriomote Islands, Okinawa Prefecture) and *Nypa fruticans* (Iriomote Island). The distribution of these species is interesting, because the islands are located almost exactly in a North-South direction, but how much of this is natural distribution related to latitude, or how much was caused by man's interference, is not known. The mangroves of Okinawa, the larger island, have been to some extent sacrificed to urban expansion and road building. At present, due to the promotion for the conservation of coastal ecosystems and enhancement of the importance of mangroves, there has been successful replanting along river estuaries, and as protection to river embankments, coastal roads and other engineering structures, species of Rhizophoraceae are mainly used on the seaward side. There are some nurseries on Okinawa and other islands; however, planting material such as propagules and seeds are easily collected from natural mangrove forests. These provide seedlings and saplings for decoration and cultivation in the intertidal zone. Most islands have some mangrove stands as remnants of earlier forest cover, some of them remain untouched as reserve forest and some are partially preserved as sacred forests. There are protected mangroves in the southernmost islands of Ishigaki and Iriomote; species variety is greatest in the latter where ecotourism is highly organised. These natural forests are occasionally damaged by cyclonic storms and high-speed destructive winds that limit the height of the trees and change their configuration and branching. At Okinawa, air temperature varies from an average of 32 °C in summer, down to a minimum of 12–13 °C in winter; mean annual precipitation is 2 037 mm at Naha, the capital city of Okinawa Prefecture, and spring tidal amplitude is 2 m. The area under mangrove cover is at present 4 to 5 km^2.

New Zealand. In the North Island of New Zealand, especially around Auckland, there are some 194 km^2 (or 287 km^2 by another estimate) stands of short *Avicennia marina* described as var. *resinifera* (Forsk.) under annual precipitation of 775 mm, tidal amplitude 0.3–3.2 m and sea water temperature 15–20 °C. *A. marina* is the mangrove species with widest geographical distribution, and in the West Pacific it occurs in a major area that goes from Papua/New Guinea, the Solomon Islands, around Australia, New Caledonia, Vanuatu and reaches its extreme southern distribution in New Zealand. It seems reasonable to assume that dispersal of this adaptable species may be due to purposeful transport by man, rather than passive transport subject to long immersion in marine waters. It is known experimentally that species of the genus do not survive more than a few days or at most a few weeks of immersion in full salinity sea water before the *testa* rots away and the embryo sinks; *Avicennia* does not appear to be able to tolerate long transfers by sea (Baba, pers. comm). On the other hand, propagules of *Rhizophora* may survive several months in seawater without losing the capacity to germinate. On the other hand, it is well-known that people of the South Pacific travelled extensively over distances greater than seeds or propagules of mangroves could have covered. Maoris used *Avicennia* to manage the small fishing impoundments that they made along the coastline. I believe that the role played by man in the transfer of useful species of plants across the breadth of the Pacific Ocean has been underestimated; the distribution of mangrove species in the Indo-West Pacific probably reflects to a great extent purposeful rather than accidental geographical spread of several species.

Philippines. The archipelago of the Philippines is biogeographically part of SE Asia; it includes small and larger islands, which extend longitudinally from about latitude 5° N to about latitude 18° N. They are somewhat an extension northwards of the S.E. Asia archipelago, which includes Malaysia, Indonesia, Brunei, and Papua/NewGuinea and covers a vast area rich in species variety, where passive geographical dispersal of species by ocean currents and gradual colonisation of the coastlines over geological times was possible. Mangroves are widely distributed in the many islands, and thirty species have been recorded. The absence in the Philippines of the genus *Aegialitis*, of *Nypa* and some other species present in the Indonesian islands is noteworthy. The hybrid *Lumnitzera* × *rosea* is found only in the Philippines. Much of the original mangrove forest has been destroyed in the past forty years or more, for conversion to other uses, mostly to aquaculture.

3.6.5 Small Ocean Islands of the West Pacific

Both high rise volcanic islands and low-lying coral islands are numerous; only a few will be mentioned, either because some information is available in the literature or because they are representative of unique conditions. Southeast Asia and the margin of the Pacific tectonic plate are areas of intensive geophysical activity as evidenced by the frequency of violent episodic events such as earthquakes, volcanic eruptions, tsunamis as well as slow movements of the Earth's crust, such as uplifting or subsidence of continental margins or entire islands. High-rise volcanic islands are American Samoa, Fiji, some islands of Tonga, some of Micronesia and many others. Uplifted coral islands are Guam, Northern Marianna Islands, Tuvalu, Palau, Vanuatu, and some islands of Micronesia. Low-lying islands and atolls are: Kiribati, Marshall Islands, Nauru, and Solomon Islands. In atolls and coral islands, the soil is always aerated because of the porous nature of the substratum. Anaerobic conditions do not affect mangrove plants directly, only indirectly by blocking the Fe and S cycles; the histology, aerenchyma and root formation of these islands' mangroves and non mangrove shorefront species of trees should be studied from this point of view.

In summary, what can we learn from this brief overview of the mangroves of small oceanic islands in the Indo-West Pacific? What they have in common is: *1*) Small tidal amplitude; *2*) Mangroves of atolls and coral islands grow in central depressions; *3*) They grow on substrata poor in humus content, well aerated, with relatively high pH, and there is no formation of acid sulfate soils; *4*) The substratum is well washed from abundant equatorial rains, usually well distributed throughout the year even when the climate is monsoon dominated; *5*) Air temperature is high, falls below 20 °C only exceptionally; *6*) Growth of trees is restricted by paucity of nutrients rather than any other factor, but they compete favourably with other plants because of their salinity tolerance; and *7*) Mangroves have been used routinely by man who may have been the major instrument of distribution of useful species and of those easily adaptable to new environments. Table 3.12 summarises forest cover, physical conditions and species composition of mangroves in small islands of the Indo-West Pacific region.

The mangroves of low oceanic islands occur most frequently in depressions, where water from rains; land runoff and clay, sand and nutrients accumulate. The depressions are never on the seafront, but in between the shore and the elevated landmass as well as in the centre of all coral islands and atolls. They are found fringing the lagoons, which may be enclosed or open to the ocean; they are protected from high wave action by long-shore coral sand bars and coral debris deposits. Figure 3.12 shows an example of mangrove distribution across a typical low island of the Pacific. Tidal amplitude may be greater in inland bodies of water than on the oceanfront, with lag-time between ebb and flood. Underground seepage usually occurs in coral islands, due to the porous nature of the substratum; both fresh waters and marine waters are well oxygenated. The mangrove flora of the depressions receives the benefit of different types of water. The seafront flora of high-rise islands usually includes large trees that grow on gravel or coarse sand and are tolerant of high salinity, such as *Ficus*, *Terminalia*, *Hibiscus*, *Tespesia*, and *Erythrina*. Compared to the soil of muddy mangrove swamps, the soil along the margin of atoll lagoons and coral islands is always poor in humus. There is next to no formation of sulfides because the amount of organic matter is reduced; the layer of humus is not more than a few centimetres thick and may become totally depleted, thus even a minimum amount of iron and sulfur aerobic bacteria are sufficient to mineralise all the organic content of the substratum. In Western Samoa, behind the sea shore we found stands dominated or exclusively formed by *Rhizophora* up to 2 m tall, where the substratum of mixed gravel, sand and humus was not thicker than some 10 cm. Accordingly growth is slow and a closed canopy was not formed, but the plants were apparently healthy, flowering and bearing fruit. On such substrata, propagules do not take root easily, unless they are planted. Those stands received seawater by overwash during high tide (personal observation). Apparently, the species is adapted to abrupt salinity fluctuations as during exposure during the monsoon and tropical rainstorms. The limited growth of the trees on small coral islands is due to scarcity of humus, which may become totally used up, with consequent death of the stand. High temperature favours growth but may cause death of the trees due to increased evapo-transpiration and high soil salinity if rains fail. The usual limiting factor for better development of mangrove trees on coral islands is scarcity of humus, which may become totally used up with consequent death of the stand; also violent winds and typhoons are a cause of dwarfing and change of shape of the trees, because apical buds are often destroyed, and the absence of auxine-producing tissues may cause a bushy rather than elongated tree shape. For mangroves in general, it would appear that anoxic conditions are a constraint not difficult to overcome; the most forbidding condition is the accumulation of sulfides in the substratum.

The small equatorial islands harbour lineages of some species, which were selected over long periods of time and preserve hybrids and varieties. The presence of trees 20 to 30 m tall in some coral islands, for instance Kiribati or Pohnpei, suggest that they have been there long enough for the accumulation of relatively abundant humus. It should be interesting to compare the length of time of human colonisation with the development of the trees. In contrast, mangroves growing on high rise islands or along the coasts of large land masses and compacted soils have become adapted to anaerobic conditions where the iron and sulfur mineralisation cycles are blocked due to anoxic conditions of the soil. It would be interesting to test whether coral island mangroves have retained their ability to cope with acidification and anaerobic conditions of the soil. If it were not for the washing and oxygenation of the soil provided by the tides, would there be no mangroves in the coastal swamps of the tropics? It seems logical to conclude that it is not because the soil is so anaerobic, saline and acidic that plants other than the mangroves cannot adapt to the intertidal zone, but because only the mangroves were able to tolerate this combination of factors and extreme conditions varying at short intervals of time.

Mangroves can live, grow and reproduce in anoxic, acidic and saline soils if they are washed by tides, but can grow, live and reproduce even without appreciable tides, if the soil is well aerated. If the pH is relatively low, tall trees can grow in humus rich pockets in coral islands if the soil is well ventilated. The greatest abundance of animal species which aerate the soil by digging galleries in the sediments are large Brachyura and Anomura: crabs, mud-lobsters and similar. The amount of soil removed and brought to the surface by the mud-lobster *Thalassina anomala* was quantified (Kawaguchi et al. 2000) and found to be over 20 t of dry soil ha^{-1} yr^{-1} in a *Rhizophora apiculata* dominated forest near Ranong on the west coast of peninsular Thailand. Where the underground layer is very rich in biogenic pyrite and sulfides, the displacement of soil from depth to top layers could become harmful because of the spreading of potentially acid sulfates when the mangrove cover is removed.

The role played by man in the geographic dispersal of mangroves has not received sufficient attention. Questions arise such as: is New Zealand really the 'natural' southern limit of distribution of *Avicennia marina*? How did small, fragile seeds with short viability of *Sonneratia caseolaris* or *S. alba* or their hybrid *S. alba* × *ovata* find their way to the Maldives and distant places such as Vanuatu or Solomon Islands? The spe-

Table 3.12. Mangrove areal cover, environmental characteristics and species composition of West Pacific Ocean small islands

Island	Cover (ha)	Air temp. (°C)	Annual rainfall (mm)	Tidal amplitude (m)	Species present	Observations
American Samoa	50.2	21 – 32	5 080	<1	*Bruguiera gymnorrhiza, Rhizophora mucronata, R. samoensis, Xylocarpus mekongensis*	
Fiji	39 – 52	20 – 30	2 974	1.3 (Spring tide)	*Bruguiera gymnorrhiza, Excoecaria agallocha, Heritiera littoralis, Lumnitzera littorea, Rhizophora samoensis, R. stylosa, R.* × *selala, Xylocarpus mekongensis*	High rise islands canopy up to 12 m
Guam	70 – 94	25 – 31	2 362	0.75 – 2.9	*Avicennia alba, A. marina, Bruguiera gymnorrhiza, Excoecaria agallocha, Heritiera littoralis, Lumnitzera littorea, Nypa fruticans, Rhizophora apiculata, R. mucronata, R. stylosa, Xylocarpus mekongensis*	Trees 4–8 m tall
Kiribati	?	24 – 30	Variable, in general high		*Bruguiera gymnorrhiza, Lumnitzera racemosa, Rhizophora mucronata, Sonneratia alba*	Trees up to 30 m tall
Marianas	?	25 – 30	2 000 – 2 500		*Bruguiera gymnorrhiza, Pemphis acidula, Rhizophora* sp*	* Planted in 1970s
Marshall Islands	?	27.8	–		*Avicennia alba, Bruguiera gymnorrhiza, Lumnitzera littorea, Rhizophora mucronata, Sonneratia alba*	29 atolls 5 coral islands
Nauru	2	27.8	–		*Bruguiera gymnorrhiza*	Ccoral islands
Micronesia	8 258	26 – 27	4 855	up to 1.8	*Bruguiera gymnorrhiza, Lumnitzera littorea, Rhizophora apiculata, R. mucronata, R. stylosa, Sonneratia alba, Scyphiphora hydrophyllacea, Xylocarpus granatum, X. mekongensis, Ceriops tagal , Excoecaria agallocha, Heritiera littoralis, Nypa fruticans, Avicennia alba*	
Tonga	330 – 1 000	24 – 27	875	0.3 – 1.4	*Bruguiera gymnorrhiza, Excoecaria agallocha, Heritiera littoralis, Lumnitzera littorea, Rhizophora samoensis, R. stylosa, Xylocarpus granatum, X. mekongensis*	Natural forest estimated to be 7000 years old

Table 3.12. *Continued*

Island	Cover (ha)	Air temp. (°C)	Annual rainfall (mm)	Tidal amplitude (m)	Species present	Observations
New Caledonia	20250	19 – 26	1020 – 2687	0.4 – 1.5	*Avicennia marina, Bruguiera gymnorrhiza, B. sexangula, Ceriops tagal, Heritiera littoralis, Lumnitzera littorea, Rhizophora apiculata, R. samoensis, R. stylosa, R. × lamarckii, R. × selala, Scyphiphora hydrophyllacea, Sonneratia alba, Xylocarpus granatum, X. mekongensis, Acanthus* sp., *Derris* sp.	Trees up to 20 m tall
Palau	?	27	3730		*Avicennia alba, Bruguiera gymnorrhiza, Ceriops tagal, Excoecaria agallocha, Heritiera littoralis, Lumnitzera littorea, Nypa fruticans, Rhizophora mucronata, R. apiculata, Sonneratia alba, Scyphiphora hydrophyllacea, Xylocarpus granatum, X. mekongensis*	Archipelago area 367 km^2
Solomon Islands	64000	27 – 31	3048	0.4 – 0.9	*Avicennia alba, A. marina, Aegiceras corniculatum, Bruguiera gymnorrhiza, B. parviflora, B. sexangula, Ceriops tagal, Excoecaria agallocha, Heritiera littoralis, Lumnitzera littorea, Nypa fruticans, Osbornia octodonta, Rhizophora apiculata, R. mucronata, R. stylosa, R. × lamarckii, Sonneratia alba, S. caseolaris, S. × gulngai Scyphiphora hydrophyllacea, Xylocarpus granatum, X. mekongensis*	
W. Samoa	700	25 – 27	2800	<0.5	*Rhizophora* sp.	–
Tuvalu (atolls)	≈520	28			*Lumnitzera littorea, Rhizophora stylosa, Pemphis acidula*	Some in offshore lagoons
Vanuatu	160 – 230	22 – 27	2130	0.3 – 1.6	*Avicennia marina, Bruguiera gymnorrhiza, B. Parviflora, Ceriops tagal, Excoecaria agallocha, Heritiera littoralis, Lumnitzera littorea, Rhizophora apiculata, R. mucronata, R. stylosa, Sonneratia alba, S. caseolaris, S. × gulngai*	80 high rise volcanic and coral uplift islands

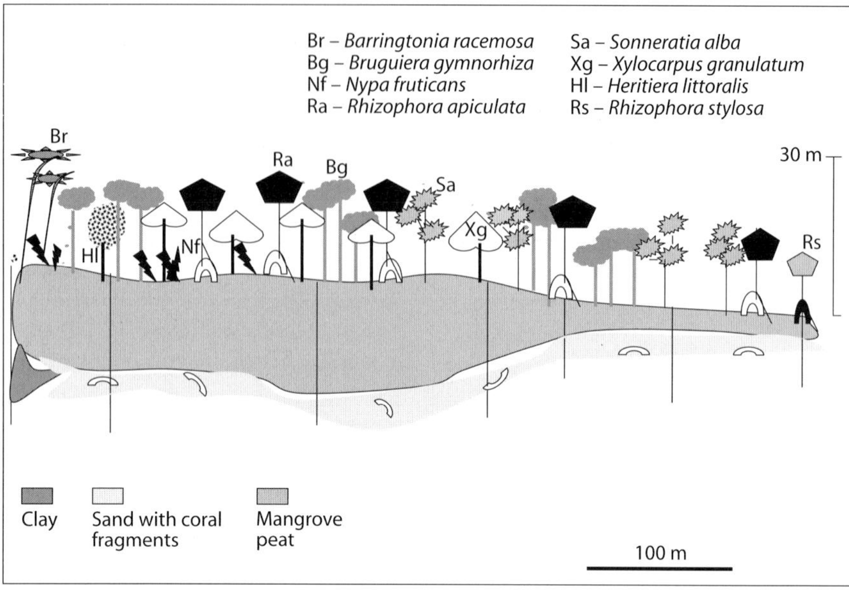

Fig. 3.12. Vegetation, landform and geologic profile of the tidal flat type on Pohnpei Island, Micronesia

cies of the genus *Sonneratia* are pollinated by bats who are known to fly swiftly over great distances, up to 50 km; in fact there are frugivorous bats in the Comoro Islands, the Maldives, Seychelles, Micronesia, and they must also be present in other islands where species of *Sonneratia* are established, such as Solomon, Vanuatu, Palau (*Sonneratia alba*), New Caledonia and probably others. How did bats come there, or do they come and go nightly to other places? Questions are still unanswered. The most widespread species of mangroves are also edible or otherwise more useful species, like *Rhizophora* and *Bruguiera*. The propagules of *B. cylindrica* are edible; the wood of the two commonest species of the genus is short lived when in contact with humid earth, though it is useful for the production of catechu, or cutch (for tannins) and as timber for making planks for boats; *Rhizophora* makes good masts. The propagules of the species of these two genera survive for long periods in seawater and could have survived relatively long periods of immersion. On the other hand, Pacific ocean islanders, especially the Polynesians are well-known for their navigational expertise and for having embarked in long distance expeditions, taking their bearings on the stars and guided by the movements of the waves. They carried live plants for future needs on the new islands they would colonise and took live embers in pots made of hardwood. I searched for details, but could not find out what purpose for which the wood was used, but it could well have been mangrove hardwood. On the other hand, mangrove species must also have spread naturally all over SE Asia, up to the Bay of Bengal and on to the Philippines and around Australia. Papua New Guinea has all the species known to the region, except Avicennia *integra* (exclusive to NW Australia), *R. samoensis* and *Rhizophora* × *selala*, and the island is placed in a strategic position for the dispersal of the species, as far as ocean currents are concerned.

On the Northern Mariana Islands, *Rhizophora* was planted in the 1970s. On the Marshall Islands, the islanders planted mangroves for the trees, and *R. mangle* must have also been introduced there by man. There are several documented records of purposeful transfer of mangrove species in recent history. *R. mangle*, *R. mucronata*, *B. parviflora*, and *B. sexangula* were introduced in Hawaii around the beginning of the twentieth century. *Nypa fruticans* was introduced in West Africa in 1906 and again in 1912. *Nypa* had been widespread in Africa, especially in the Niger River delta, where it disappeared in the Miocene.

The final remark is to point out that the threat of mean sea level rise is particularly serious for small oceanic islands. Mangrove ecosystem growth seawards may probably cope with mean sea level rise, where rivers carry sufficient amounts of silt and other suspended matter that may be deposited at the land /sea interface. Small island rivers are short and mostly do not have as much silt as would be required to out-balance intrusion by the sea. On the other hand, high rise islands usually have a narrow insular shelf and a precipitous fall down to great depths near to the shore line. Mangrove seeds and propagules may be lost to the sea before taking root.

3.7 Systems Dominated by Socio-Economic Factors. Management and Phenology

The use of mangrove resources is surely as old as the first contact of *Homo sapiens* with the ecosystem, as far back as prehistoric times. Most probably the first use of mangroves was as a source of food and materials to make fire and rafts on which to go fishing. The aborigines of the Northern Territories of Australia until recently transported goods and displaced themselves on rafts made of mangrove woods. Inshore waters of protected bays, lagoons, and flooded coastal plains are easier to cross over by water than by travelling overland, and mangrove poles make useful plying poles. Some mangroves have lightwood that can easily be used as logs on which to sit, propelling one with hands and arms. Certain Apocynaceae, like those of the genus *Cerbera* and others could well be used for this purpose. The origin of the use of mangrove-derived substances as medicines is lost far back in time. Early man would have learnt very quickly how to control diarrhea, disinfect infected wounds, make poultices or compresses with leaves or decoction of bark of some species to cure or soothe skin diseases caused by leprosy or mycoses and tooth aches. The milky sap of Euphorbiaceae and Apocynaceae contains different alkaloids, some of which are poisonous and some curative. Extracts from some species of the genus *Derris* are used as fish poison. The tradition is alive in the north of Brazil on what fishes, mollusks and crustaceans should be eaten, for what purposes, at what season or age of the person and which are suitable for puerperal and nursing mothers. They know the use of various chemicals, like tannins, colours and dyes for different purposes. The widespread use of *Excoecaria agallocha* as a cure for leprosy is well-known and already recorded by Roman authors.

Generation after generation, man became either a seasonal visitor or, where convenient, a permanent resident well integrated in the mangroves. To this day, fishermen from as far as the Chittagong area seasonally come to the mouth of the distributaries of the delta in the western Sundarbans to catch the highly praised anadromous fish *Ilsa ilisha* on its seasonal migrations. Shell mounds or coastal kitchen middens are a testimony to the use of the mangroves by prehistoric man, extending to the present. More recently, a few centuries ago, the magnificent Sun Temple at Konarak, on the east coast of peninsular India was built adjoining a mangrove area were devotees came to be cured of leprosy. In prehistoric and ancient historic times, diffuse and varied use of mangrove resources was not great enough to cause degradation of the system. Even at present, for instance in dense littoral and supralittoral forests, when the sagu palm (*Metroxylon sagu*) and other palms become depleted but not extinguished in an area, the group moves elsewhere to allow for the young plants to grow to maturity. Practices of "shifting extraction" are applied empirically and are comparable to practices of shifting agriculture; they are wise practices, provided enough time is allowed for regeneration of the many species of the original forest. Some attention is now given to the traditional medicines obtained from mangroves, as Linnaeus knew well. For this reason he coined the genus name *Avicennia*, which is the latinised version of the name of a famous Arab physician: Abu-bin-Sinnah. The species name *officinalis* indicates the medicinal properties of the plant.

Practices of "shifting extraction" are ecologically valid where the human population density is low and has wandering habits, which ensure ample time for the mangrove forest's spontaneous recovery, but they become damaging where the mangrove forest species diversity is low, as in dry coastal areas where only two or three species of trees were harvested to almost complete extinction. This happened around the Arabian Peninsula and the Island of Socotra where *Rhizophora* and *Avicennia* were used as timber and firewood until timber had to be brought in from as far away as present-day Tanzania.

This was the picture until a few centuries ago: some areas had been almost totally depleted of mangroves, but most of the tropical belt still had pristine flourishing mangrove forests and supra littoral varied flora of mangrove associates in the transition zone to dry land vegetation. At present, there are only a very few places in the Indo-West Pacific region where mangrove forests are still in a pristine condition. As mentioned above, the Nicobar and much of the Andaman Islands in the Bay of Bengal may be the only original mangrove forests of the Indian Ocean. Probably they are a refuge of the species of the eastern Tethys Sea and the only representative left of the flora of the Tethys Sea mangroves. I do not know of any palynological study carried out in the Andaman-Nicobar Islands, or parts of major islands of SE Asia and the West Pacific may also be in their original condition, but many are now being logged over before they are described and studied. Scattered undefiled "Sacred Forests" common throughout the Indo-West Pacific, besides serving their designated purposes, such as disposal of the bodies of deceased persons, or monuments to departed respected elders and leaders or as a place of worship of particular deities or simply as a place for initiation rites, meditation and respect towards Mother Nature, have played a recognisable role for the preservation of species, of the ecosystem and for the integrity of the coastal zone.

The flora, including microorganisms of the mangroves of the world is far from well-known, apart from the vascular plants that create the ecosystem. Non-vascular plants, symbioses like the very common lichens, higher and lower fungi, yeasts, multicellular and unicellular algae, and other microphytes, are imperfectly known, and their role in the system is almost ignored; for instance, how much do the Cyanophyta (blue-green algae) contribute to nitrification of the soils? We know that the number of species may not be greater than in other ecosystems, but their individual numbers may be astronomical; witness to this the large number of browsers and filter feeders that live on and off the microorganisms present in the sediments, in the pore and surface waters and in the humid film of mucilage that coats most surfaces bathed by tidal waters. Colonies of mudskippers and giant gastropods like *Terebralia* spp. and *Telescopium telescopium* may blacken solitary unspoiled shores or carpet muddy mangroves soils where they feast on organic debris and rotting leaves. Burrowing bivalves and crustaceans or swimming crabs may grow to very large sizes and reproduce to such an extent that at some seasons only crab and barnacle larvae are found in the plankton and the coelenterates that feed on them; giant burrowing bivalves like *Anadara granosa* or *Geloina coaxans* are common enough to become regular food for the people of some places, for instance in the lagoons of the west coast of Sri Lanka, but their interaction with the environment, their growth and reproductive rate, their population dynamics and general physiology are very poorly known if at all. Kawaguchi et al. (2000) have recently measured the amount of minerals and organic matter recycled by the Anomuran crustacean *Thalassina anomala* of the mangroves of peninsular Thailand (Table 3.13 and 3.14).

Among the plants, *Acrostichum* ferns are usually considered to be pests; they establish themselves in logged over areas, especially where tidal inundation is less frequent; they grow so fast that in six months or one year they may totally cover the ground, preventing propagules to take root. This situation is common in Matang and other places where total felling followed by replanting is the rule. The fern has to be eradicated mechanically or has to be sprayed with herbicides, which later contaminate the waters.

Table 3.13. Particle size distribution of ejected fresh subsoil, old mound slope and soil surrounding *Thalassina anomala* mounds (Kawaguchi et al. 2000)

Soil	Particle size analysis (%)						
	Gravel (>2 mm)	Coarse sand	Fine sand	Sand	Silt	Clay	Texture
Fresh subsoil	3.0	6.6	25.1	31.7	27.6	40.7	Lic
Old mound slope soil	3.1	6.8	24.8	31.6	28.8	39.6	Lic
Surrounding soil	10.0	15.8	33.7	49.6	18.2	32.3	Lic

Table 3.14. Chemical composition (% dry weight) of fresh subsoil, old mound slope soil and soil surrounding *Thalassina anomala* mounds (Kawaguchi et al. 2000)

Soil	Total-C	Total-N	Total-S	Total-P_2O_5	Bray2-P_2O_5
Fresh subsoil	3.96	0.192	0.543	0.0674	0.00401
Old mound slope soil	3.85	0.189	0.540	0.0663	0.00529
Surrounding soil	3.29	0.165	0.354	0.0497	0.00459

The Fauna of the mangrove and of brackish and coastal waters in particular is better known and studied because it comes under various fields of interest, such as: taxonomic variety (almost all phyla are represented), seasonal, reproductive and metagenetic cycles, physiological adaptations, relationships with other ecosystems, osmotic balance, migrations, food, large animals and materials, and others. Marine Biology, wildlife and ornithological textbooks and journals always have a fair representation of the fauna of mangroves and brackish waters, often mangrove bordered. However, this is not a textbook and will unavoidably disappoint many readers who will not find here answers to specific questions. The appended bibliography is an attempt to make up for this gap. What, however, is important to be mentioned here is that the list of vertebrate animals of the mangroves of different parts of the world is usually given in country by country reports; especially well-known are birds and migratory birds in particular. Also mammals are included, because many of them are endangered species, like bats and their role as pollinators of Sonneratiaceae, and others, like the Royal Bengal Tiger, which, apart from being a magnificent animal species is a disagreeable animal that has frequent preferences for human flesh, as the hundred odd humans, mainly honey collectors killed yearly in the Indian Sundarbans alone may testify. Reptiles and amphibians are few in the Indo West Pacific, and the decline in numbers of the *Gavialis gangeticus* is a boon for the fishes on which it feeds and for its competitors, the dolphins. Small invertebrates exist by the hundreds of thousands in waters and soils, are free living or sessile, and air borne. Most of them are poorly known, including insects, some of them vectors of diseases and others are pests. The soil infauna may be extremely rich; however for most taxa, even the taxonomy is poorly known; Protozoans, Nematodes, Amphipods, Collembola and other lower insects may occur by the millions, while Phoronidea, Oligochaetes and Polychaetes may not be as numerous and together with Coelenterates are somewhat better known. The large bottom fauna, mainly crabs and gastropod or bivalve mollusks are the best known. However the physiology, and the exchange of matter and energy or the ecological role of the mangrove fauna as a whole is far from quantitatively known. The most common pest are Sesarmid crabs, especially species of the genus *Sesarma* which ring the propagules or young seedlings of mangroves to munch on the inner tissues, thus causing the death of the plant. Boring larvae of Scolytid beetles, especially *Poecilips fallax* perforate propagules and may cause their wilting and death. Occasionally hordes of monkeys, *Macaca fascicularis* in SE Asia may devastate newly planted sites, though the Proboscis monkey, *Nasalis larvatus* is a non-damaging regular inhabitant of the mangroves of Borneo.

In addition to these, there are several animal speciesthat can be classed as pests, and others that may be vectors of plant and animal or human diseases. Mention was made of insects like anopheline mosquitoes or other Diptera, which are blood sucking and who could be vectors of diseases. Schistosomiasis, however, does not occur in the mangroves, because the trematode parasites are intolerant of brackish waters. *Ancylostoma duodenalis* and other round worms like *Ascaris* and *Filaria*, the latter transmitted by insects, are frequent in areas polluted by untreated urban sewage. Microbial pollution from brackish water-resistant strains of *Shigella* and *Vibrio* are also to be found.

The use and management that Man has always made of the mangroves initially varied in time and place. Undoubtedly, the longest standing association he has had with the mangroves is in South and SE Asia. There was no replanting of mangroves in the past centuries, even though at places, extraction was greater than natural regrowth. Where the mangroves were finished, timber and poles were sought for elsewhere. Much of the navigational expertise of people bordering the northern Indian Ocean was due to trade, already at the time of Sumer and Babylonia, and much of the trade had something to do directly or indirectly with the mangroves. Planks made of mangrove timber were used to make dhows, which transported goods and traders. Would mangrove wood have been used when the technology of construction evolved from sun-dried bricks to kiln-burnt bricks? We know that wood from present-day Rajasthan was used for the construction of the cities of the Harappan civilisation, as for instance Moenjo-Daro, but how much of this was mangrove wood? Palynological studies applied to archalaeology could throw some light on past life-styles. Be that as it may, let us return to what we already know for sure.

Small, family or village level management must have started very early, for instance planting propagules in the central depressions of coral islands, or on the small islands or in raised ground around shallow lagoons, or on river banks and mud banks, as is still done all over the region. Throughout South and South-east Asia, small ponds at family or village level are bordered with mangroves or have some mangroves growing in them. Methodical management of larger areas of which there are historical records started a few centuries ago; but degradation of mangroves due to over-exploitation has been growing since the nineteenth century.

The mangrove ecosystem, as all marginal ecosystems (Vannucci 2000), is fragile; biodiversity is low at any given place at any time, though it may be high in the region as a whole as in the Sundarbans, or Malaysia and Indonesia. Once destroyed, the mangrove forest is usually unable to regenerate spontaneously. In such cases, assisted regeneration is provided by planting seedlings of the same or different species and by organising nurseries for the supply of seeds and propagules to be later out-planted at the appropriate place and time. The natural or planted mangroves of small islands of the Pacific Ocean, whether high-rise or low lying islands, were undoubtedly utilised from the moment man sat foot on them. As the resources of small islands were no longer sufficient for growing numbers of the human population, man migrated to other islands, carrying with him seeds, cuttings, tubers or other parts of edible or otherwise useful plants. Mangroves would not be an exception. The discontinuous geographic distribution of *Sonneratia caseolaris* is a case in point, since this species, which is pollinated by bats, occurs in many oceanic islands where the edible fruits, which make excellent refreshment, may have been most welcome where potable water is often scarce. The wilful transport by man may have facilitated the hybridisation between species and the formation of varieties through geographic isolation. Genetic isolation is far from complete in the most common genera: *Rhizophora*, *Avicennia*, *Bruguiera*, and *Sonneratia*. On the other hand, along the sinuous, strongly indented coastlines of the larger landmasses and many islands of S.E. Asia, species may have spread by gradual natural dispersal. In flourishing coastal mangrove areas, hybrids and varieties are not so common, and impoverishment of the ecosystem takes much longer to take place, if at all.

Any change, small or big, once introduced into an ecosystem, becomes a form of management, positive or negative. The mere presence of man, whether temporary or permanent, induces changes; the removal or addition of any component of the system may cause a chain of further unpredictable changes. All of these are forms of management, whether they are appropriate or not to the place. I will focus here on three areas of the Indo-West Pacific region that have been managed for a well-known long time. The first is the largest of the three: the Sundarbans, on which we have accurate, though sparse information dating from Greco-Roman authors, well over 2 000 years ago and managed scientifically for over 120 years. The second, managed empirically since over 400 to 500 years ago on which we have precise and detailed information are the Kerala backwaters. The third is the area that has been most recently put under methodical management for specific purposes: the Matang forest in peninsular Malaysia.

The first example is from the Sundarbans of India and Bangladesh. It covers an area of about 12 000 km^2, a figure that does not include the accretion areas of the eastern part, nor the Chakoria Sundarbans. In earlier days the Sundarbans extended from where the city of Calcutta is at present to the seafront. The city of Calcutta grew from three villages bought by the East India Company as the nucleus for the new metropolis. This is one of the first documented contemporary examples of massive destruction of mangroves for urbanisation purposes. The first thing that happened when the mangroves were cut was the outbreak of malaria epidemics, which killed people by the hundreds. Malaria set in because the common anopheline mosquitoes, vectors of the *Plasmodium* spp. protozoan, which grow into and destroy red blood cells, develop only in fresh water, while the pools of water left in the mangrove forest during ebb tide are brackish. However a few species, the commonest being *Anopheles sundaicus* tolerate brackish water conditions, provided there is enough sunlight. This condition arises when the earlier closed canopy of the mangrove is torn open by felling of trees, optimal conditions for breeding brackish water tolerant anopheline mosquitoes were thus created when the mangroves were felled for building the new city of Calcutta. Thus started the mismanagement of the Sundarbans, after centuries of peaceful coexistence between Man and his environment. At present, the total population of the Sundarbans includes about 10 million people, and human population density is greater than ever before due to intrinsic population growth and due to migration from elsewhere to the Sundarbans, as for instance the considerable migration of Santhals from Bihar to the 24 Parganas South District, in West Bengal. The local people lived in and of the mangroves and traded with people who visited them coming both from inland and the sea. By and large this immense area was self-sufficient for the human population; forest, waters and above tidal areas provided for daily needs, and trade with outsiders was flourishing. To this day, the practice is alive to plant vegetables in between the pneumatophores of *Heritiera fomes* in the intermonsoon period, when tidal waters do not flood the higher reaches of the forest. Fishery products, honey and wax were used then, as at present, for local consumption and trade. Felling the trees of islands kept for this purpose produces to this day firewood and coal, and extraction is on a rotational basis. Selective felling of only mature trees is a traditional practice still in use. The Sundarbans were put under methodical management some 120 years ago, and part of the vast area was set aside as a Reserved Forest for production and follow-up studies. Theoretically, all the products of the forest are noted down for control purposes, but illegal and unrecorded uses of the forest and of the waters are unavoidable and uncontrollable in such an extensive area where the maze of waterways eludes all surveillance. The Sundarbans continue to be a multiple use forest as originally; the problems that have become acute are an apparent decrease in productivity, which may be partially due to poaching, an unrecorded but well utilised resource of the forest. The main problem however, is due do socio-economic pressure: the conversion of much of the mangrove forest to agriculture and the conversion of some mangrove areas to intensive aquaculture. The latter is the major problem facing the Sundarbans, also because women and old men who collect the shrimp larvae by dragging nets in the creeks are forced to spend hours in sea water which is particularly noxious to the women's reproductive system. Another serious problem is a condition that affects the trees of *Heritiera fomes* known as "top dying", which is the wilting of the trees from top to bottom; it kills the plant over a period of few years. The condition is probably caused by increased salinity of surface and pore waters due to diversion of river water at dams upstream across River Ganga, at Farakka. In conclusion, the Sundarbans have been managed as a multiple use system that would cater for all the needs of the people, including ports for trade and trans-shipment of goods towards in land, now at Khulna, Port Canning and elsewhere. Scientific methods applied to traditional knowledge have maintained the productivity of this large system over the ages. Most probably this was possible because of the immensity of the area, the variety of its resources and the absence until recently of intensive exploitation of single resources. Above all, this was possible due to by the system of selective felling of mature trees, rather than clear felling of small patches. The traditional management of the Sundarbans was inevitably an adaptation to a rich but variable and threatened environment that could and still can provide all basic needs for a large resident human population.

The second example I should like to bring forth here is that of the Kerala lowlands, one part of them briefly described above as the Cochin backwaters. The Kerala backwaters cover an area smaller than the Sundarbans of Bengal, however they are also densely populated "since ever", as people say. Better said, they are populated since prehistoric times. Legends are the embodiment of the collective memory of a people. They are a poetical glorified manner of preserving the knowledge of the past history to be taught to the new generations. Myths introduce the supernatural in legends; major telluric events like volcanic eruptions, earthquakes or tidal waves often find their way into myths as personalised acts of supernatural beings capable of performing feats beyond man's power. The myth about the origin of the Kerala waterways tells us that Parashurama, an Avatar of Vishnu, threw down his battle-axe and opened up the land for the people. Water invaded the troughs opened by the mighty axe and became the waterways open to travel and trade along the length of the land; these waters could be travelled over at all seasons, by any weather, without running the risks of heavy winds and violent swells of the ocean. The land was then managed ever after as a multiple use system to provide for all the needs of the people. Present-day people will say that this is of course a myth, but they add: "if you look at the map of Kerala, it really looks like a giant axe, the "*parashu*" had opened the land ajar between the hills (the Western Ghats) and the Ocean, creating the low salinity backwater system of lagoons." Would there have been in aeons past an earthquake or a gigantic landslide or a series of such landslides from the Western Ghats into the Sea? Whatever the prehistory, the fact remains that the waterways and the coastal plain of Kerala, from roughly over latitude 8° N to about latitude 13° N have been densely populated since prehistoric times, and people have used the waterways as means of communication and transport of people and goods ever since. Records of trade between Drawidian South Indian and the Arab merchants and mariners of the western Indian Ocean are common knowledge since B.C. times. The entire area must have been originally a large mangrove swamp, and as mentioned, mangrove stands even now flourish here and there. One of the major temples, dedicated to Lord Rama, attracts pilgrims who ritually feed fish that come for a treat up to the steps that lead to the temple ground. The temple stands guard over a small patch of islands. One of the oldest Royal Houses of Kerala, the Varma dynasty traces its ancestry, recorded in writing and dated, to well over 2 000 years ago. As witnessed in the records, already in early days, basketful by basketful of mud was shaped into a highly productive checkerboard of fields and ponds as I have described above. The system was ecologically sound as well as socioeconomically satisfactory and productive; it provided for internal population growth, which over the centuries absorbed several streams of immigrants: Jews, Syrian Orthodox, Arabs and later Europeans of different nations and creeds. Imbalance grew only in the twentieth century and with it increasing destruction of mangrove areas. Eutrophication of waters increased due to the large amount of organic effluents; waves from motorised craft caused erosion of the earthen banks, as they do also on the marble palaces of Venice in Italy. Urbanisation and industrialisation grew in an explosive manner and added to the destruction of the original vegetation and agricultural areas, with inevitable increased pollution, further exacerbated by the Ernakulam oil-terminal. Though the backwaters still function as a multiple use system, the life style of people has changed and so have their daily needs. The latter have grown in kind and quantity; the system is no longer self-sufficient and the mangroves have lost their material useful life span due to modernisation but not due to biological-ecological degradation and loss of productivity. The Rama fish-temple continues to be a centre of pilgrimage; it should perhaps be looked upon as a standing warning of ecological costs to be paid in the name of development. Perhaps, if the Kerala backwaters were as large as the Ganga-Brahmaputra delta, they would survive as a balanced multiple-use sustainable productive mangrove ecosystem as it is; they are at present a totally different ecosystem. The traditional management of the Kerala backwater was necessarily a multiple use system which was imperative for survival of a dense human population confined between the ghats and the sea, and to provide adequate living conditions favoured by an equatorial monsoon climate.

The third example of mangrove systems dominated by socioeconomic factors that I would like to mention here is the Matang mangrove forest in the State of Perak, west coast of peninsular Malaysia, facing the Straits of Malacca. It is by far the smallest of the three systems and more recently put under regular management. The entire forest covers only 40 711 ha, about 85% of which are managed as a production forest. The remaining 15% are an accreting zone with 3 089 ha dominated by *Avicennia* and over 2 205 ha dry land forest. The average annual rainfall is 2 500 mm with a relatively dry season from February to June–July. Tidal waters are semi-diurnal and inundate most of the area, except the higher inland border, which may be inundated only at spring tides. The maximum and minimum air temperatures are 35 °C and 22 °C, respectively. The forest is sited and they are sited at nearly latitude 5° N. Like the Kerala backwaters and the other mangrove forests of peninsular Malaysia and of Sabah and Sarawak, they are typically equatorial ecosystems.

Recently Chan (1996) wrote a concise Case Study of the Matang Mangrove, which was first declared a Reserved Forest in 1902. The Matang mangrove has been managed since 1902 mainly for poles, charcoal and firewood. Quoting from Chan (1996), silvicultural operations presently adopted at Matang include pre-felling inventory, final felling, enrichment planting and stand thinning. Systematic clear felling at 30-year rotation is the harvesting system practised. Intensive reforestation follows after felling, which on an average amounts to replanting just over 700 ha yr^{-1}. The species chosen for replanting are *Rhizophora apiculata* and *R. mucronata*, which are the two most important commercially, because of their straight bole and high calorific power. The forest management is basically based on clear felling in strips, leaving a prescribed number of mother trees standing to provide for natural regeneration, and by supplementing it with nursery grown saplings. Tables 3.15, 3.16 and 3.17 from Chan (1996) give

Table 3.15.
Summary of silvicultural operations in Matang (Chan 1996)

Year	Silvicultural activities
1	Eradication of *Acrostichum* ferns and estimation of areas that require enrichment planting
2	Planting of *Rhizophora* in blank areas
3	Assessment of planted areas to determine survival and beating up if necessary
15	Thinning I, using a 1.2 m stick
20	Thinning II, using a 1.8 m stick
29	2% enumeration of trees greater than 7.5 cm diameter to determine expected yield and premium to be imposed
30	Final felling

Table 3.16.
Annual mangrove reforestation efforts in Matang (Chan 1996)

Year	Planted area (ha)	Cost at US$34 per ha
1987	736	25 020
1988	690	23 460
1989	585	19 890
1990	189	6 430
1991	1 158	39 370
1992	926	31 480
Total	4 284	145 650

Table 3.17.
Annual site preparation efforts (eradication of *Acrostichum* ferns) prior to planting in Matang (Chan 1996)

Year	Site preparation (ha)	Cost at US$280 per ha
1987	234	65 520
1988	607	169 960
1989	54	15 120
1990	217	60 760
1991	105	29 400
1992	91	25 480
Total	1 308	366 240

an idea of the replanting effort, which is probably one of the main factors for its success. The first working plan was drawn by Noakes (1952) for the period 1950–1959. The forest is managed on 10-year plans aiming primarily at the production of quality wood for firewood, charcoal, timber and poles, on a sustained yield basis. The emphasis of the working plans is intensive reforestation following clear felling, but other objectives are the maintenance of the ecosystem, which supports a valuable fishing industry and a variety of wildlife species. In the five years from 1987 to 1992, about 4300 ha of mangroves that had been clear-felled for wood were reforested essentially with *Rhizophora mucronata* and *R. apiculata*. Chan (1996) considers that "the success story of the sustainable forest management in Matang is mainly due to intensive reforestation efforts." The management of Matang forest mainly consists of assisted regeneration by intensive replanting and thinning and in the eradication of *Acrostichum*; management costs are related to these basic activities, including the labour involved.

Other mangrove forests in Malaysia are not regularly managed. In comparison, the intensive total felling of mangroves in Sabah and Sarawak (insular Malaysia) for the wood chip industry yielded only 2.5% of the Matang yield/ha, based on economic consideration and provision of employment (Spalding et al. 1997).

Quoting from Chan (1996), the total annual revenue from forestry, including timber and other forest products, is about US$9 million and provides employment for a direct workforce of about 1400 workers and an indirect workforce of 1000. The fishing industries, on the other hand, provide a revenue of about US$33 million and direct employment for 2600 workers and indirect for about 7500. Summing up, the 40000 ha Matang mangroves provide employment for about 12500 people and an annual revenue of about US$42 million, or a monetary return of just over US$1000 $ha^{-1} yr^{-1}$. What to my knowledge has not been included in these estimates are the ecological costs of reducing a multi-species association into one with reduced numbers of key species. During the recent past, the Matang mangrove has had to compete with economically invading oilpalm plantations. The question arises about the long-term ecological and economic sustainability of a near monoculture situation where costs of maintenance inevitably will grow at a pace faster than spontaneous regeneration. Should the forest become degraded, the fisheries will also decline, probably in an unpredicted manner.

The main hazard for the mangrove forests of the Indo-West Pacific as elsewhere, is the conversion to other uses, especially to intensive aquaculture for shrimp farming. Conversion to agriculture, especially paddy rice and sugar cane has not been sustainable over long periods of time. The only exception is the area turned over to agriculture, including paddy rice, in the accretion areas of the eastern Sundarbans where agriculture is practised successfully. The reason for the unusual success of agriculture on the islands reclaimed from the sea through mangrove afforestation in Bengal are: low salinity of the tidal waters, abundant fresh water input by rivers and monsoon precipitation that wash away sulphides and deposit terrigenous sediments. The mangrove forest at Matang is under entirely different ecological conditions. Both systems, at Matang and in the Sundarbans are good examples of the wisdom of appropriate rational management of mangroves. First of all, it must be pointed out that at Matang, by safeguarding the productivity of the forest, the fishing industry kept and increased its returns to a level two to five times higher than the revenue from the forest itself (Spalding 1997); fisheries also have an employment capacity five times higher than silviculture; secondly, production capacity can be maintained only if natural ecological processes that are essential for the sustenance of the mangrove ecosystem are respected. Altering these processes "will inevitably be accompanied by adverse environmental consequences that are extremely difficult to resolve. Often, the ecological price to be paid surpasses the socioeconomic benefits of converting mangroves to other land uses" (Chan 1996). The type of management of Matang mangroves is comparable to the modern approach of systems managed having in view long sustained economic returns.

The example of the management plans for the Matang mangrove forest cannot be extrapolated as a model to be adopted everywhere, as also the pattern of management of the Sundarbans cannot be extrapolated to other areas without considerable adaptations. The main reasons are: *1*) The nature and magnitude of ecological factors vary from place to place; there being no two environments exactly alike, there cannot be a single pattern of management; *2*) The needs of the people and socioeconomic conditions also vary; for instance, life-styles and needs of coastal dwellers of the three examples brought up here had always been linked to their particular environment, but each was different from the others; *3*) Species composition of plants and animals are unique to each place; and *4*) The Matang success is to a great extent due to ecological impoverishment in terms of a decrease in species diversity: From a large number of species, to only two of the same genus, only *Rhizophora apiculata* and *R. mucronata* are now cultivated. Monoculture or reduced species biodiversity is always an ecological hazard to be considered when the ecological cost is assessed. Probably the management plan, which is nearly perfect for Matang could not be extended even to other mangroves of the same coastal area without appropriate changes. In fact, the accreting area of Matang, which is under *Avicennia*, does not undergo the same pattern of management as the rest of the forest. In addition, there are also 42 ha left untouched as "virgin jungle forest", which serves the function of a "sacred forest"; there is also, at Matang, a small portion of dry land forest that is seldom tidally inundated.

The Matang forest is well-known as one of the best – if not the best managed mangrove forest in the world in terms of defined production, monetary revenue and socioeconomic sustainability. Intensive assisted regeneration and regular harvesting achieve maximum returns. Major expenditure is incurred in the eradication of *Acrostichum*. Losses are due to poaching or other diverted gains. The major ecological loss is a fall in biodiversity. The major ecological improvement is the enhanced returns from fisheries, which besides their monetary value can also be considered a measure of the hidden benefits that a well managed mangrove ecosystem provides free of charge. It could be said that Matang is an outcome of the modern era and an example of the rational application of the principles and strategies of the industrial age to nature management.

The great success of the Matang Mangrove Forest Management system is vastly due to the wisdom of the management plans which take advantage of the exceptionally favourable environmental conditions of the coastal zone facing the Straits of Malacca at that point, and most importantly are revised at regular intervals of ten years. The same model, however, could not be directly transferred even to other mangrove stands further south on the same coast, where marine currents are much stronger.

However, there is one feature which is common to the three cases: this is the preservation of some parts of the forest in their pristine condition, including the sacred forests of the Cochin backwaters, parts of the Reserved Forests in general in the

Sundarbans and the "Tiger Reserve", as well as the "primary forest" in Matang. Though the three systems have their merits, none can be advocated as a model of general validity that could be transferred unchanged anywhere in the world. The reasons why they were kept intact varies; in the Sundarbans and Kerala they are traditionally maintained as "sacred forests", while in Matang, the stated reason is that they have been conserved primarily because they are of no commercial value, though they serve as a biodiversity reserve which includes mangroves and upper-tidal species.

A system of traditional management that originated in East Java and is at present widely used in S.E.-Asia as far as Vietnam and was brought by the Portuguese to Mozambique, Angola and north-east Brazil (the *viveiros* of Pernambuco State and elsewhere) is the *tambak* and *tumpangsari* systems. The basic principle of these systems is similar to that of the Kerala backwaters and other monsoon systems, which is to utilise tidal forces to alternately inundate and partially drain plots sunk down to a level somewhat below mean ebb tide. The system is applicable wherever the tides are semi-diurnal, as in the equatorial zone and much of the tropical belt. The mangrove area is divided into square plots roughly one hectare in size, with bunds raised over the entire perimeter. The bunds are made up of the mud taken from a moat dug down to a depth of about one meter below mean ebb tide. In the centre of the plot, there is a slightly raised platform where mangroves, often a *Rhizophora* species, are planted. The surface area of the mangrove stand is about one third of the entire surface of the plot. The peripheral moat or ditch occupies about two thirds of the entire surface of the plot, and is left uncovered. A narrow canal not wider than about 50 cm traverses the bund and is provided with a small sluice gate; this canal is linked to a major tidal canal that brings and takes back tidal waters to and from the sea. The topography and gradient of the coastal plain and the amplitude of the tides over the entire lunar month determine the extent of the area worked into a checkerboard of plots, divided by bunds one from the others and each one with its central mangrove island, peripheral ditch and link canal with sluice gate. Each major feeding canal usually waters two rows of plots, one on either margin; in large areas, there are several feeding canals perpendicular to the coastline. Where tidal amplitude is sufficiently great, the feeding canals may be branched and may feed more than two rows of plots. Attempts at different topographical designs are not always successful. Waters from incoming tides bring with them fish fingerlings and Penaeid shrimp larvae, which are prevented from returning back to sea by nets placed across the sluice gates. The system functions at sea level and uses only tidal forces with no need for costly equipment to pump sea water and aerate the ponds. Management consists of accurate levelling of the plots, precisely tidally regulated and manual opening and closing of the sluice gates, and weeding out carnivorous species of fishes. Finally, management culminates in harvesting the crop with push nets two or rarely more times annually, when fish and shrimps accumulate in the peripheral ditch at low tide. Management also requires a good knowledge of the different habits of the fish and shrimp, knowledge of their seasonal and breeding cycles and an accurate estimate of optimum population density in the ponds. Further, the plots must be allowed to go fallow at specified intervals of time, according to different types of soils, which may be two or three seasons or years; the purpose is to aerate the bottom sediments and occasionally replant the densely packed plants of the central stand. I have also occasionally seen mangrove species planted on the higher parts of the banks and bunds, which are not inundated daily. The principle is simple, but it requires experience, and

I have seen in places where the elders have passed away, and subsequently the new generation had run into trouble and finally abandoned the plots or turned them over to intensive or semi-intensive shrimp farming. Evidently, details on how to operate the system are site-specific. The main factors to be considered are tides, topography, reproductive and seasonal cycles of the species to be cultured, knowledge of the most suitable mangrove species, of the sediments, pests and parasites, maximum possible density of fish and shrimp stocks, thinning of both mangrove and animal stocks, mesh size of the net at different seasons, and so on. Transfer of the system to other places has often resulted in failure, which is a general rule when any bio-ecological system of management is transferred to other places without appraisal of different environmental constraints. Within Indonesia itself, there are many variants of the system in regards to design, mangrove species used, the size of the plots, and their management (Fig. 3.13).

In East Java, there are at present about 6 000 ha of *tambaks* operated by recent traditional methods, with *Avicennia* or *Acanthus ilicifolius* planted on the banks; villagers use the leaves of these species to adjust the pH and the salinity of the waters; they also consider *Avicennia* preferable to *Rhizophora* to consolidate accretion areas. The rotational period of these species is about 20 years. Other traditional methods used by local people include the periodical use of green and blue-green algae at certain stages of the succession of the operations for the maintenance of the pH and nutrient availability in the water of the ponds (Inoue 1999).

The silvi-aquacultural tambak and *tampangsari* system evidently has many similarities with the *chemmin-kettu*, *pokkali*, *bheri*, and other systems of silvi-agri-aquaculture practised around the world. The species of mangrove used may vary; for instance local species are used in Brazil, mainly *R. mangle*. In other places, the system and species used may be different where mean sea level variations are seasonal and monsoon regulated rather than daily regulated. In such cases, the most common species of mangrove used are *Excoecaria agallocha*, *Heritiera* or *Nypa* and others which tolerate weeks and months above inundation level or, like *Nypa*, tolerate near fresh water conditions or other extremes of the environmental parameters. By and large, the *tambak* system is the most suitable for long-term sustainability of mixed mangrove ecosystems in the equatorial and tropical belts of the world.

The obvious merit of such systems are *1*) preservation of the integrity and diversification of the coastal zone; *2*) sustainability; *3*) preservation of the hidden benefits offered by the mangrove; *4*) no requirement of electricity, costly equipment, additional feed, antibiotics, soil additives and hatcheries; *5*) immunity from bacterial, fungal and other diseases caused or favoured by crowding, non renewal of waters, absence of natural terrigenous antibiotics, and the like; *6*) intensive aquaculture ponds are liable to become degraded after a short period of few years, while *tumpang-sari* and other sea-level systems of silvi-aquaculture, if adequately managed with traditional wisdom may have a life-span of decades and centuries and in addition preserve most of the hidden benefits provided by the mangrove ecosystem, including enrichment of coastal marine waters. The only marked disadvantages are when the yearly return or the return/season is not as great, or may be just about one third of the intensive aquaculture ponds. The latter, however, have a short life span of few years and in addition cause degradation of the environment, often irretrievable. *Tambak*, *tumpang-sari* and similar systems of silvi-brackish water aquaculture are the nearest possible approximation of the very human wish better expressed as: "to have one's cake and eat it, too." The system

Fig. 3.13. Mangrove replanting in Bali, Indonesia (*photo:* courtesy by Dr. S. Baba)

has stood the test of time of uncounted skilled generations. During the second half of the twentieth century, the system expanded from Sumatra (mainly Aceh Province), to Java, Sulawesi, Kalimantan, Maluku and Irian Jaya. It is recently introduced in many other countries in Asia and elsewhere. The system, however, has collapsed in some other parts of the world when productivity is pushed to extremes by intensive culture, for the sake of quick and large monetary returns.

Fig. 3.14. Narrow mangrove fringes along the Gulf of Thailand surrounded by intensive shrimp farming (*photo:* courtesy by Dr. S. Baba)

In contrast, a review of the running 1990 Management Plan of Matang that is not under intensive farming shows that the profit from the fisheries is five to eight times greater than the profit from timber production. The figure includes all mangrove-dependent species of shrimp and fish. There is quantified information to show that the decrease of the forest area was accompanied by a decrease in offshore sea fisheries production, especially in the catch-per-unit effort of the wild stock of shrimp at sea. The reason for the decrease is that the numerically more frequent and economically more rewarding species of shrimp are those of the family Penaeidae, whose larval stages are mangrove-dependent for shelter and food. The nearly mature larvae return to coastal marine waters for final growth and reproduction. This is the wild stock of adult Penaeid shrimp, which are the main revenue from coastal marine fisheries.

The application of the industrial ideal of maximum possible production in the shortest possible time and smallest cost when transferred to mangrove ecosystems has proved to be unsustainable. In Thailand, large mangrove areas have been converted into shrimp farms, leaving narrow fringes of native mangroves unable to protect the shoreline and play their important ecological role as nurseries and nutrient suppliers for coastal waters (Fig. 3.14). In the Chokoria Sundarbans, shrimp aquaculture industry has converted mangrove forests at fast rate decreasing mangrove forest cover from 7 500 ha in 1976 to 973 ha in 1988 (Fig. 3.15). Another example that has happened is the "improved" contemporary management of the *tambaks* of Indonesia, which all belong to the State by force of law, but can be licensed for exploitation by other concerns. *Tambaks* were pushed into maximum production in the shortest possible time. The total area of *tambaks* in 1975 was 180 000 ha with a production of 9 600 t. In 1991, the

Fig. 3.15.
Conversion of mangrove forests into shrimp ponds in the Chokoria Sundarbans (adapted from Choudhury et al. 1990)

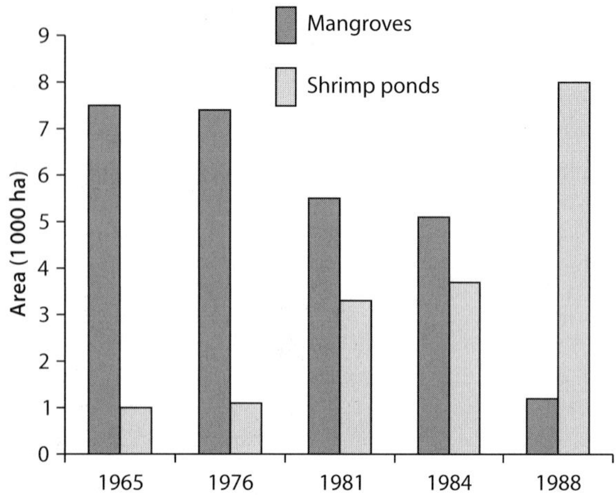

area had jumped to 290 000 ha and the production to 140 000 t yr^{-1}. In other words, the increase of one and a half times the area managed as *tambaks* produced about 15 times more fish, mainly *Chanos chanos* (milk fish). Further conversion of *tambaks* to fish and shrimp aquaculture is envisaged as the Fishery Department of the Ministry of Agriculture has declared 840 000 ha of mangrove forest suitable for conversion to *tambaks*, with a mangrove/pond area = 4:1 instead of about 3:1 as is traditional. The higher return in 1991 was due to intensive aquaculture methods, such as a high density of stocking. However the tambaks under such stressing conditions become degraded by the acidification of soils and waters, and become subject to viral diseases of the fish and shrimp stock and other evils due to crowding.

The management of *tambaks* or *tumpang-saris* requires the expertise and "clinical eye" of local people. In fact, when more and more tambaks were put under intensive cultivation, there was an explosive spread of viral diseases, and by 1990, 15 000 of them were abandoned. Diseases are not the only cause for abandoning *tambaks*; the soil and waters themselves as mentioned above become acidic and unfit for culture. The net result was that more virgin mangrove land was clear-felled for tambaks to be managed under intensive aquaculture, which in turn have a short life of just three to five years. But the process continues. Basically, the causes of the disaster are the eagerness for large monetary returns in a short time and also that new systems have abruptly displaced old methods and old, experienced people. The most harmful system is that of intensive farming that requires high stocking, artificial feeding, aeration, soil additives and significant capital. Only traditional or semi-intensive and moderate systems are sustainable practices in the long run. The recuperation of such degraded *tambaks* takes a long time for the reconditioning of soils, though gradually they may be brought back to the original condition. Semi-intensive aquaculture may be viable for longer periods of time; it combines traditional with experimental practice.

Management of mangroves varies according to the purpose for which each one is managed and varies also with the environmental conditions of the area. For instance, in Indonesia, high growth-rate forests are managed on a rotational basis of 30 years for timber and charcoal production, while slow growth-rate forests, which are managed mainly for wood chips, are on a 20-year rotational cycle. The Matang forest, which is managed mainly for timber and charcoal, is on a 30-year rotational cycle. These decisions were taken after various management plans had been tried (see Chan 1996; Inoue 1999). In Indonesia, experience has shown that extensive aquaculture can be highly profitable if *Avicennia* is used instead of *Rhizophora*. *Avicennia* is usually planted on the banks of ponds and for consolidating the mud banks of accretion areas. In contrast, in the accretion areas of the Sundarbans, *Sonneratia* plantations have given the best results. In accretion areas and in the green belt of Bangladesh, indirect benefits offered by the mangroves, where they are used for reclamation and prevention of erosion and as green fuel wood, may in the long run be greater than the direct benefits such as timber and charcoal production.

There are other examples of different types of mangrove management elsewhere. In East Java, 6 000 ha of *tambaks* are operated by recent extensive traditional methods, with *Avicennia* or *Acanthus ilicifolius* planted on the banks. *Avicennia* is considered to be preferable to *Rhizophora* to consolidate accretion areas. The rotational period of the trees is about 20 years. Some other traditional methods used by local people include the cultivation of green algae as one of the steps for *tambak* maintenance. Details of different methods currently used in the Indonesian islands and a detailed socioeconomic analysis of the recent operational changes are given in Inoue et al. (1999).

In Tamil Nadu, on the East coast of South Indian, "canal fishing" was developed about 270 years ago. The system continues to flourish, thanks to the active involvement of the community. Out of 79 canals, 71 are managed by the families traditionally charged with the maintenance and management of the system. In the area of Muthupet, the canals are distributed over three Reserve Forests. The canals are parallel to each other and dug at a right angle to the sea. Each one is opened at the seaward end by a wide mouth. The canals are seasonally closed at the landward end after water is poured into the saltpans placed behind the mangrove forest. The length of each canal varies from 1.5 to 2.0 km, the width varies from 1.2 to 2.5 m; the average depth is about 1.2 m. The canals cross the mangrove forest and were originally dug out to supply seawater to the saltpans, which were from the very beginning placed behind the mangroves. The mangrove forest is at present degraded due to over-exploitation, but fish and shrimp continue to be trapped in the canals during the late monsoon season and are collected regularly up to the end of the post-monsoon. Meanwhile, water evaporates from the saltpans, leaving behind it good quality salt. The system is beneficial to the mangrove forest as regular tidal flush is maintained, and there is neither accumulation of stagnant water nor increase of water salinity in the canals. Mangrove regeneration and growth can thus take place. It should also be noticed that *Prosopis* sp. is entering and becoming adapted to the mangrove ecosystem (Selvam et al. 1999).

3.8 Concluding Remarks

A correct, complete discussion on the mangrove ecosystem should include the description and discussion of all abiotic and biotic components as well as of renewable and non-renewable resources. Relationships with capture and captive fisheries as well as related ecosystems such as coral reefs, sea-grass beds and seaweed cultivation should also be part of the studies. As there is neither space nor time to do this for the Indo-West Pacific in this book, I will leave those subjects to another opportunity and let this be just a general overview. Subjects such as hydrology, geology, pedology, palaeontology and palynology, physics and chemistry of sediments and waters, flora, fauna, microbiology and others have so far either been dealt with exhaustively elsewhere as specialised subjects in their own right, or are published as chapters or research papers scattered in scientific books and journals. In the mangrove context, hydrology, geology and marine biology are among the oldest subjects that have been and are still dealt with methodically since Linnean and pre-Linnean times. Each of these would be useful as comprehensive review books to collate and update our knowledge on mangroves. The taxonomy of the mangrove flora was updated by P. B. Tomlinson in his book "The Botany of Mangroves" published in 1986 by the Cambridge University Press. The only subjects that are dealt with in some detail here are the study on Mangrove Phenologies, by Norman D. Duke, and a general framework for mangrove ecosystems management by Padma N. Lal. I take phenology of mangroves as an expression of both the variability and adaptability of mangrove species and mangrove ecosystems to various environmental conditions. A framework for management is the ultimate purpose of the book.

Most readers will be disappointed because they will not find answers to many of their questions. I am well aware of the many gaps in the text. The lacunae are intentional, partly because short mention to those topics would not do justice to the complexity of the subject, or worse still, would convey at best vague or biased information. Partly the omissions point to the need for formulating integrated research projects, of which the Ranong Program, already 14 years old, was a pioneer. The need for integrated and appropriate site-specific management plans is obvious. It is more and more the consensus of all those interested in a rational development of the coastal zone itself and in the development of the permanent dwellers of the coastal zone, that no serious long-term management programme can be achieved without the wholehearted participation of the people whose homes are the mangroves, the coastal lagoons, the estuaries, the delta islands and open coasts of the tropical and equatorial belts of the world.

Community participation is essential for the success of any program involving rational management and development of mangroves and people. This was one of the most important conclusions reached at a Symposium on "Significance of Mangrove ecosystems for Coastal People" held at Hat Yai, Songkla Province, Thailand, from 19–21 August 1996. The conviction was reiterated again in the Concluding Remarks of a Symposium on Mangrove Ecosystems held at Bali, Indonesia, from 6–9 September 1999. Both were ISME activities, at the UNU Workshop on Asia-Pacific Cooperation on Research for Conservation of Mangroves, held at Okinawa, 26–30 March 2000, and at the "Mangrove 2000" Conference on Sustainable Use of Estuaries and Mangroves: Challenges and Prospects" held in Recife, Brazil, 22–28 May 2000.

The developed world is now living in what has been called "post-modernism" times. I would like to consider this undefined term as the recognition that the "modernism" of the past century and a half is now gone and buried. But is it? Ruthless exploitation of renewable and non-renewable resources and numerical growth of the human population would point in another direction. Have we not embarked on an ostrich policy of "let's go while the going is good?" While it is true that there is at present some awareness of the need for conservation and recycling, the major part of the world's population is indifferent to the pollution and desertification it contributes to build up. People at large are still unaware of the radical changes that "modernism" has brought about, except for what the media choose to show as personal advantages and benefits. When talking of mangroves and other renewable resources, the word "renewable" brings with it a curse, because many understand the word as "unlimited". One example is "intensive" shrimp farming methods or the clear felling of forests in "faraway" parts of the planet, as for instance the forests of Maisin in Papua New Guinea, or the areas clear-felled of mangroves in Irian Jaya and other places. But, those "faraway places" are the here and now for many people who have not gone to places "faraway" for them, to demolish our cities.

Management of mangroves cannot be dissociated from that of other tropical coastal systems upland and coastal marine waters downstream. Upland systems include natural supra-tidal vegetation tolerant of occasional inundation by tidal waters and of relatively saline soils and saline pore water conditions. Below tidal systems include coastal marine currents, coral reefs, seagrass beds and coastal fisheries.

Management plans should also consider:

- Forestry practices, including raising nurseries wherever natural recruitment of seedlings appears not to be sufficient for natural replenishment
- Consolidation of accretion areas and establishment of green belts on the seafront and along river embankments as a protection against erosion of embankments and engineering structures, as well as and protection against cyclones and strong winds
- Rationalising the exploitation of wild stocks of fisheries and aquaculture in the intertidal zone
- Developing non-destructive silvi-aquaculture or silvi-agri-aquaculture
- Destruction of mangroves causes decreases in wild stock fisheries
- Return from fisheries in mangroves and coastal tropical marine waters is greater by a factor of five or more than the return from the mangrove forest. In other words, return in terms of local or export dollars in well managed controlled mangrove areas is far greater than the return from timber, charcoal, fuel wood and minor forest products put together. The word "fisheries" includes fish, mollusks, crustaceans, marine algae and others.

Here, I wish to repeat again that it is a matter of global interest to manage mangroves and other natural ecosystems in a rational, sustainable manner. These general conclusions refer only to salient points to be considered for a wise long-term management of mangrove ecosystems. Since correct management implies the understanding of the natural and man-made forces that act on a natural ecosystem, these "Conclusions" implicitly presume the knowledge of functioning of the system in all its aspects. The word "Function" in the title of the Book includes all relevant knowledge derived from traditional knowledge, local customs and laws, as well as scientific research. It involves scientists, policymakers, administrators and, above all, the coastal dwellers themselves, who have acquired and developed much know-how.

Hopefully, our collective knowledge of the function of mangrove ecosystems will lead to wise management. Some important aspects of mangrove ecosystems emerged while this book was being written; they are especially relevant for management purposes. Surprisingly, some of these aspects are precisely those that have been disregarded or ignored when management of the tropical coastal zone was discussed, planned and implemented. Even when stressing forces were identified, they were seldom quantified. The conclusions warranted at present can be summarised:

- Damage to mangroves and coastal areas at large are caused by natural and man-made stressing local and distant forces, present or past. Some of these are: tectonic movements, climate changes, damaging episodic events such as earthquakes, landslides and natural barrages formed over feeding rivers, cyclones, dams, impoundment and diversion of water upstream, changes in fresh water flow, building sea walls, changing course and flow of rivers and creeks in the coastal plain, felling of mangrove trees for aquaculture ponds in the intertidal belt, reclamation and conversion of mangrove areas for agriculture, urbanisation, industrial complexes, airports and the like. These factors should be identified and quantified when plans are made for the management of the tropical coastal zone and mangroves in particular. There are at present many examples and models of how this can be done.
- The direct and indirect ecologic, economic and social benefits offered by the mangroves must be identified and quantified to define locally appropriate mangrove management plans. Management plans must be revised periodically. The mangrove ecosystem's hidden benefits having greater relevance are: filtering of silt and suspended matter in general, control of solid polluting matter and immobilisation of heavy metals, and consolidation of coastal sediment deposits. Mangroves also provide protection against erosion, promote land accretion, enrich coastal marine and brackish waters with nutrients, and finally, mangroves are a complex ecosystem with a very high recycling power. Mangrove ecosystems are a highly productive transition zone between land and sea, which benefits other upstream and downstream ecosystems. Soils and brackish waters are key components of the mangroves. Material products vary among mangroves – management should promote optimal production of goods and safeguard and preserve hidden benefits. Appropriate long-term management should be based on the quantified knowledge of acting forces.
- There are quantified numerical data (Malaysia, Indonesia) to show that several coastal water fisheries are mangrove dependent; recruitment of juveniles to the wild stocks draws from mangrove area nurseries; destruction of mangroves causes a decrease in the catch from wild stocks.
- Return from fisheries in mangroves and in coastal tropical marine areas can be greater by a factor of five or more than the return from the forest. In other words, returns in terms of local or export dollars in well managed, controlled mangrove areas is far greater than the return from timber, charcoal, fuel wood and minor forest products taken together. The word "fisheries" here include fish, crustaceans (mainly shrimp and crabs) and mollusks, algae and other resources.

It has been said: "development is increase of production, while conservation is the maintenance of the means of production." To preserve mangroves is to preserve the means of much of the production of the tropical coastal zone.

Acknowledgements

I should mention that a large number of persons have contributed to this text. The classics on mangroves, such as Chapman, MacNae, Tomlinson and others were the obvious bases on which to build. The actual starting points were: the UNESCO Bibliography (1975); Mangroves of Asia and the Pacific: Status and Management (1987), which is the final report of the UNDP/UNESCO Research and Training Pilot Programme on Mangrove Ecosystems (RAS/79/002). The ISME Publications: The Economic and Environmental Values of Mangrove Forests and their Present State of Conservation in the South-east Asia/Pacific Region (1993), by B. F. Clough and others; the World Mangrove Atlas (1997); and the Manual for Restoration of Mangrove Ecosystems (1996). The task of writing this chapter was given to me probably for no better reason than that I have lived in the region for over 30 years, and know firsthand most of the places mentioned in the chapter; I have actually lived in many of them for shorter or longer periods. Since I cannot take a bird's eye-view of the mangroves of half the Globe, I will leave that task to those who study them by Remote Sensing Techniques whose co-operation is gratefully acknowledged; I will take the more prosaic approach of discussing them from a crab's eye-view.

To all the people and coastal communities with whom I have spent much time and from whom I learned so much, I wish a better understanding by those in power to give them a better future. To all of them my simple thank you. Though they may never see this book, they know my thoughts and feelings. This chapter was written during my stay at ISME's Headquarters in Okinawa during the summer/autumn of year 1999. The constant support and facilities offered by the Secretariat are gratefully acknowledged. To Prof. Shigeyuki Baba I am especially indebted for many valuable inputs and remarks and unfailing help. To Ms. Nozomi Oshiro, a decade-long devoted colleague, I owe the critical reading of the manuscript and prompt help as a mediator of the frequent misunderstandings between the computer and me. To Dr. Marc Steyaert, Vice President of ISME, grateful thoughts, in the first place for persuading me to become the Chief Technical Advisor of the UNDP/UNESCO Mangrove Projects of Asia and the Pacific from 1982 onwards, for critical reading of this manuscript and above all I am indebted to him for an ever present, decades-long encouraging friendship, often badly needed. To each and every one of those who contributed their knowledge to the text I am grateful for complementing a task I could never have even thought of doing by myself. I am ever grateful to Dr. Denise Navas Pereira for her unfailing follow-up and reminders to all my endeavours. Among the many whose names do not appear as direct contributors to the text, I wish to express my gratitude, especially to: A. D. Agate; A. K. Bandyopadhyay; H. T. Chan; M. Jaffar; Padma N. Lal; M. M. T. Qureshi; F. Rahim; V. Selvam; A. D. Untawale; and M. Wafar. Last but not least, I am indebted to Bridget Ganguly for her patience while revising the text in spite of my repetitive mistakes in the use of the English language. Finally I am greatly indebted to ISME, the International Society for mangrove Ecosystems, to CNPq the National Research Council of Brazil, and JICA, the Japan International Co-operation Agency for financial support for travel and subsistence.

Chapter 4

Mangrove Phenologies and the Factors Influencing Them in the Australasian Region

Norman C. Duke

4.1 Introduction

Phenology is the study of seasonal cycles in plants, focusing on leafing (vegetative) or reproductive cycles, in particular. These studies, however, logically follow into longer-term cycles, which change both the magnitude of phenological events and their timing from year to year. In all cases, the effects on the reproductive cycles have a profound influence on the adaptability, survival and distribution of species.

Phenological patterns in mangrove plants follow cycles and trends which depend largely on a range of factors from seasonal and year to year differences, to features of particular geographic locations at different latitudes and for different climatic conditions, to aspects of genetic variability including differences in species type. For these reasons, it is not possible to broadly generalise about phenological patterns in mangrove plants without specifying the species in question and its location.

Mangroves normally occupy a particularly narrow niche in marine and estuarine environments, especially in regard to their restricted elevational range between mean sea level and the high tide mark. This restricts their ecological range, but it also restricts their wider geographic occurrence to an often thin strip of coastal vegetation bordering landmasses and continental margins. This is punctuated by some wider stands in larger embayments and at the mouths of rivers, and further upstream in some cases as thin riparian borders. Species preferences for particular parts of this range further restrict the connectiveness of mangrove genetic entities and population groupings. To cope with these conditions, mangroves have highly developed strategies, of which the most obvious are their buoyant, often large propagules. It also appears that flowering is developed to maximise dispersal of genetic material. It therefore follows that the factors, which influence flowering and propagule maturation, would also have a profound effect on species presence and their sustainability.

Several studies have attempted to seek influencing factors from data collected at single locations, but this does not take into consideration how each species copes with different conditions throughout its range. For this reason, such single-location studies have failed to advance our overall understanding of factors influencing phenologies except in their contribution to the larger data pool.

In this review, I re-assess our current understanding of mangrove phenologies and the factors which influence them, drawing on available observations as much as on data at hand from personal studies within the Australasian region. Where appropriate, examples were taken of events and circumstances, which best illustrate particular influencing factors.

4.2
Definitions

In this treatment, I use several terms, which need to be defined in this context. I list these along with a brief explanation and definition of each:

- *Initiation*: the event which marks the beginning of the reproductive cycle. This is thought to be triggered by exogenous factors.
- *Phenoevent*: also called a phenological event, is an event of peak production for particular component stages of growth. Notable examples of common events include: leaf fall, leaf flushing, flowering and propagule maturation.
- *Phenophase*: the period of growth development between two particular phenoevents. Examples of phenophases include: initiation of the reproductive cycle up to peak flowering and flowering to peak propagule maturation.
- *Propagules*: mature end products of the reproductive cycle. Depending on species, propagules of mangrove plants might be fruits, encapsulated seeds, spores, or viviparous hypocotyls. In all cases, propagules are responsible for the dispersal and re-establishment of species.

4.3
Factors Affecting Mangrove Phenologies

The factors influencing mangrove plant phenologies are many and varied. The chief cycles of floral and vegetative phenologies are also likely to interact to create the appearance in some locations of uniqueness for major phenoevents and their timing, shown as flowering and fruiting, or leaf fall. Several instances of phenological case studies have been presented in literature and many authors have found correlations with certain climatic factors, such as rainfall and temperature. However, consistent factors which broadly influence phenologies have proven illusive, and only one study has shown patterns in phenological cycles over a wide geographical area. I draw on this earlier study, in combination with others, to enhance this review of chief factors influencing phenological events of mangrove plants.

4.3.1
Seasonal Annual Cycles and Species Differences

4.3.1.1
Floral Development

Factors affecting phenologies of individual plants depend primarily on species type and season. In all species, regardless of growing location, the major phenological events of flowering and propagule maturation, etc., mostly occur seasonally irrespective of latitude. The timing and significance of these events varies from place to place, but species assemblages are characterised by the regular and common occurrence of major phenological events each year. It is fully expected that these patterns are dependant on location; nevertheless, it is useful to review the breadth of variation at any single location. A brief summary of the dates of flowering and propagule maturation of 41 mangrove species is shown in Table 4.1, based on sites in northern Australia and southern New Guinea.

Table 4.1. Phenologies of 41 mangrove plants in tropical Australasia, showing months of peak flowering and propagule maturation with approximate months of development between events (modified and updated from Duke et al. 1984)

Mangrove Species – northern Australia and southern New Guinea	Months of peak flowering (*) and propagule maturation (?)												Dev. period
	J	A	S	O	N	D	J	F	M	A	M	J	
Acanthus ilicifolius L.			*	*	*	*	?	?					2 – 4
Aegialitis annulata R. Br.						*		?					2
Aegiceras corniculatum (L.) Blanco	*	*	*				?	?				*	6 – 7
Avicennia alba Blume						*	*		?				3
Avicennia integra N.C. Duke			*	*	*	?	?						3
Avicennia marina (Forsk.) Vierh. *sensu lato*					*	*			?	?			3
Avicennia officinalis L.			*	*	*		?	?					4
Avicennia rumphiana Hallier f.				*	*	?							2
Bruguiera cylindrica (L.) Bl.					?							*	5
Bruguiera exaristata Ding Hou			*	*				?	?				5
Bruguiera gymnorrhiza (L.) Lam.	*	*					?	?		*	*	*	5 – 9
Bruguiera parviflora (Roxb.) Wight & Arn. ex Griff.		*	*	*			?	?					4 – 5
Bruguiera sexangula (Lour.) Poir.		*	*?	?	?	?							4
Camptostemon schultzii Mast.							*	*	?	?			3
Ceriops australis (White) Ballment, Good. & Sm.					*	?	?	?					13 – 15
Ceriops decandra (Roxb.) Ding Hou			*	*	*	*?							12 – 15
Ceriops tagal (Perr.) C. B. Rob.							*	*	*?	?	?		14 – 16
Cynometra iripa Kostel	*	*				?	?		*	*	*	*	6 – 9
Diospyros littoralis R. Br.			?		*								?
Dolichandrone spathacea (L.f.) K. Sch.			*		?								3
Excoecaria agallocha L.						*	?	?					1 – 2
Heritiera littoralis Ait.			?	?	?	?		*	*				7 – 9
Lumnitzera littorea (Jack) Voight.					*			?	?				3 – 4
Lumnitzera racemosa Willd.						*		?	?				2 – 3
Lumnitzera X *rosea* (Gaud.) Presl.					*			?	?				3
Nypa fruticans Wurumb.					*			?	?				3 – 4
Osbornia octodonta F. Muell.							*	*?	?				1 – 2
Pemphis acidula J.R. & G. Forst.						?					*	*	6
Rhizophora apiculata Bl.								?		*	*		9 – 10
Rhizophora X *lamarckii* Montr.		*	*	*	*	*	?	?	?				3 – 9
Rhizophora mucronata Lam.				*	*	*	?	?	?				15
Rhizophora stylosa Griff.							?	?*	*				11 – 12
Scyphiphora hydrophyllacea Gaertn.				*				?	?				4 – 5
Sonneratia alba Sm.				*			?						3
Sonneratia caseolaris (L.) Engl.	?						*					?	5 – 6
Sonneratia X *gulngai* N.C. Duke		?				*			*?				3 – 5
Sonneratia lanceolata Blume			*	*			?	?					4
Sonneratia ovata Backer				?							*		5
Sonneratia X *urama* N.C. Duke						?					*	*	6
Xylocarpus granatum König	?	?	?			*	*					?	6 – 8
Xylocarpus mekongensis Pierre			*	*			?	?	?				3 – 4

The overall patterns and the differences were discussed in detail by Duke et al. (1984). Briefly, the majority of flowering was observed in warm-dry months (September to December), and propagule maturation in warm-wet months (November to March). However, despite this, there are some species that defy the common pattern. One unexpected major difference is for *Sonneratia caseolaris*. This species, in contrast with its congener in sympatry (see Table 4.1), has peak fruit maturation in June–July, while *S. alba* fruits in January. Curiously also, individuals of the putative hybrid between this species and *S. alba*, called *S. × gulngai*, showed a combination of the two different phenologies (Duke 1988). Another pair of congeners with similarly different seasonal timing are *Xylocarpus granatum* and *X. mekongensis* (also see Table 4.1). The reasons for such major differences seem profound and likely to be based on aspects of genetic isolation in the evolution of these taxa. If so, this further emphasises the great importance of phenologies in plants where they clearly control the transfer and exchange of genetic material, whether it be via pollen from flowers or as propagules.

Phenoevents might range from the more obvious, such as leaf fall, flowering and fruiting, but clearly this is based on the particular stage being monitored. For example, in Fig. 4.1, eight stages of reproductive development are shown for *Rhizophora stylosa* and *R. apiculata* in litter fall collected monthly over a three year period in NE Australia. These events are those periods shown as peaks in graphs for each of the components shown.

In each case, the three peaks mark the annual occurrence of respective events for each of the stages of development. One great benefit in presenting the data in this way is that particular cohorts can be traced in time as reproductive units develop.

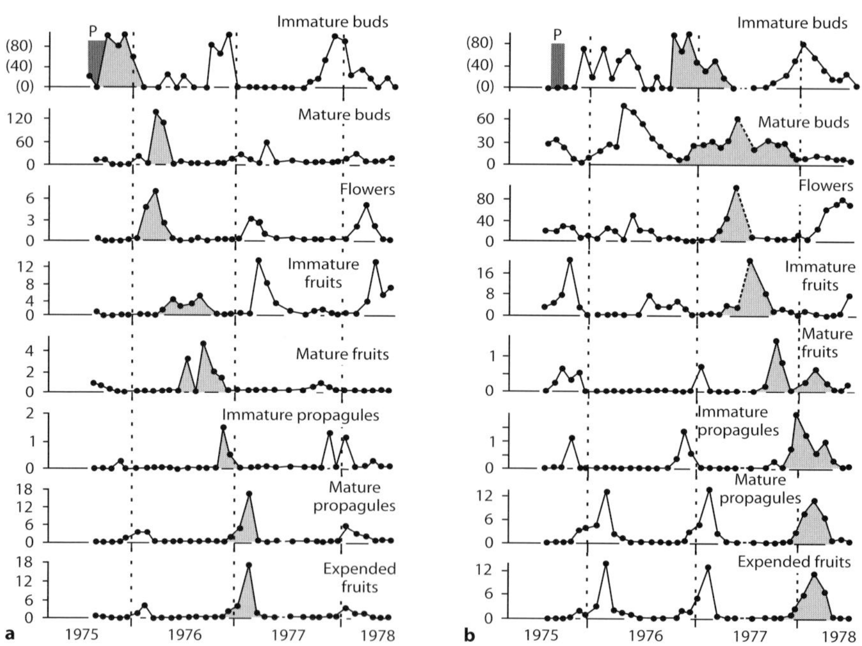

Fig. 4.1. Floral development over three years (1975–1978) for **a** *Rhizophora stylosa* and **b** *R. apiculata* in north-eastern Australia (Duke et al. 1984)

This figure as such provides a schedule of reproductive development up to propagule maturation.

Initiation of the cycle is indicated best with the first appearance of bud primordia or immature buds. However, while this was the most likely case for *R. stylosa*, it was not for *R. apiculata*. For the latter species, primordia stand dormant for approximately one year and delay the reproductive development cycle accordingly. Development periods for other species follow similar development progress, although rates of development differ considerably.

In Fig. 4.2, five comparable stages of reproductive development are shown for *Avicennia marina* (var. *eucalyptifolia* (Val.) N. C. Duke) growing in Australia and Papua New Guinea (Duke 1990). In each site, initiation may have occurred around September, since this marks the earliest appearance of immature buds in litter fall. However, flowering and propagule maturation shift appreciably for sites in higher latitude locations. The time from flowering to propagule maturation was around 3 months in *A. marina*, while for *Rhizophora*; it was around 11–12 months to 9–10 months, respectively.

4.3.1.2
Leafing Phenologies

Species are also characterised by their quite different vegetative phenologies, although there are few data available to properly characterise events such as leaf flushing and leaf fall. These occurrences are sometimes less well-defined as an event, but there are

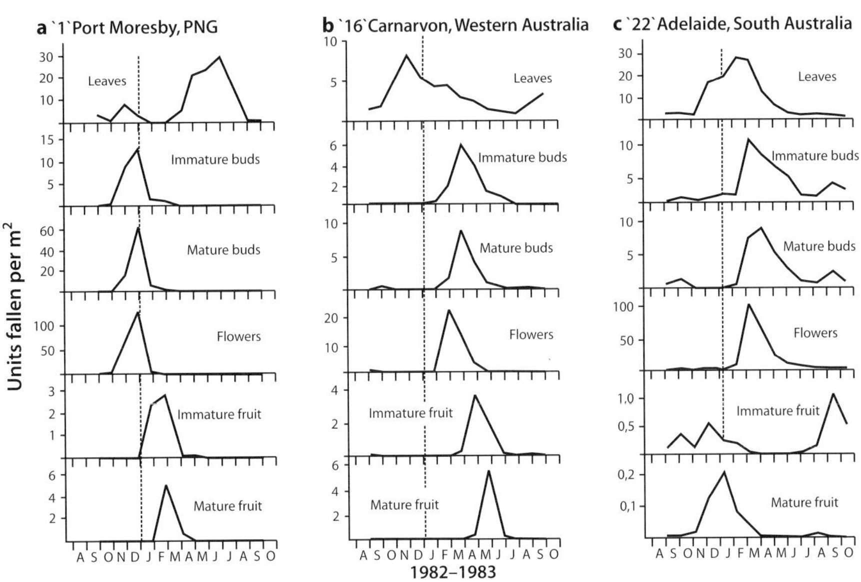

Fig. 4.2. Floral development over one year (1982–1983) for *Avicennia marina* in **a** Port Moresby in Papua New Guinea, **b** Carnarvon in Western Australia, and **c** Adelaide in South Australia (Duke 1990; also see Fig. 4.3. for locations)

at least two species with quite notable leaf fall, namely *Xylocarpus mekongensis* and *Excoecaria agallocha*. For *X. mekongensis* in particular, leaf fall is distinctly deciduous, where leaves become reddened and drop altogether, leaving trees bare for a few weeks to a month each year. In northern Australia, this occurs around August, and shortly afterwards flowers and new leaves are produced on the bare branches. Leaf turnover in these species is clearly annual, or less, but in other mangroves, leaf longevity is longer, with 19–27 months noted for *Rhizophora*, and even longer (up to 42 months) in other Rhizophoraceae (Duke et al. 1984). Essentially, while leaf flushing and leaf fall are annual events and leaf longevity commonly exceeds one year, these and most mangrove species are considered evergreen (Tomlinson 1986). However, despite this image of foliage continuity, canopy leaf densities in such species do pulse and flux in a seasonal way in response to the net effect of leaf flushing and leaf fall, which in themselves are each seasonal. The data for *Rhizophora* are better than for most species simply because of the occurrence of leaf stipules. These stipules enclose young leaves, and they are dropped as leaves emerge. In this way, litter collection is sufficient to monitor both leaf fall and leaf production (Duke et al. 1984). The difference between these two rates is net leaf production in the canopy. Clearly, in normal circumstances, this might fluctuate between positive and negative values each month, but after a year it would be expected to add up to zero in normal conditions where canopy density is maintained from year to year. But, when there are year to year differences, this technique reveals whether canopies have greater or lower leaf densities than normal and whether trees are relatively stressed or not.

4.4 The Influence of Floral Phenologies on Genetic Variation

The previous section identified phenological variation across a range of different species. On closer examination of this information (referring to Table 4.1.), however, it was evident that the greater amounts of difference in occurrences of phenoevents were shown in congeneric species pairs. Could this be evidence of phenological differences driving genetic separation, or something else? Species pairs, or closely related sister species, are common in mangrove assemblages world-wide (Duke et al. 1998). In northern Australia, there are at least eight such pairs growing in sympatry. In order to assess this further, these closely related species were ranked by closeness of occurrence of phenoevents, as shown in Table 4.2. Each species pair was broadly classified according to its morphological and genetic differences.

Morphological classification was based on general characteristics but related ultimately to attributes of flower parts. Genetic difference was broadly classified from field observations of hybrid occurrence in locations where putative parents occurred in sympatry. This relatively simple assessment identified two quite different strategies in genetic isolation of mangrove taxa. The eight species pairs can be divided evenly into two distinct groupings, each having comparable but independent and overlapping rank orders of phenologically difference:

a Morphologically different – genetically similar. This group of four species pairs is characterised by the occurrence of hybrid taxa between sister species in sympatric stands (see Table 4.2), and a relatively strong difference in morphological charac-

Table 4.2. Closely related species pairs from Table 4.1 are ranked by differences in occurrence of phenoevents and compared with degrees of morphological and genetic difference. Classification of morphological character is general, but relates ultimately to floral attributes. Two groupings of species pairs (*A*) and (*B*) are discussed in the text. Genetic difference was broadly classified using field observations of hybrid occurrence where putative parents occurred in sympatry. Observations were chiefly made in northern Australia (Duke and Bunt 1979; Duke et al. 1984; Tomlinson 1986; Duke and Jackes 1987; Ballment et al. 1989; Duke 1994; pers. observ.)

Species pairs (ranked by occurrence of phenoevents from close to increasingly different)		Morphological character	Hybrid occurrence
Lumnitzera littorea	– *L. racemosa*	Different (A)	Rare
Sonneratia alba	– *S. lanceolata*	Different (A)	Uncommon
Rhizophora mucronata	– *R. stylosa*	Virtually indistinguishable (B)	None
Rhizophora apiculata	– *R. stylosa*	Different (A)	Common
Bruguiera gymnorrhiza	– *B. sexangula*	Virtually indistinguishable (B)	None
Ceriops australis	– *C. tagal*	Virtually indistinguishable (B)	None
Xylocarpus granatum	– *X. mekongensis*	Different, except for flowers (B)	None
Sonneratia alba	– *S. caseolaris*	Different (A)	Common

ters, particularly among flowers. The pair showing the greatest amount of difference in this grouping was *S. alba-S. caseolaris*. Genetic isolation of this pair and others in this group were shown mostly by the notable differences in flower and fruit attributes (Duke and Jackes 1987). This, however, contrasts with the closeness of genetic makeup where the hybrid, *S.* × *gulngai*, for example, was observed in every estuary where the parents co-exist in northern Australia. The occurrence of hybrids (a measure of genetic similarity) appears to increase in species pairs with greater differences in phenoevents.

b Genetically different – morphologically similar. This group of four species pairs is characterised by a lack of hybrids in sympatric stands (see Table 4.2), and a lack of distinctive morphological characters by which to separate the species based on floral characters. The pair showing the greatest amount of difference in phenoevent occurrence in this grouping was *X. granatum-X. mekongensis*. Genetic isolation of this pair and others in this group were shown mostly by the lack of hybrids and the many notable differences in tree form, bark character and fruit (Tomlinson 1986). This, however, contrasts with the lack of difference in flower characters. Morphological differences appear to increase in species pairs with greater differences in phenoevents.

In these cases of species pairs, genetic isolation was supported by differences in flowering times and morphological characteristics. The occurrence of significant overlap in the rankings of these groups (Table 4.2) also implied that each of the two strategies for genetic isolation were equally important and independent. These points must however be re-assessed after genetic character comparisons for each species pair have been better described and quantified using molecular techniques. But for the moment, it appears the closer some species pairs are genetically, the more different will be their respective phenologies. Also, divergent phenologies of other species pairs reflect greater morphological differences.

4.5
Year to Year Differences in Floral Development

In the previous section, the importance of seasonal annual events were emphasised; however, it was also evident that these events vary from year to year in both quantity and timing. The changes in quantity from year to year create the impression with observers of 'good' and 'bad' years. This applies to the different major phenoevents, whether it be leaf flushing, flowering or propagule production. This view depends on the observers interest; for example, those people monitoring general forest condition would follow leaf density in the canopy, while beekeepers would follow flowering, and those wanting to collect propagules for planting projects would follow propagule maturation. This diversity of interests reflects the diversity of ways in which phenological cycles are manifest. It also demonstrates the flux and dynamics of mangrove forest systems adapting and coping with fluctuations and changes in environmental conditions. We may not know yet the factors most responsible for changes in phenoevents each year, but we can appreciate that these fluctuations will effect the wider biotic inter-relationships of mangrove ecosystems, not to mention the character and survival of these unique forest systems.

Unfortunately, most evidence of year to year differences are anecdotal, and as such they reflect the interests of the variety of observers' interests, as noted above. Few general studies of mangrove phenologies have run over more than one or two years in any single location for a group of individual trees. In the Australasian region, the studies around Hinchinbrook Island in northeastern Australia have been perhaps the most comprehensive to date (Duke et al. 1984). At least five common species were monitored from litter fall collected monthly over a three year period from 1975 to 1978. The results from this study alone cannot show year-to-year patterns, but the results do show substantive differences in the phenoevent quantities of these trees. The differences also, as might be expected, are species specific. Some data from this study were presented in Fig. 4.1 for *R. stylosa* and *R. apiculata*.

For *R. stylosa*, relatively large peaks of mature propagules and expended fruits were observed in early 1977, while smaller peaks were evident in corresponding months in 1976 and 1978. Because of the detail of these data and the large amounts of aborted material, noting the earlier reproductive stages, it is possible to trace the progress of successive annual cohorts from the appearance of bud primordia through to propagule maturation. These studies were supported with concurrent shoot observations. The 1977 mature propagule peak was thus recognised as derived from the reproductive crop commencing around mid 1975, noting the primordia around August 1975, mature buds and flowers in March 1976, mature fruits in September 1976, and completing development in February 1977. Similarly, we might trace the cohort initiated in 1976 through to maturation in January 1978, and final development of the cohort beginning in 1974 and the early development of the cohort beginning in 1977. Clearly, these cycles overlap in *R. stylosa* so that at any time, this species has two distinctly different stages of reproductive development in its canopy at any time. It is also notable that flowering and propagule maturation were synchronous in this species.

For *R. apiculata*, peaks of mature propagules and expended fruits were roughly the same, contrasting with *R. stylosa*. Also differing from *R. stylosa*, bud primordia in *R. apiculata* were produced in mid 1975, and lay dormant for one year in the leafy shoot before developing further into immature buds, to mature buds and flowers around May 1977, to mature fruits around October 1977, and to mature propagules and expended fruits in February 1978. The early dormancy phase of the primordia extends the development period of this species by one year, and there are up to three stages of reproductive development in the canopy of this species at any time. Flowering followed propagule maturation by about three months in this species. Despite this difference, propagule maturation was synchronous between *R. stylosa* and *R. apiculata*.

In view of the 'waves' of successive reproductive cohorts, it is clear for *R. stylosa* that the peak of maximal development was notable through most of the earlier reproductive stages as aborted units. This is presumed to be indicative of a comparable rate of the success at each developmental stage. For this reason, it seems likely that the 1979 crop of mature propagules might also be relatively large, based on the relatively large peak of flowers in early 1978. By contrast, annual quantities of *R. apiculata* reproductive components appeared relatively consistent over the three-year study period.

For *A. marina*, particularly in southern Australia, the quantity of fruit production is also notably different from year to year. There is some indication that large fruit crops in Moreton Bay, eastern Australia, occur every second year, and particularly in recent times for odd numbered years; for instance, fruiting was high in July 1999 and virtually non-existent in 1998 (pers. obser.). Clearly, longer term monitoring is required to substantiate this observation. This changeable occurrence may be the result of overlapping reproductive cycles (Duke 1990), where plants that have used resources to produce a one-year crop then are unable to support the next year's crop but can then support the following year's crop, and so on. As such, irregular annual fruiting might reflect year-to-year resource allocations in the plants.

Year to year differences may also reflect effects from El Niño events and cycling of the SOI. Recent evidence from rainforests in Panama and other places (e.g., Wright et al. 1999) show the importance of the longer-term influences. Unfortunately, there are no long-term data sets on phenologies of mangrove forests from which to make comparable observations.

Given the relationship between phenologies and temperature, however, there are likely to be profound effects on mangrove habitats, particularly in subtropical and temperate latitudes, with increasing global air temperatures. As noted, the influence on mangrove development will be dramatic where temperature appears to affect reproductive development by increasing growth rates by a factor of two or three, for each 10 °C rise. The result will be a possibly dramatic shift in several aspects. First, there will be a shift in dates of flowering and propagule maturation causing the disruption of the organisms that depend on this resource. The shift might be dramatic in some instances, notably for places where development periods currently run through winter, like around Moreton Bay, southern Queensland. Or, those further south, like in southern Victoria, where the cycle extends through winter and into the next summer. If temperatures rise in Victoria, then propagules will shift their maturation from summer to winter, making them unavailable to animals that depend on early to midsummer food resources. Second, the amount of material produced each year will be altered. This is shown particularly again in higher latitude sites where annual cycles clash and appear to induce plants to alter the amount of floral production considerably from year to year.

4.6 Latitudinal Influences on Floral and Leafing Phenologies

Major trends with latitude were observed in leaf fall and reproductive phenologies of *Avicennia marina* assessed during 1982–83 using litter fall collections from 25 sites in Australia, Papua New Guinea and New Zealand (Fig. 4.3). Latitudinal variation of collection locations ranged from 9°32' S to 38°21' S.

Flowering shifted from November–December in northern tropical sites, to May–June in southern temperate sites. Periods between flowering and fruiting increased from 2–3 months in tropical sites to 10 months in southernmost sites. Leaf fall was more variable with uni-modal annual peaks in temperate sites and often multi-modal patterns in the tropics. In Fig. 4.4, these observed patterns in Australasian sites are shown. Sites are ranked from low to high latitude locations combining hot arid sites in Western Australia with variable climate sites in eastern Australia. Despite the variations in climate, there are clear overall trends that describe the predominant influence of a few factors.

Correlative evaluation of simple models suggested that initiation of the reproductive cycle occurred when day length exceeded 12 h, followed by rates of development to flowering and fruit maturation given by a linear function of mean daily air temperature (Duke 1990). The best model explained around 92% of variance in total cycle duration and timing. Temperature appeared to affect reproductive development by increasing growth rates by a factor of two or three, for each 10 °C rise. The model was verified from independent observations of reproductive events in sites not used in its derivation.

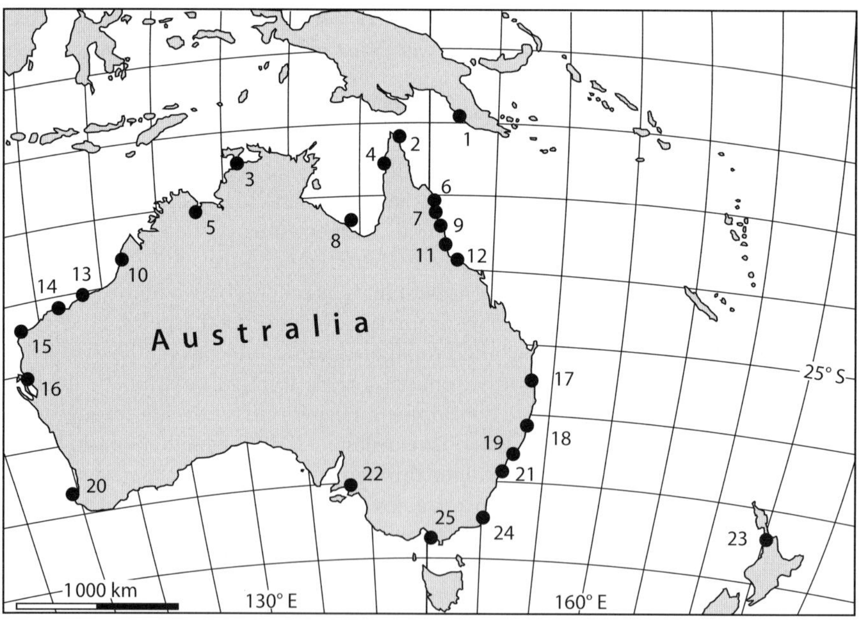

Fig. 4.3. Sites in the Australasian region comparing *Avicennia marina* phenologies (reported in Duke 1990)

Fig. 4.4.
Seasonal leaf fall of *Avicennia marina* from 25 Australasian sites ordered by latitude south. Reproductive events included for the same sites also are putative initiation (*dashed line*), flowering (*open circle*) and propagule maturation (*solid circle*)

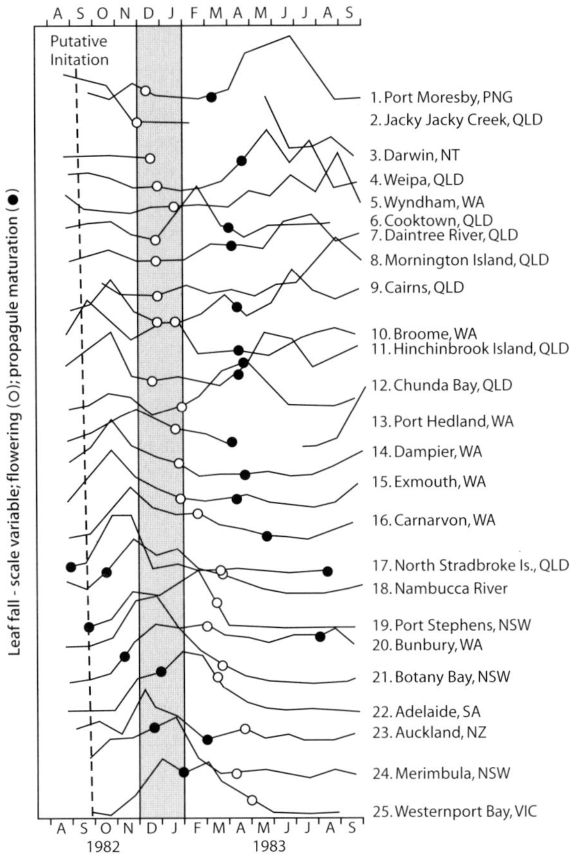

4.6.1 "Continuous" Flowering in Equatorial Sites

Avicennia trees in equatorial sites are sometimes characterised as having 'continuous' flowering and a lack of seasonality. However, this may only be an impression based on presence-absence observations, rather than on quantitative measures of peaks in phenoevents.

Continuous flowering might also be explained in the model by taking into consideration two additional points. Firstly, in equatorial sites, day length fluctuates throughout the year at just over 12 h d^{-1}. It is conceivable that this might repeat the trigger for initiation several times a year, depending on amounts of photoperiod and the extent of cloud cover from day to day. The response of the plant will therefore depend on these factors, along with rainfall and nutrient availability. The second point relates to the trees' ability to support two crops in one year. In some circumstances, *A. marina* trees may be able to initiate two cycles in one year. And, depending on climatic conditions and nutrient availability, one or both of these cycles might result in propagule maturation. Clearly, flowering may not always result in propagule maturation where resources are limited or the climate is unsuitable. It is also of interest that in sites at latitudes less than 10° (equatorial sites), the floral development cycle is completed within six months, leaving ample time for a second cycle, where conditions might be suitable (see Fig. 4.4). It is of further interest that floral development can also overlap from year to year, as noted in the high latitude sites in Australasia.

To resolve the questions surrounding continuous flowering, it is recommended that litter fall and specific phenological studies are required to quantify monthly amounts of inflorescence material. Such studies need to be conducted preferably over two or more years to further isolate and identify the variables.

4.7 Links between Floral and Leafing Phenologies in *Avicennia marina*

The relationships described in the model all relate to the floral cycle. The question, therefore, is how does this relate to leafing phenologies? And, what are the factors which appear to drive leaf appearance and leaf fall?

While floral cycles were relatively simple to interpret from their progressive development through different reproductive stages, interpretation of leafing was complicated by the lack of information on when leaves are produced on trees. Direct shoot observations have been rarely quantified, and leaf appearance may not be evident in litter fall studies, particularly for *A. marina*. However, for this species, it was observed that leaf appearance was closely correlated with leaf fall. Therefore, the relationship between appearance of new leaves on canopy shoots and leaf fall was investigated closely at one site in north Queensland (Duke 1990). Seasonal highs in leaf appearance were observed to be followed by peak falls a month later, and leaf longevity was estimated to be around thirteen months on average. Furthermore, the timing of new leaf appearance and fall compared favourably with observations from comparable studies conducted in the region (Attiwill and Clough 1978; Clarke and Myerscough 1991; Goulter and Allaway 1979; Davie 1984; Duke et al. 1984; May 1999).

Based on such findings, leaf appearance in *A. marina* was interpreted from leaf fall data collected throughout the region (see Fig. 4.4). Leafing was characterised by variable multi-modality. This occurrence was taken to reflect the secondary importance of vegetative phenologies, where they apparently defer to the ordered progress of reproductive cycle events (Duke 1990). There are three overall trends or patterns evident in these observations, and the findings are briefly summarised here based on two kinds of tropical patterns and another pattern in temperate climates.

1. In the tropics, where freshwater was not obviously limited, notably in sites of higher rainfall, leafing occurred between floral cycles, namely after propagule maturation and before initiation of the next floral cycle. The leafing cycle appeared to be independent.
2. In the tropics, where moisture was limited, leafing occurred at both the appearance of inflorescence material, termed 'inflorescence leafing', and between floral cycles, the 'independent leafing'. The respective amounts of leafing (i.e., comparing the respective peak heights) appeared to depend on moisture and temperature conditions, such that where moisture was less and evapotranspiration high, there was a greater tendency for inflorescence leafing. In extreme circumstances, such as in severe seasonally dry, hot tropical sites, there might be no leafing observed between floral cycles. The timing of floral cycles appeared generally unaffected by moisture conditions.

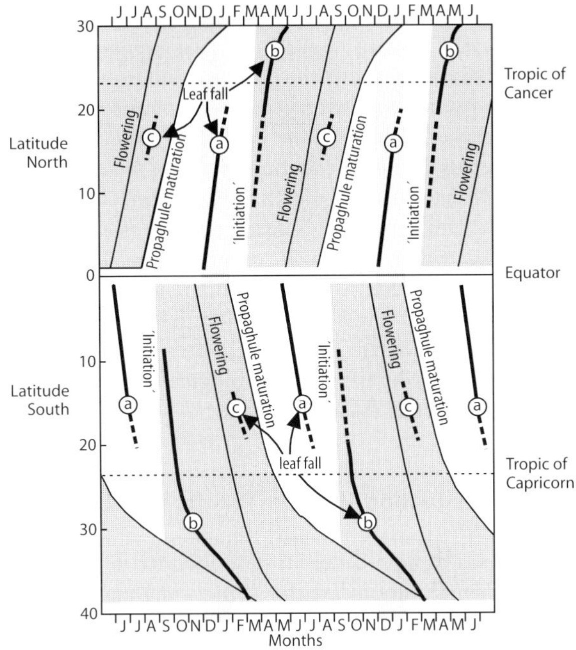

Fig. 4.5. Predicted mean occurrence of floral and leafing phenoevents, including initiation, flowering, propagule maturation, and leaf fall, as reported in Duke (1990). Note that leaf appearance occurs one month earlier than leaf fall

3. In temperate climates, irrespective of moisture conditions, and where cycles were characteristically much longer in duration, leafing occurred in the summer and notably after reproductive initiation, and before peak flowering. These peaks might be the same as the 'inflorescence leafing' noted in tropical sites, but direct shoot observations are required to confirm this view since there is a lag of 4–6 months from putative initiation to leafing (see Fig. 4.4).

The factors controlling leafing therefore appear to be strongly linked with floral development, temperature and moisture conditions. Predictions of these patterns for the time being must be general, however, and be based on the patterns shown in Fig. 4.5. Specific differences in sites affected by abnormal conditions might be explained in respect to the three patterns described above.

4.8 Predicting Floral Phenologies

4.8.1 Predictions Based on Observed Patterns in *Avicennia marina*

Prediction of floral phenologies of *A. marina* can be relatively precise. The prior assessment of *A. marina* demonstrated that phenologies were primarily influenced by a few factors, and that the annual appearance and timing of events is predictable. This study has been used to predict the duration of floral development and the timing of the key events of initiation, flowering and propagule maturation. These predictions are based on two factors: day length and air temperature. Therefore, future predictions can be made in either of two ways: The dates for particular peak events can be extrapolated from the diagram in Fig. 4.5. This figure shows the observed and expected patterns in southern and northern hemispheres based on averaged observations for respective latitudes of occurrence and years. It also provides a view of the overall pattern of phenological events and their interactions across the full distributional range of *Avicennia marina*. Or, the dates of peak events can be calculated using equation models based on the putative initiation date and air temperatures (Duke 1990). This latter method, as expected, provides a means of deducing phenologies where site climatic conditions are not average, and it is particularly useful in quantifying expected year-to-year differences.

Distributional limits of *A. marina* in higher latitudes coincide with trends toward zero reproductive success (notably seen in flowering). This apparently was brought about by a convergence of phenological events within the more limited growth period of summer in these latitudes.

4.8.2 Using the Equation Model to Predict Floral Phenologies of *Avicennia marina*

In this section, a procedure is described for determination of expected flowering and propagule maturation dates for *A. marina*. Data required are site specific, and comprise the site latitude and respective values of mean daily temperature during floral development. The temperature values ideally should be averaged values at the site measured daily during the actual development period(*T*m). But, in the case of predictive assessments, averaged mean daily values from previous years are appropriate. Values can also be monthly averages, if daily values are not available to average for specific periods of development, but estimates based on less precise data will be less accurate.

The calculations for determination of floral development duration and dates of the chief phenoevents, initiation, flowering and propagule maturation are made in three or seven steps, depending on what is required:

Estimate putative initiation date. In Fig. 4.6, the site latitude (nearest whole degree) is used to extrapolate the approximate initiation date in the relevant hemisphere. If you only want to estimate propagule maturation date, proceed to (6), otherwise continue to (2).

Calculate the duration of the initial phenophase – the period from initiation to flowering, or pre-anthesis development. Use the estimated initiation date above (1) to average the

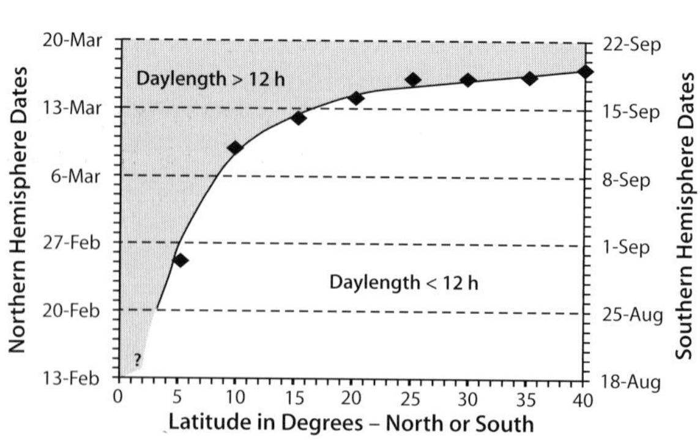

Fig. 4.6. Putative initiation dates for the reproductive cycle in *Avicennia marina* in both northern and southern hemispheres. Predictions are based on day length becoming greater than 12 h. Sites at the equator may never have days less than 12 h, so the occurrence of initiation might be less predictable in equatorial sites

mean daily temperatures (*Tm*) during the three month period after initiation. Use this mean value, in the first iteration, to calculate the duration of the phase using the equation:

$$\text{Initiation to flowering – Duration} = 888.6 + (-55.86(Tm)) + 1.010(Tm)^2$$

Refine the calculation using the averaged mean daily temperature for the period covered by the duration estimate from this equation. Depending on the amount of difference with the earlier estimate, further iterations may be necessary.

Estimate putative date of flowering, or anthesis. Based on the estimate of initiation date, (1) use the estimate of duration (2) to derive the date of peak flowering. If you want to also estimate propagule maturation date, continue to (4), otherwise you have completed the steps.

Calculate the duration of the last phenophase – the period from flowering to propagule maturation, or post-anthesis development. Use the estimated flowering date above (3) to average the mean daily temperatures (*Tm*) during the three month period after flowering. Use this mean value, in the first iteration, to calculate the duration of the phase using the equation:

$$\text{Flowering to propagule maturation – Duration} = 738.9 + (-43.92(Tm)) + 0.725(Tm)^2$$

Refine the calculation using the averaged mean daily temperature for the period covered by the duration estimate from this equation. Depending on the amount of difference with the earlier estimate, further iterations may be necessary.

Estimate putative propagule maturation date. Based on the estimate of flowering date, (3) use the estimate of duration (4) to derive the date of peak propagule maturation. If you want to affirm your calculation using the model for the full cycle, continue to (6), otherwise you have completed the steps.

Calculate the duration of the total phenophase – the period from initiation to propagule maturation, or total floral development. This step is useful if you require an estimate of propagule maturation without considering flowering. Use the estimated initiation date above (1) to average the mean daily temperatures (*Tm*) during the six month period after initiation. Use this mean value in the first iteration to calculate the duration of the phase using the equation:

$$\text{Initiation to propagule maturation – Duration} = 1518.4 + (-89.63(Tm)) + 1.512(Tm)^2$$

Refine the calculation using the averaged mean daily temperature for the period covered by the duration estimate from this equation. Depending on the amount of difference with the earlier estimate, further iterations may be necessary.

Estimate putative propagule maturation date. Based on the estimate of initiation date (1) use the estimate of duration (6) to derive the date of peak propagule maturation.

4.8.3
Testing the Model in a Special Case – a Near Equatorial Site at Singapore

Mathews (1992) conducted a detailed study of *Avicennia marina* in Singapore during 1990–91, noting the dates of particular phenoevents from field observations. The cycle progressed over five months from April to September. Bud primordia first appeared in April, flowering occurred 18 days later and continued for 32 days, and propagule maturation occurred 63–77 days afterwards. The last propagules were observed in September.

Testing the model proposed by Duke (1990) for these observations has special value since Singapore, at around 1° N latitude, is appreciably outside the study range used to develop the model, namely for latitudes 10–38° S. It is important to note that Singapore is not only equatorial, at much less than 10° latitude, but also that it is in the northern hemisphere. The behaviour of phenological cycles at these latitudes have been largely unknown, so such additional observations greatly extend and enhance our understanding of factors influencing phenological development.

In consideration of the field observations made in Singapore, the steps outlined in the previous section were followed, using a mean daily temperature estimate of 29–32 °C, in order to calculate the expected dates of peak occurrence of chief pheno-events. The predicted dates of initiation, flowering, and propagule maturation are shown in Fig. 4.7, comparing them with the observed events described above from the Singapore study. There is clearly close agreement between observed and predicted occurrences of phenovents.

This result is remarkable for several reasons, but the chief reason is that the putative day length trigger appeared to be quite sensitive and distinct even where day length might never be less than 12 hours, as they are in sites near the equator. It is therefore possible that day length may be slightly greater than the 12 hours suggested by Duke (1990). More study is required to work this out. Another important observation has been that predicted events provide a better description of resource commitments for the trees, in view of the predicted occurrence of peak events, as well as the more complete view of floral development from early growth and putative initiation, to propagule maturation.

These observations indicate how *A. marina* might readily commence a second cycle of floral development, and possibly complete it. As noted earlier, this might also explain observations of continuous flowering in such locations.

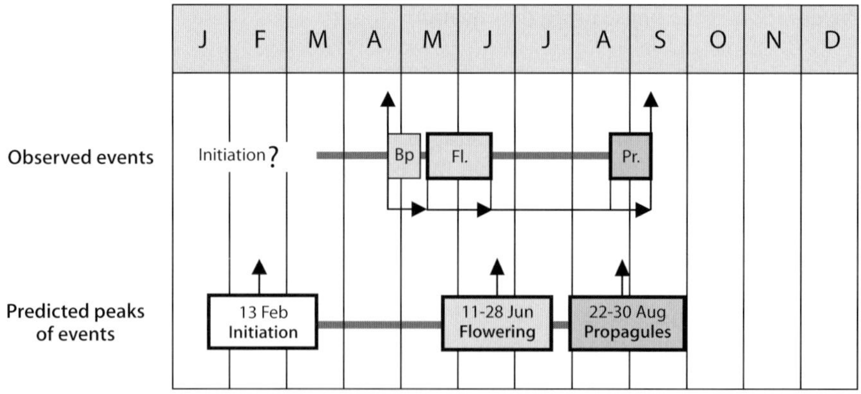

Fig. 4.7. Comparison of observed and predicted occurrence of phenoevents from Singapore (Mathews 1992)

4.9 Conclusions

This review has drawn together observations of mangrove phenologies with the intention of compiling a list of chief factors responsible for influencing the occurrence and timing of floral and leafing cycles. The assessment of all species in north-eastern Australia identified the extent of variation in the region, noting that some of the more important differences were observed in congeneric species. The reason for this appears to be related to genetic isolation through different schedules of pollen transfer and availability, notably with regard to flowering. Clearly, this might also be expected to impact subsequent development through to propagule maturation. These factors might therefore also effect the colonisation and replacement of species throughout the region, as well as contribute to the overall latitudinal range restrictions of each species.

The annual re-occurrence and quantity of maturing propagules is notably more variable in temperate locations, but it is not restricted to these latitudes. In tropical north-eastern Australia, there is also apparent year to year periodicity of propagule production in *Rhizophora stylosa*, for example. This cannot be quantified, however, without longer-term study projects. Therefore, despite recent evidence of El Niño influences on phenologies (Wright et al. 1999) and the likely implications for mangrove phenologies, we are not in a position currently to assess the importance of year to year differences in phenologies.

By contrast, the study of phenologies of *Avicennia marina* throughout Australasia has provided a considerable understanding of seasonal cycles and how they vary with latitude. This understanding has led to the first tangible evidence of large-scale factors influencing floral development, in particular. Thus, it has been demonstrated that reproductive cycle initiation is possibly triggered by day length exceeding 12 hours, and subsequent development to flowering and propagule maturation is controlled by temperature. Leafing events, although seasonally regular, occur notably between the stronger floral cycle patterns which as noted, depends very much on latitude.

Particular seasonal peaks of leafing in *A. marina* were related chiefly to either the reproductive cycle during initiation of inflorescence development, or independent growth. Respective heights of these peaks for particular sites appeared dependent on environmental factors of temperature and moisture. Thus 'inflorescence leafing' was predominant in low moisture and low temperature sites during early summer, and 'independent leafing' occurred mostly in wetter tropical sites during winter months.

Despite these important insights, there is still much to be learnt about mangrove phenologies. For instance, how might phenologies of other species change with latitude? The occurrence of distinct patterns in *A. marina* are not expected to be unique, and there are likely to be important differences in other species. Such observations might also help explain the remarkable differences in ecological attributes of each species, including their often distinct distributional ranges. It would also be particularly interesting to compare phenologies of congeneric species pairs over their latitudinal ranges. There are many questions regarding factors forcing genetic variation in mangroves.

Chapter 5

Integrated and Adaptive Mangrove Management Framework – an Action Oriented Option for the New Millennium

Padma Narsey Lal

Much has been written about mangrove management
but there is still too little action
Peter R. Bacon, March 1997
Intercoast Network: International Newsletter for Coastal Management

5.1 Introduction

Mangrove losses have been occurring at an alarming rate despite general acceptance of the role that mangroves play in maintaining the productivity of the coastal ecosystem, and despite many concerted efforts to curb the destruction. It is generally accepted that mangrove ecosystems are highly productive, dynamic and complex. They provide four types of ecological services – (i) converting solar energy, nutrients and water through primary and secondary productivity processes and food chain interactions into fauna and flora; (ii) providing physical habitat and refuge for various fishes and non-fishes at different stages of their life cycle; (iii) storing and filtering sediment and nutrients, recycling nutrients and maintaining the quality of the aquatic system; and (iv) maintaining the biological order in the system (Hamilton and Snedaker 1984; Twilley et al. 1993; Gilbert and Janssen 1998; Costanza et al. 1997).

These ecological services support various environmental functions producing many goods and services directly or indirectly valued by humans (see Table 5.1). The quality of mangrove cover drives the ecological processes, which determine the performance of the environmental functions producing goods and services (Gilbert and Janssen 1998).

Globally, more than 50% of mangroves have been lost in recent times (World Resources Institute 1996). They continue to be depleted in places at an alarming rate. In the Asia-Pacific region, for example, an average of about 1% per annum in mangrove area loss has been reported. Amongst the causes of mangrove destruction, conversion to agriculture, aquaculture and housing development have been most significant. Internationally, agriculture has been the principal cause of wetland loss (Finlayson and Spiers 1999).

By 1985, about 56–65% of the European wetlands had been drained for agriculture. In Asia, about 27% of the wetlands were reclaimed for agriculture. Even in small island nations, such as Fiji, where the extent of mangrove wetlands is small by most country standards, 86% of all mangrove reclamation since the turn of the century has been for sugar cane and rice farming (Lal 1990). In some cases, commercial logging, cutting of wood for domestic purposes, and illegal fishing and hunting have also led to the destruction of the wetlands in the Asia-Pacific (Dugan 1990). In many Pacific islands, subsistence and semi-subsistence timber and fuel wood harvest, have led to significant losses and degradation (Lal 1990b). Other threats to wetlands in developing countries include pollution, urban development, and recreational overuse (Table 5.2).

Table 5.1. Ecological processes, environmental function and goods and services supported by mangrove ecosystems (adapted from Gilbert and Janssen 1998; Costanza 1997)

Ecological process	Environmental function	Goods	Services
A. Primary and secondary productivity and food chain interaction			
– Fixation of solar energy and primary production	Production of plant materials	Leaves, wood, tannin, propugules, charcoal for fuel and energy. Timber for building and industries. Nipa shingles and leaves for roof thatch; fruit and propugules for food and drink and as alcohol source; Medicinal resources	Preserving biodiversity
– Secondary production	Fish and non-fish fauna production in estuaries and exported to coastal areas	On site and offsite harvesting of fish, shrimps and other animals for food	Preserving biodiveristy
B. Mangroves as a physical habitat	Preserving safe and live habitat; Buffering of storm and dissipation of energy	Production of hell and bone fish, and other animals	Shoreline and land based activities protection
C. Ecosystem as storage, filters and recyclers	Maintenance of water quality by filtering sediments, trapping nutrients; Maintenance of the quality of habitat		Sediment and flood control; Waste removal; Preserving the natural system dynamics
D. Biological order	Preserving of food web interactions and biological quality of the system; Maintenance of the physicochemical quality of the ecosystem		Preserving biodiversity, natural system dynamics and cultural and aesthetic value; Future option values

More recently, losses in mangrove areas have occurred because of the boom in the shrimp demand. Globally, as much as 50% of the reclamation of mangrove areas in the last two-three decades has been due to the sitting of shrimp ponds in mangrove areas (Bacon 1997). In the Philippines, over 70% of the mangrove forest's deforestation has been due to shrimp or fish aquaculture. Between 1970 and 1993, over 260 000 ha of mangroves were lost to shrimp farming, and the deforestation continues despite governments passing legislation prohibiting mangrove deforestation (Janssen and Padilla 1997). In Vietnam, too, large areas of mangroves have been cleared for shrimp farming, despite government regulations banning such clearings. During the 1982–87 period, over 4 600 $ha^{-1}\,yr^{-1}$, or about 2 291 ha of mangrove forests in the Minh Hai Province alone have been cleared, mainly for shrimp farming (Hong and San 1993). The destruction of mangroves for lucrative shrimp farming continues, despite many government directives prohibiting forest destruction (Vietnam Government 1999).

Table 5.2. Incidence of threats to tropical wetlands, expressed as percentage of the total number of wetland sites (*source:* Barbier 1997)

Causes	Asia	Latin America and the Caribbean
Hunting and associated disturbance	32	31
Human settlement	27	–
Drainage for agriculture	23	19
Reclamation for urban and industrial development		11
Pollution	20	31
Fishing related	19	10
Commercial logging ad forestry	17	10
Wood cutting for domestic use	16	–
Conversion to aquaculture ponds or salt pans	11	–
Diversion of water	9	–
Over grazing by domestic stock	9	–

In small countries such as Fiji and Honduras, mangrove areas have also been lost to shrimp farming. In Fiji, mangrove reclamation took place for shrimp farming which never came into production because of acid sulphate conditions that developed once the soils were exposed to the air (Lal 1990). In Honduras, much of the shrimp industry, which is a profitable venture generating over US$140 million, is situated on mangrove cleared areas (Smith 1998). The reclamation in Honduras is continuing, and with the current rate of 2 000–4 000 ha yr^{-1} of recent mangrove deforestation, all the mangroves are expected to be lost by year 2020.

5.2 Causes of Destruction and Wetland Loss

One of the key reasons for the continued loss of mangroves, despite concerted efforts made by governments and international agencies, is the mismatch between the mangrove ecosystem and human governance institutions. Governance, as defined by Hanna (1999), is the interaction between property rights, the institutional environment, and individual behaviour. The property rights could be about use and ownership as well as management.

5.2.1 Property Rights – Use and Ownership Rights

Property rights regimes are bundles of entitlements that define owners' rights and duties, and rules under which those rights are exercisable and enforceable (Bromley 1989; Randall 1987). Individuals may have ownership rights. Alternatively, users may only have use rights with the ownership resting either with the state or non-users. A clearly specified property rights regime identifies the resource over which the rights are held, the nature of uses, and transactions permitted. Well-defined rights are enforceable, and enforced, and non-right owners are excluded. There is also certainty over the rights during the tenure period.

Property rights are a subset of broader institutional rules and regulations that determine individual incentives and constraints within which human interactions occur (North 1990). Regulatory policies and institutional rules determine conditions of access and utilisation (Freeman III 1991), which also influence economic value of the goods and services. Outcomes of human activity are dependent on the interaction between the economic processes and the underlying ecological processes within a given social context (Barbier 1987). The interaction between these institutions and the underlying ecological processes determine the net value of environmental functions and goods and services. The necessary conditions for market mechanisms to provide an efficient use of a resource are that the good be privately owned (or at least the use rights are held privately), the rights are transferable from one person to another, and are protected from encroachment and use by others (Young 1992). In an efficiently operating market system, all costs and benefits of a resource use are adequately reflected, including externality costs of some human activity.

5.2.2
Property Rights and Mangrove Ecosystems

Being found on land-water interface, mangrove wetlands challenge the definition of property rights about them. The rights over the terrestrial component of the system can be, and often are, easily demarcated, fenced, enforceable and enforced. As a result, in many countries, particularly in former British colonies in the tropics, wetlands above spring high water mark are often included in private land titles. The aquatic system, on the other hand, does not lend itself to easy division, demarcation and enforcement. Consequently, wetland areas and resources below high water mark often remain as public good and defined as being owned by the state (Munro-Faure 1991). In these circumstances, the market's 'invisible hand' and market-based transaction are not likely to lead to a socially optimal use of a resource, and by implication, a sustainable outcome. The main reason is that in a complex ecosystem such as the mangroves, any land based activity may generate externality costs on environmental functions in the aquatic subsystem transmitted by the underlying ecological processes in the wetlands (see Fig. 4.1). These externality costs on others will not be borne by the individuals using the resource in the absence of property rights over them. When all costs are not fully borne by those using the resource, it is undervalued (Freeman III 1991), and generally overexploited, degraded and abused, a phenomenon known as market failure.

Market failure is a common cause of mangrove loss. In the absence of adequate property rights, mangrove resources are subject to pressures for reclamation and conversion to alternative uses without adequate consideration being given to the value of goods and services, and/or the underlying ecological processes that may be lost. This results in excessive conversions and loss of the mangrove resources (Lal 1990; Turner and Jones 1990; Dixon and Lal 1994).

In some countries where remnants of indigenous rights can be found, no distinction is often made about the land, aquatic components and the underlying ecological processes (Ruddle and Johannes 1993; Batibasaqa et al. 1994; Waqaraitu 1994). In Fiji, for example, a traditional clan, or mataqali, communally has claim over a vanua, or the physical resources and the environment, which together encapsulate 'nature', including the mangroves. However, there has been confusion about the nature of mataqali rights (Lal 1983; Waqaraitu 1994). Traditional clans are recognised communal owners of the coastal areas, but the government also declared that these rights were usufructs only, and were not recompensable (Lal 1990). Such an ambiguity affected the entitlements the mataqalis could claim. The traditional owners could not adequately exercise their 'ownership' rights and demand adequate compensation for the loss of mangroves due to reclamation or waste disposal. One could argue that much of the reclamation in Fiji might not have occurred if the communal rights were recognised as compensable rights.

During a brief period in the early 1980s, the rate of reclamation was reduced considerably once the government shifted its stance about the nature of the indigenous claims over the coastal resources. The political pressure led to partial acceptance of the traditional claims over the aquatic resources, just as the nature of the indigenous ownership of the land was undisputed. While the state was still declared as the rightful owners, the mataqali were seen as custodians. With this recognition came large claims about the value of the expected loss of fisheries resources and their source of livelihood as a result of mangrove reclamation. At one stage, the engineer in charge of the reclamation on behalf of the government exclaimed, in response to the mataqali demands, that it (reclamation) was no longer 'worth the hassle' (Ernest pers comm 1983). Despite government owning the mangrove areas as well as the adjacent coastal waters and resources within them, and thus having the ultimate rights over their use and management, mangrove losses continued.

5.2.3
Government Failures and Management Rights

Governments themselves have been responsible for much of the losses because of their attitude, and lack of appreciation of the importance of mangrove forests' role in supporting and maintaining the dynamics of the ecosystem. Throughout the world, governments, until recently, have regarded mangroves as wastelands, harbouring undesirable vermin and a breeding ground for mosquitoes, thus encouraging their use for solid waste disposal, draining the areas for agricultural purposes and reclaiming for aquaculture, industrial and urban developments (Hamilton and Snedaker 1984). In Fiji, the government was responsible for over 70% of the reclamation since the turn of the last century (Lal 1990). About 90% of this had been for agriculture. In many of the South East Asian countries, shrimp farming expanded with the encouragement and blessing of many governments (Earthsummit Watch 1999).

Much of the wetland management to date has focused on individual sectors/ resources. The overall pattern of management in different countries is very similar. In most countries, the Fisheries Department is concerned with 'regulating' commercial harvest of fish and fish products, using command and control strategies of licenses and fishing permits. The Department of Agriculture or Drainage and Irrigation regulates drainage and agriculture development on mangrove-reclaimed land. The Forestry Department controls the commercial harvest of mangrove timber and wood for charcoal production by controlling the number of concessions issued, volume of timber harvested for timber, and firewood and charcoal production (see, for example, Lal 1983; Chan 1987; Aksornkoae 1987; Philippine National Mangrove Management Committee 1987; Said, Othman and Khan 1992).

Often there is no or very little communication and co-ordination among the plethora of agencies. In Malaysia, there are about 10 different agencies, with each authority dealing with only one aspect of wetlands (Said, Othman and Khan 1992) and controlling mainly commercial activities. In Fiji, there are seven government agencies at the national level, in addition to the local town and city councils, which have jurisdiction over mangroves within city boundaries (Lal and Minerbi 1985).

Not only are there many agencies with overlapping jurisdictions, there is also a lack of co-ordination. In many countries, different arms and levels of government often also pursue conflicting policies. In Fiji, local city councils, under the Ministry of Local Government and Housing, have treated wetlands as wastelands and used these as household garbage disposal sites. On the other hand, the government's policy, as stated by the Department of Fisheries, clearly stipulates conservation of the mangroves as a fish habitat and nursery grounds for many species of fishes harvested for commercial and subsistence purposes. Even within the same ministry, different departments have pursued policies in direct conflict with each other (Lal 1983). In the Ministry of Agriculture, Forestry and Fisheries, the Department of Irrigation and Drainage encouraged drainage and reclamation of mangrove areas while the Fisheries Department often objected to this on the grounds of its importance as nursery and habitat for many valuable food fish species.

Such conflicting policies stem from the government's lack of recognition of spatial and temporal interconnectedness of the different components of the mangrove ecosystem and the effects of one activity on another dimension of the same system. In many instances, governments failed to enforce their own rules and regulations, even where mangroves were declared protected areas or acknowledged as areas that should be conserved because of their importance as nursery grounds.

Governments have often justified development projects producing short-term economic returns, while ignoring the spatial and temporal linkages within the mangroves. Land and land based resources have generally been treated in isolation from aquatic ones. The harvest of forest products are regulated without due regard for the dependent fisheries, and mangrove land is reclaimed without any regard for its effects on the fisheries resource. This lack of recognition of the interconnectedness between the terrestrial and the aquatic subsystems has been the main cause of wetland loss, overharvest of mangrove timber, over exploitation of the fisheries' resources, and the unsustainable use of mangroves as industrial and urban waste disposal sites.

Conversion of mangroves has often been justified using partial economic analysis focusing on the value of land based activities (Dixon and Lal 1994; Hamilton and Snedaker 1984; Barbier 1994). Generally, only the marketed goods produced on-site have been valued while the subsistence uses of the same goods and the value of ecological services have been ignored (Table 5.3).

Non-marketed values and the value of off-site fisheries are often found to be at least equal to, if not more than, the on-site value of the forest products (Christensen 1982; Lal 1990; Ruitenbeck 1992; Ronback 1999). By ignoring the value of such insitu resources, governments have often justified mangrove reclamation, resulting in non-optimal loss of mangroves.

Table 5.3. Marketed and non-marketed goods and services and their valuations (adapted from Hamilton and Snedaker 1984)

Location of goods and services	Values of goods and services	
	Marketed	Non-marketed
Onsite	Usually included Commercially harvested timber and wood for construction; Commercial harvest of wood for charcoal and energy; Commercial harvest of fish caught in mangrove areas	Seldom included Subsistence use of timber and firewood; Medicinal uses; Nursery area for juvenile fish; Feeding grounds for fish; Nutrients and sediment filtration; Storm buffer; Biodiversity maintenance; Preserving biological order
Offsite	May be included Commercial harvest of fish or shellfish dependent on mangroves but caught in adjacent waters	Seldom included Nutrient flows to adjacent waters

5.3 Diversity in the Mangrove System – Spatial-Temporal Variation in Wetlands

Continued destruction of mangroves is also partly due to earlier approaches that focused on declaring protected areas and encouraging the protection of all areas regardless of the quality and the nature of the particular mangrove. The Ramsar Convention, discussed below, encouraged preservation of mangroves. All mangrove ecosystems were considered to be similar (Bacon 1997), and thus subject to generic mangrove management guidelines.

Environmental production by mangrove ecosystems shows large spatial-temporal variations throughout the tropics. Reported finfish production varies from 257 kg ha^{-1} yr^{-1} in Fiji (Lal 1990) to 5 840 kg ha^{-1} yr^{-1} in Queensland, Australia (Morton 1990). Shrimp production world-wide varies from 13 kg ha^{-1} yr^{-1} to 765 kg ha^{-1} yr^{-1}, with a direct correlation between latitude and penaed catch (Turner 1977). Even within a single country, variations can be large; shrimp production in Indonesia varies from 16–165 kg ha^{-1} yr^{-1} in Indonesia (Martosubroto and Naamin 1977).

Environmental functions and goods and services supported by mangroves depend on a complex interplay of factors, including the area of mangrove forest, the level of carbon fixed by the mangroves and the extent of organic carbon and nutrients exported from mangroves (Robertson et al. 1992). Gilbert and Janssen (1998) suggest that the quality of the mangrove cover also has a direct influence on the productivity and physical structure of wetlands. The better the mangrove cover; the better is the performance of ecological process and environmental functions. The ability of the mangroves to maintain coastal water quality also is an important determinant of the systems' productivity (Marten and Polovina 1982; Pauly and Ingles 1986; Ronback 1999).

In addition to the spatial variations in the ecological processes and the level of primary and secondary productivity, there is also a difference in the economic value. As

Table 5.4. Diversity in economic value of goods and services supported by mangroves in selected countries (US\$ $ha^{-1} yr^{-1}$)

	Fiji[a]	Indonesia[b]	Philippines[c]	Kosrae[d]
Forestry	6	67	251	178
Fisheries	100	117	60	461
Biodiversity		15		
Erosion		3		
Waste disposal	5820			
Total				426 – 640

[a] Lal (1990).
[b] Ruitenbeek (1992).
[c] Gilbert and Janssen (1998).
[d] Naylor and Drew (1998).

discussed above, local market conditions and regulatory policies and rules determine the values of resources. The values of fisheries products reported (Table 5.4) include US\$60 in Philippines (Gilbert 1998); US\$100 in Fiji (Lal 1990), US\$117 in Indonesia (Ruitenbeck 1992) and \$461 in Kosrae (Naylor and Drew 1998). The value of forest products showed even greater variations from \$6 $ha^{-1} yr^{-1}$ in Fiji (Lal 1990) to \$251 $ha^{-1} yr^{-1}$ in Pagbilao, Philippines (Gilbert and Janssen 1998).

Despite these variations, there has been a tendency to generalise about the structure, function and value of mangroves when advocating wetland management. Environmentalists often call for mangrove preservation/ conservation without considering the ecological characteristics of the particular site or the relative economic merits of the alternative options. It is possible that some of the losses in wetland areas could be justified. In Fiji, reclamation of mangroves close to urban areas could have been justified on purely economic grounds, since the economic value of foreshore land was at least three times the value of in situ uses (Lal 1990). These mangroves were also so degraded and being lost due to sedimentation, that the mangroves would have been 'lost', anyway. Thus, not conserving all mangrove ecosystems will always be in the best interest of a society.

The real issue is under what circumstances an alternative use may be more desirable and how best to manage the mangroves for non-destructive alternative uses. Many different approaches have been advocated and/or used with different degrees of effectiveness.

5.4 National Mangrove Policies, Strategies and Plans

In recent years, many countries have attempted to develop wetland policies, wetland management strategies and wetland action plans, with the encouragement of international initiatives, UNESCO/UNDP, ISME, IUCN and the Ramsar Bureau, established after the 1971 Convention of Wetlands of International Importance held in Ramsar, Italy.

Initially, the focus of much of the international initiatives was on preservation. Under the Ramsar Convention, the contracting countries were obliged to identify and declare at least one site of international importance based on representative or unique wetlands, plants, animals and waterfowl, fish and fisheries values. However, realising the inadequacy and ineffectiveness of merely establishing protected areas as a strategy for wetland conservation and management, as was the focus of during the 1970s and 1980s, the Contracting parties to the Ramsar Convention in late 1980s changed their focus.

Since 1987, various conferences of the contracting parties to the Ramsar Convention have accepted 'wise use, which involves the maintenance of their ecological characteristics as a basis not only for nature conservation but for sustainable development' as the central theme for the implementation of the convention. The signatories to the Ramsar Convention, apart from declaring sites of international importance, now also agree to promote the conservation of the designated site and other wetland areas so as to 'promote the conservation of the wetlands' (Article 3.1). They are also encouraged to develop and implement national mangrove policies, and wetland strategies and plans (Rubec 1999). The management policies and plans are expected to also reflect other international and national obligations for nature conservation, including the conservation of biodiversity, climate change and pollution control as adopted by the UN Conference of Environment and Development (UNCED, Rio 1992), and other international conventions, including the 1992 Convention on Biological Diversity.

As of December 1999, 117 countries have joined the Convention (Rubec 1999) and have declared at least one site of international importance under Article 2.1. So far, the Contracting Parties have designated 858 sites for inclusion in the Ramsar List of Wetlands of International Importance covering over 55 million ha of different wetland types. Of this about a third, or about 18 100 000 ha of mangroves are declared as Ramsar wetlands. Ramsar sites include, for example, 596 000 ha of Sunderbands in Bangladesh, 35 042 ha of Manglares Churute site in Ecuador and 54 000 ha in the Caicos Islands (Bacon 1997). However, this approach of acting locally for the global good has not been totally successful in curbing the destruction. Many mangrove areas, including much of the Ramsar sites, are under constant threat of degradation and total destruction. In 1999, about 84% of the Ramsar listed wetlands have either been degraded or are threatened by ecological changes caused by human activities (Finlayson and Spiers 1999).

Even with the declaration of protected sites, not many countries have adopted a more holistic view of the management challenge. Of the 117 countries that are Contracting Party to the Ramsar Convention of Wetlands, as of May 1999, only 12 countries have adopted a national wetland policy, and 50 have some form of national wetland strategy/ national wetland action plan. But what is more important to note is that over 50% of the countries have no action reported in relation to the development of national wetland policy or local wetland strategy or action plan (Table 5.5).

Where wetland policies and action plans have been developed, these have in some cases tended to be very generic in nature, listing key actions/strategies that countries should adopt. For example, the Action Plan developed for the Wetland Conservation in the South Pacific Island Nations (SPREP 1994) lists eight themes for action. Most of the points reflect the guiding principles adopted by Ramsar (Rubec 1999), calling for the establishment of national wetland committees, national wetland legislation, the development of effective legislative framework and linkages between all levels of government, etc. (see Table 5.6).

Table 5.5. Status of National Wetland Policy, National Wetland Strategy or National Action (Rubec 1999)

	National Wetland Policy	National Wetland Strategy/ National Wetland Action Plan
Adopted	12	50
In draft form	65	12
Development under consideration or proposed	26	39
No action reported	70	13

Table 5.6. Examples of generic action points listed in the Action Plan for Wetland Conservation in the South Pacific (*source:* SPREP 1994)

Theme 1: Policy and legislation

- National Wetland Committees should be established, to develop national wetland policies and/or legislations for wise use
- Development of an effective legislative framework will require appropriate linkages to local and traditional laws and existing enforcement mechanisms, implementation of an awareness program
- Facilitation of the development of village level regulations
- Private sector should be encouraged and assisted to develop industry-specific standards for avoiding and minimizing impact
- Development of code of ethics

Theme: Management

- A wetland checklist for the region should be developed
- National government should use existing environmental assessment procedures
- National environmental management strategies, zoning controls and projected area protocols should be used and appropriate strategies for maintenance should be developed
- Local communities should be empowered to manage their own resources
- Incentives for customary land and resource owners should be developed and supported by national government to encourage wise use
- Ecotourism should be developed as income generating activity
- Review for wetland conservation and rehabilitation should be imposed on resource users
- Downstream effects of government and development actions should be taken into account
- EIAs and site action plan must incorporate monitoring programs
- Rehabilitation of degraded lands should be carried out where feasible and cost effective

Such an institutional approach is only as effective as its implementation and enforcement. In Fiji, for example, the environmental impact assessment has been required for large development projects since the early 1980s. National Mangrove Management Committee, comprising of key government departments, including Forestry, Fisheries, Lands, and Drainage and Irrigation, and the National Trust, had the opportunity to review the merits of every project involving mangroves. However, in practice, the

Committee had no legislative backing and had very little say in the actual decision-making process. The Committee often found that despite objections, many of the developments were approved, with the government justifying projects on the basis of partial economic evaluation.

Some of the strategies outlined in the action plan above need to be put in operation. For example, calling for an inventory of the resource will not be sufficient unless this information is used to define specific actions, such as producing resource use zones. Nor will the formation of National Mangrove Committees and community participation be effective unless they become an active player in the actual decision-making processes. Levies for wetland conservation and rehabilitation cannot be imposed without a clear idea about what the values will be and what are they expected to reflect. For this, economic valuation of the externality costs or the cost of rehabilitation is required. The underlying assumption is that if users were made to internalise the externality costs, the rate of conversion will be reduced.

5.5 Economic Valuation

Economists use cost-benefit analysis as a tool to choose between competing uses. In BCA, comparison is made between the net benefits of in situ uses of the wetlands and the benefits, net of costs – including the costs of off-site impacts – of converting it for other uses. BCA is a static partial equilibrium assessment. Moreover, the assumption underpinning such an approach is that a resource will be most efficiently utilised, – social (economic) benefits will be maximised – when the resource is allocated to that use which generates maximum benefits, net of all costs. Temporal considerations can be taken into account by estimating net present value of alternative uses; however, the debate about the choice appropriate rate used to discount the value of future generations can not be disregarded (see for example, Sen 1967, 1982). In the past and as discussed above, only partial economic analysis has been the norm, resulting in non-optimal conversion of mangroves. Consequently, to address this problem of partial or incomplete cost-benefit analysis, many international initiatives have targeted economic valuation of wetlands, as one of the areas that need particular attention.

The Ramsar Bureau, for example, advocates that one must include amongst other things economic values of wetlands in the background paper and the national 'issues statement' or 'vision statement' that is used to underpin the development of national wetland policy (Rubec 1999). The underlying assumption is that economic valuation will provide information needed to make strong arguments about conservation of the in situ values of the wetlands, particularly when faced with demands for conversions of mangrove areas for non-compatible uses largely on economic grounds.

In situ economic values of mangroves is specific to a biophysical environment, the cultural context, social values and institutional setting. The economic values of goods and services supported by a natural ecosystem is a product, as summarised in Fig. 5.1, of the interaction of three sets of functional relationships: policy environment, which determines the institutional rules and regulations stipulating the nature of use acceptable, biophysical processes, and production economic links (Freeman III 1991). Government policies, for example, could define a particular environmental or resource attribute such as mangrove cover, which determines the level of nursery area and the

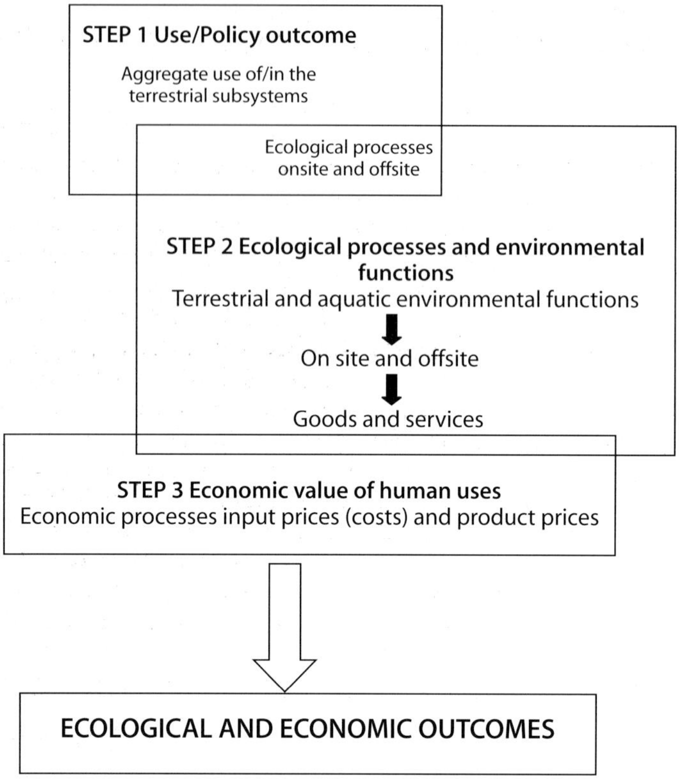

Fig. 5.1. Value of goods and services supported by mangroves as a product of policy, biophysical processes and economic production process (Gilbert and Janssen 1998)

amount of detritus exported into the coastal areas. These factors as discussed above determine the primary and secondary productivity of wetland and the quantity of goods and services produced by the system. Ultimately, the economic welfare of the dependent community will be affected.

Properly conducted economic valuation will capture the total marginal economic value. Economic values will be captured in the economic valuation of goods and services for which a market exists, a proxy market can be determined, or for which hypothetical markets can be created. In neo-classical economic valuation, such estimates are based on the assumption of marginal (small) change in the amount of goods and services, or there is a fixed relationship between mangrove area and products and ecological services. Such assumptions hold when the development area is small relative to the existing mangrove cover. However, mangrove ecosystems exhibit non-linearity in relationship between mangroves and goods and services they support (Pauly and Ingles 1986). There are thresholds and discontinuities in the relationship, and it is possible that loss of mangroves beyond a certain threshold can lead to a collapse in the whole system. In such a situation, the value of the products of ecological processes will not be captured by the standard valuations techniques.

5.6 Uncertainty about Spatial-Temporal Interconnectedness

Much has been written about ecological relationships (see for example Robertson and Blaber 1992; Matthes and Kapetsky 1988; Larsson et al. 1994; and Beveridge et al. 1997; Robertson and Duke 1987; Morton 1990; and Odum and Heald 1972). There exists, however, a great degree of uncertainty and a lack of understanding about spatial and functional relationships between mangrove areas and associated ecological and environmental functions (Ronback 1999; Gilbert and Janssen 1998). The relationship between mangrove area and the quality of mangroves, the underlying ecological processes, and primary and secondary productivity is not known for most mangroves (that is between steps 1 and 2 in Fig. 5.1). There is also uncertainty about the effect of human activities on the ecological functions, and on the quantity of fisheries and other wetland dependent goods and services supported by the mangroves (between steps 2 and 2 in Fig. 5.1).

Many different approaches have been used to determine the functional relationship between the wetlands, the goods and services produced by the underlying ecological processes and the environmental functions. But the underlying relationships remain unclear. Firstly, statistical analysis of observed yield levels has been used to determine the relationship between areas of mangrove and fish yield. Pauly and Ingles (1986), for example, find a logarithmic relationship between the area of mangrove cover and shrimp output, implying that as mangrove areas decrease, the corresponding decrease in shrimp output increases.

A systematisation of observed yield levels and an assumed functional relationships have commonly also been used. Nickerson (1999), Barbier and Strand (1997) and Lal (1990), for example, all assume a proportionate relationship between the area of mangroves and the mangrove dependent species harvested. Lal (1990) assumes a linear relationship between mangroves found in major estuaries and coastal areas in Fiji and the weight of mangrove-dependent fishes and non-fish species sold through market and other outlets and subsistence catch. Nickerson (1999), on the other hand, assumes a proportional relationship between mangrove area and carrying capacity of fisheries production and determines the catch estimates using Schaefer fisheries biological model. Barbier and Strand (1997)also assume a proportional relationship between mangrove area and carrying capacity of fisheries production, but uses Schaefer-Gordon bioeconomic fishery model to estimate the impact of mangrove on shrimp catches. Others such as Sathirathai (1998) uses a static optimisation Cobb-Douglas model, assuming a direct proportionate (non-linear) relationship between quantity of crabs harvested and level of fishing effort, keeping the area of mangroves constant. Sathirathai (1988), assuming specific type of management regime, simulates the level of quantity and price associated with determining the profit and cost functions. Using marginal product approach of Freeman III (1991), Sathirathai (1988) estimates the relationship between mangrove area and the flow of fisheries output. The exact relationship is complex and would no doubt be context specific. It thus cannot be known without spending a huge effort. Even then, it is unlikely that it will be known with any great degree of certainty, given the complex ecological system that mangroves are.

Where uncertainties and incomplete information exist, some economists have advocated a more 'precautionary approach' of stipulating safe minimum standards

(Bishop 1978), ecological threshold effects, or adopting a macro environmental standards approach (Lal 1990). The need to adopt a precautionary approach is emphasised particularly where the development is irreversible. Barbier (1994) suggests that an economic marginal valuation approach and the use of a cost-benefit analysis as a decision-making tool may be relevant only when a development is likely to affect the environment within the bounds of the 'safe minimum standard' criteria. Moreover, purely economic arguments, as often advocated by neo-classical economists, cannot be used in circumstances where subsistence, semi subsistence and commercial economies CO-exist, and their respective economic values are not known. Lal (1990) suggests that properly conducted cost-benefit analysis be used to choose between alternative uses only as a second tier criterion. That is once broad ecobiomes have been defined outlining the binding biophysical constraints for each use zone.

5.7 Land Use Zoning

Land use planning has also been used to manage mangroves in some countries. This approach recognises multiple demands, which may be complementary/ competing/ and or conflicting. In Fiji, for example, mangroves have been zoned primarily as a resource base for capture fisheries (Watling 1985), and secondarily as the renewable source of products which supports underlying ecological processes determining the quality of life of associated coastal communities. In Kosrae, the process was applied to define the mangrove management plan (Lal 1989). Lal (1990), using published detailed inventory reports, first delineated the mangroves into key plant species alliances according to the dominant species, the characteristics of underlying soils and the biophysical characteristics of adjacent aquatic habitats. These biomes then formed the basic resource management units and were ranked according to their ecological suitability for alternative uses. The alternative uses considered included from one extreme a sanctuary allowing no extractive use, to the other extreme of irreversible conversion. The biomes were then ranked according to the uniqueness of their biological characteristics and landscape values, and needs and aspirations of the local communities and the compatibility of alternative uses within a district. The overlapping characteristics of the biomes and the ecological and economic compatibility of uses within the biomes were then used to define the resource districts, or the ecobiomes. The district defines the overall binding constraints of acceptable quality of mangroves, acceptable level of change and overall biophysical constraints and safe minimum standard. Such a hierarchical designation of zones allows a degree of flexibility within at least the 'managed' and developed zones.

The national resource reserve zone also recognises the importance of conserving mangrove areas for the role it plays in maintaining the underlying ecological processes that support coastal fisheries and the biological order of the ecosystem. However, this cannot be used as the sole criteria, as was initially promoted under Ramsar Convention. Access to countries' resources for traditional subsistence uses are important as is the need for commercial harvest of forest and fisheries products. Similarly, the demand for land for urban and periurban development cannot be disregarded, when the social pressure on existing land is acute. Thus in the presence of incomplete information and uncertainty, and concerns about equity, a resource use zonation approach could be relevant. This approach adopted has, however, been only one part of a 'blue print' and remains static, with little scope for changes if socioeconomic circumstances, which changed the relative value placed by society on the different uses.

5.8 Community Participation

In the presence of uncertainties and incomplete information, an ethical judgement is required about the desirability of a change, particularly when it may be irreversible and the decision may incur potential costs to future generations. Governments, as guardians of the interests of future generations, and communities, who stand to lose or benefit today from such a decision, collectively would need to make decisions. Even where such value judgements are not required, without community participation, development and management decisions made by government agencies or external donors can be inappropriate. A lack of participation in a number of World Bank rural development projects, for example, resulted in the choice of inappropriate technologies and the setting of unrealistic agricultural production targets (Upoff 1991).

Community participation has been widely advocated (see in the context of mangrove management, for example, IUCN 1991 and Kushiro Statement of the Contracting Parties to the Ramsar Convention 1993, and other Ramsar Conferences). The nature of community participation has varied considerably (Pretty 1995), but all stress the importance of incorporating local views, knowledge and needs in decision-making processes (Chambers 1997; Thrupp et al. 1994). The ideal form of community participation, according to Pretty, is "consensus" participation, where stakeholders are involved as equal partners and active participants in the entire decision-making process. The stakeholders should be involved in determining a vision of development for themselves and thus determining the categories of current uses they endorse, the level of risks they wish to take and the safe minimum standard they are comfortable with, and the level of future options they wish to maintain. Such an approach was used by Lal (1989) in Kosrae. Lal asked the local communities to specify their visions, needs and goals as well as the types of uses of mangroves in their vicinity and the nature of developments they wanted for their children. On the basis of this information, and data on the ecological characteristics of the mangroves and the compatibility of uses, she zoned the mangroves into use districts, which defined the acceptable levels of changes in the biome quality. Communities should also be asked to make decisions about the nature of management strategies and instrument options they wish to adopt in wetland management and the extent and the nature of external enforcement mechanisms required.

A mere involvement of communities is not an answer, as there are both benefits and costs associated with community participation. The process is often time consuming. Vocal and organised groups may dominate; differences in opinion may cause conflicts to develop, causing disharmony amongst the community members, as summarised in Table 5.7.

It is also worth noting that community participation has not always led to desired goals. In practice, varying degrees of success have been achieved, depending on what form of community participation took place, and how involved communities were in the entire project cycle (Whyte et al. 1998). The greater the community involvement was from the concept stage to implementation and management of a project, the more successful the project was found to be.

Table 5.7. Benefits and costs of participation from a policy-making perspective (Tacconi 1998; Whyte et al. 1998)

Advantages	Costs or difficulties
Identification of those affected is facilitated	Decision making process may be prolonged and slow, which may result in loss of inertia
Stakeholders learn about the dimensions of the issues/problems	Administrative costs may be increased
Policy is better matched to the needs through local feedback	Opposition to policy may develop; conflicts may arise and thus require conflict resolution mechanisms
Increasing sense of ownership of policy outcomes and compliance may be increased	Expectations of increased power in decision making created
Development of consensus on major policy choices	Inaccurate information may have strong influence
Public issues, which many not have received attention, are brought into focus	Vocal and organized groups may dominate and overshadow other views
Accountability of public institutions increased	Those participating may not represent the wider public

5.9 Integrated Adaptive Mangrove Management Framework (IAMMF)

The discussions so far suggest the desirability of adopting a holistic management approach that reflects amongst other things the need for:

- A recognition of complex ownership and use rights and management rights found in a country and the respective roles of the holders of these rights;
- An acknowledgement of the need for and active participation of local communities dependent on the resources and many different government agencies, at national and local levels, responsible for the management of different aspects of the mangrove system;
- A recognition of spatial and inter-temporal interconnectedness of the wetland systems and on-site and off-site effects of human actions;
- An acceptance of the reality of imperfect knowledge and information and the uncertainties about the functional relationship between mangrove forest areas and the underlying ecological processes, environmental functions and the effects of human activities and ecosystem's reaction;
- A combination of institutions, including the traditional command and control instruments as well as economic instruments, co-management and voluntary measures; and
- A dynamic and adaptive management system (Holing 1989).

An integrated and adaptive mangrove management framework (IAMMF) that reflects these considerations is proposed below.

The IAMMF is made up of a three-tiered decision-making process – the reflective stage (Fig. 5.2), the broad policy options stage; and the action management stage, with a built in mechanism for regular feedback and revisions, and involving the ac-

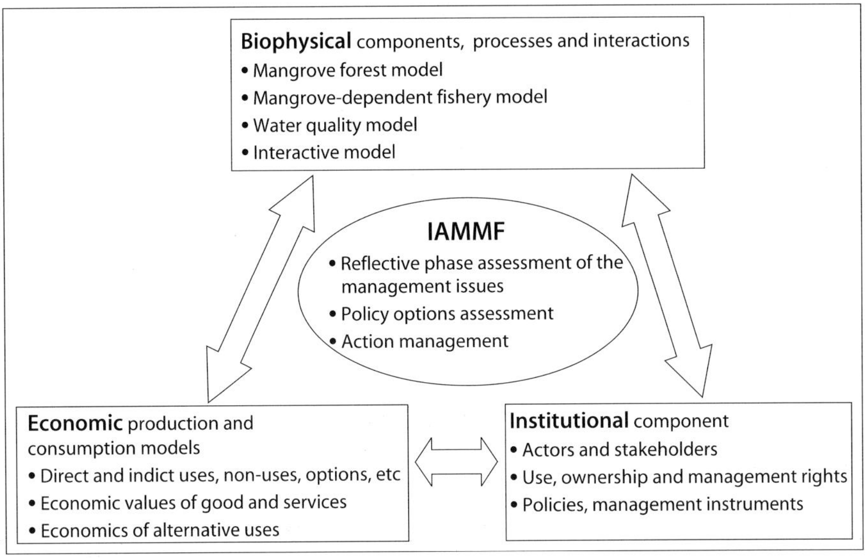

Fig. 5.2. Three tiered Integrated, Adaptive Mangrove Management Framework (IAMMF)

tive participation of all stakeholders at every stage. The three-tiered process is an extension of the decision-making processes proposed by Norton et al. (1995) and the two tiered macro environmental standards approach (MESA) adopted by Lal (1989) in Kosrae.

5.9.1 Reflective Stage

The objective of this phase is to define the current status of the environmental condition, actors and stakeholders and their use and management rights. This stage comprises of four components:

- A resource inventory, an assessment of natural processes and their dynamics which underpin the current status of the environmental quality and resources at an appropriate spatial and temporal scale;
- Stakeholder assessment, which defines who the main actors are, what their rights and responsibilities are, and whose decisions and which decisions need to be influenced;
- Inventory of rules and regulations which underpin activities within and elsewhere but which affect mangrove ecosystems; and
- Assessment of the interaction between natural and human processes to identify the specific problem or the management issue that needs to be addressed.

The information generated and descriptive and causative inferences drawn collaboratively by researchers and stakeholders will then help define specific management problems, issues or concerns. The management issue identified in this stage would then become the focus of research and analysis. For this, as a minimum, an interdisciplinary team, including at least an ecologist, an economist and a social anthropologist, with active participation of owners, users, and policymakers/managers, should be involved.

Ecological and biophysical modelling together with bioeconomic modelling and geographic information systems (GIS) can be used to understand spatial and inter-temporal relationships within the mangrove ecosystem, and the relationship between human activities and the reaction of the ecological system. Policy simulation analysis, using these models and computer-based tools, can help identify the combination of policy and management instruments that may be most cost effective and socially acceptable as well. The application of this stage will not always be easy because of the complexities (Antunes and Santos 1999) caused by:

- Several activities producing a single outcome;
- Multiple impacts of a single activity;
- Inter-connectedness within and between land based and aquatic components of the ecosystem;
- Indirect and synergistic or cumulative effects; and non-linearity in relationships.

5.9.2 Policy Options Stage

There are two main policy issues associated with this stage – what is the appropriate decision-making process, and what are the suitable institutional arrangements? The answers to these two questions will depend on the nature of management decision that needs to be made. Two main types of management issues pertaining to mangroves are commonly found in developing countries. Allocation of mangrove land between competing uses – conservation and wise use of the mangroves and alternative uses after cutting down and converting mangrove forests – is one of the most important issues. The other is about the appropriate rate of use, such as the rate of harvest of forest and fisheries products; or the use of an aquatic system for filtering nutrients. The management decision-making process and management strategies for these two categories of issues will be different.

5.9.2.1 *Allocation Between Competing Uses*

If there is perfect information about the effects of human activities on the ecological processes and economic values, and if the integrity of the underlying ecological processes is not threatened, then market mechanisms can provide an efficient and optimal allocation between competing uses. This, as discussed above, assumes that market values reflect all the costs and benefits, and everyone has perfect information to engage in appropriate market transactions. For the policymaker, a cost-benefit analysis can be used to decide between alternative uses.

However, as discussed above, often there is imperfect information and uncertainties over actual outcomes. In such a situation, market failures are the norm, and the outcome is neither efficient nor ecologically sound. Moreover, BCA as a decision-making tool is not suitable. Choice about the best use of resources would need to be made using other decision-making processes and using a set of criteria, including economic net returns and adopting a precautionary approach.

A combination of administrative and market-based processes could be more suitable (Lal 1990; Barbier 1994; and Norton et al. 1998). The criteria could include some safe minimum standard of an acceptable minimum condition of the mangrove ecosystem, inter and intra generational equity considerations, and a set of binding constraints about the quality of the mangrove ecosystem and the level of acceptable changes.

Such criteria will help define the broad policy options, including identifying resource use suitability categories, as well as the resource use and management districts or ecobiomes (Lal 1989). A resource district, or ecobiome, is an area of a natural system where organisms, including humans exist, and their biophysical environment interacts dynamically, within which a balance between inflow and outflow of materials and energy is sustained (Cassells 1991). Or it can be a sub area, which can be distinguishable as a subsystem, with a definable boundary. Ecological principles and processes guide the delineation of the boundary, since they determine the extent and the level of environmental impacts and outcomes of any activity. The actors, communities and government agencies help define the needs and aspirations and future use scenarios through a consultative process. The interaction between ecological, economic and social factors and processes define the actual boundary of the ecobiome (that is economic biome), and will reflect the spatial and inter-temporal interconnectedness in the system.

The designated ecobiomes define a hierarchy of characteristics, including acceptable quality of a wetland area, the stakeholders that have rights and responsibilities over use and management of the resource and whose inputs are relevant, and the binding constraints about upper limit of acceptable change. These ecobiomes form the basis of more specific action-oriented management. Within these ecobiomes, decisions can be made about the rate of harvests of mangrove-based goods and the rate of use of the ecological services.

5.9.2.2
Policy Regarding Rate of Use

Where the management's concern is about the rate of harvest of fish or forest products or the assimilative capacity of coastal waters as a pollution sink, the bioeconomic models, incorporating the ecological functional relationships could be used. The rate of forest harvest, for example, determined purely on the grounds of rotational cycle reflecting biological growth would need to be tempered to reflect the role that forests play in maintaining coastal fisheries. The optimal rate of harvest of timber and fish products will also depend on their relative economic values. Where there is a lack of information and some uncertainty about the effect of human activities and incomplete value information, the market mechanism cannot be relied on when making decisions about the rate of use. Instead a participatory approach based on an iterative process, the adoption of the precautionary principle is likely to be more suitable.

5.9.3
Action Management and Management Instrument

Action oriented management would target the users as well as the managers of the resources focusing on their decision-making process. Management instruments are used to provide incentives/disincentives for actors to engage in an activity, restrict their activities within a specified area or to meet a specific target, and/or to self regulate.

There is a plethora of management instruments (at least 16 according to Dover 1999), which can be divided into three categories of instruments. The categories reflect different degrees of user incentives, 'top down' prescriptive rules and user initiated management rules, regulations and enforcement. The command and control is the most rigid of all, stipulating strict conditions about, for example, what is permitted in a use zone and managed using restricted entry permits, licenses or fixed standards of water quality. The enforcement is usually by government agencies. Economic instruments, such as transferable fishing rights; and taxes or resource rent charges provide users with incentives to change their behaviour, without being prescriptive. It relies on market mechanisms and the pursuit of profit maximisation goals to encourage users to change their behaviours; enforcement of these instruments is by the government agencies.

The third category, co-management, is a process-oriented management tool, which explicitly recognises the diverse range of users and actors, use and management rights, and their motivating forces. Co-management is based on the mutual accommodation and sharing of management responsibility (Hyviding 1996). Enforcement is also practised by the users themselves with governments providing largely legislative and institutional support.

For the management of mangroves or other resources, too often instrument choice has either been based on a particular discipline, is ideologically driven or is simply a matter of convenience or expediency. For example, natural scientists and environmentalists tend to favour licenses and rigid standards, whereas economists prefer the use of economic instruments such as taxes and property rights. However, within the IAMMF, any combination of instruments can be used. Actors representing the users, government and non-government agencies make the choice of instruments.

In choosing the set of relevant instruments, two sets of criteria about effectiveness and implementation, can help (Dover 1999; Antunes and Santos 1999). Effectiveness criteria include: information requirement; dependability (regarding goals); corrective versus anecdotal focus; flexibility (across context and time); cost effectiveness or gross cost; efficiency; response time and cross sectoral influence. The implementation criteria would address issues such as equity impacts; political/social feasibility; legal/constitutional feasibility; institutional feasibility; monitoring requirements; enforceability /avoidability and communicability. The overall goal of adopting a combination of instruments is to change the behaviour of resource owners/ users and managers to ensure an acceptable quality and quantity of wetlands within which an ecobiome is maintained.

5.10 Application of the Three-Tiered Process

The application of this IAMMF should be through an iterative process involving resource owners/users, researchers and the government. An interdisciplinary research team drawn from key biophysical disciplines, including ecology, from social sciences including social anthropology and economics, is essential because of the complex nature of the ecosystem, spatial and ecological inter-connectedness within the mangrove ecosystem and the link between ecological productivity and economic production. Similarly, recognising that the current management is split among many different government agencies, the involvement of key government actors from the major ministries, such as forest, fisheries, lands, and marines is also critical. Such interaction could be encouraged through National Mangrove Management Committees, or other similar organisations. Whatever the nature of the organisational interaction, they collectively must be given the responsibility and legislative power to make informed decisions, and not just act in an 'advisory' capacity, as is commonly the case in many countries. They must also be provided with an appropriate level of resources. It is important to also recognise that participatory decision making often takes longer than usual and results in conflicts within and between key actor groups. Conflict resolution skills within such a group are also vital. Only then they together will be in a position to make appropriate policy decisions based on the best information available.

As more information becomes available and the understanding about the dynamics of the system improves, management policies could evolve and adjustments could be made to the mix of instruments used to manage the mangroves. This adaptive management is an essential element of the IAMMF.

5.11 Conclusions

Despite the recognition of the importance of mangrove ecosystems and despite concerted efforts both locally and internationally, loss and or degradation of mangrove areas continues even today. If the local and global goal of the 'wise use' of mangroves is to be achieved, national and international 'wetland policies', 'national strategies and actions plans' need to be translated into more specific and directly implementable intervention measures that target the root cause of the problems.

In this paper, an integrated adaptive mangrove management framework (IAMMF) and a three-tiered decision-making process is outlined. The three-tiered decision-making process comprises the reflective stage, policy options stage and action management stage. In this adaptive management, the focus of management is on providing the actors with an appropriate incentive and a sense of ownership of the outcomes and responsibility. The emphasis is on the implementation of a process in which the management strategies and combination of instruments evolves over time, as more information becomes available and the understanding of the dynamics of the system improves. Building on Lisbon principles (Costanza et al. 1998), key principles that underpin the IAMMF include:

- *Use and Management Rights and Responsibility Principle* – the recognition of the complex ownership and use rights, and overlapping management rights found in a country, and the respective roles and responsibilities;
- *Scale-matching Principle* – the recognition of spatial and inter-temporal interconnectedness of the wetland systems and on-site and off-site effects of human action and the need for management regimes to reflect this;
- *Precautionary Principle* – the acceptance of the reality of imperfect information and uncertainties about the functional relationship between mangrove forest areas and the underlying ecological processes and environmental functions; the effects of human activities and the ecosystem's reaction; and the need for the adoption of a precautionary approach;
- *Binding Constraints Principle* – the acceptance of pre-determined binding biophysical constraints guiding use and management;
- *Participatory Principle* – the acknowledgement of the need for, and active participation of, all stakeholders – local communities dependent on the resources and different government agencies at all levels responsible for the management of different aspects of the mangrove system. Active participation of all stakeholders is at all stages of the use and management decision making;
- *Instrument Mix Principle* – the importance of the use of a combination of institutions, including the traditional command and control instruments as well as economic instruments, co-management and voluntary measures;
- *Adaptive Management Principle* – the acceptance of the reality of imperfect knowledge and evolving understanding and knowledge accumulation, and the need for regular revisions to management.

References

Abele LG (1992) A review of the grapsid crab genus *Sesarma* (Crustacea: Decapoda: Grapsidae) in America, with the description of a new genus. Smithsonian Contr Zool 527:1–60

Abeysinghe PD, Triest L, De Greef B, Koedam N, Sanath H (1999) Genetic differentiation between *Bruguiera gymnorhiza* and *B. sexangula* in Sri Lanka. Hydrobiologia 413:11–16

Acosta-Fabelo JM (1974) Estudio de las comunidades vegetales de la bahía de Los Totumos. Bol Soc Venez Cienc Nat 3:79–112

Adaime RR (1987) Estrutura, producão e transporte em um manguezal. Simp. Ecossistemas da Costa Sul e Sudeste do Brasil – Sintese de Conhecimentos. Acad. Ciênc. Estado de São Paulo, São Paulo, pp 80–99

Agate AD (1988) The sulfur and iron cycles. In: Agate AD, Subramaniam CV, Vannucci M (eds) Role of microorganisms in nutrient cycling of mangrove soils and waters. UNDP/UNESCO RAS/86/120, Manilla, p 101–107

Ahamada S (1997) Situation de la mangroves des Comores. SAREC-SIDA Regional workshop on mangrove ecology, physiology and management, Zanzibar

Ahmad IU (1997) Communication from the Resource Information Management system. Forest Department, Dhaka

Aksornkoae S (1987) Country profile – Thailand. In: UNDP/UNESCO (ed) Mangroves of Asia and the Pacific: Status and management. Tech Rep UNDP/UNESCO Research and Training Pilot Program on Mangrove Ecosystems in Asia and the Pacific (RAS/79/002), Manila, pp 321–343

Aksornkoae S (1999) Timber production from mangroves. ISME Mangrove Ecosystems Proc 4:83–87

Alexander TR (1967) Effect of hurricane Betsy on the southeastern Everglades. Q J Fla Acad Sci 30:10–24

Ali MO (1998) Development of fish resource and management technique. Fishery Resource Development, Fisheries Directorate, Dhaka

Alim A (1974) Instruction manual for planting on coastal areas. Bangladesh Forest Department, Dhaka

Alió JJ (2000) Los recursos vivos del Sistema de Maracaibo. In: Rodríguez G (ed) El Sistema de Maracaibo. Instituto Venezolano de Investigaciones Científicas, Caracas, pp 153–173

Al-Muharrami M (1994) Resource management of mangroves in the arid environment in the Sultanate of Oman. MSc dissertation, University of Newcastle, UK

Alvarez-León R (1993) Mangrove ecosystems of Colombia. In: Lacerda LD (ed) Conservation and sustainable utilization of mangrove forests in Latin America and Africa regions. Inter Soc Mangrove Ecosystems, Okinawa, pp 75–114

Alvarez-León R, Hernández-Avellaneda A (1997) Variation chromatique du feuillage de *Rhizophora mangle* dans la Ciénaga Mestizo (Córdoba), sud-est de la cote caraibe de Colombie, In: Kjerfve B, Lacerda LD, Diop ES (eds) Mangrove ecosystem studies in Latin America. UNESCO/ISME/USFS, pp 108–110

Amend T (1992) Los habitantes de Los Roques en el pasado y el presente. In: Amend T (ed) Parque Nacional Archipiélago Los Roques. Fundación Polar-Instituto Nacional de Parques, Caracas, pp 1–42

Andrade J (1984) Folclore da região do Salgado. Escola de Folclore, São Paulo

Andrews TJ, Clough BF, Muller GJ (1984) Photosynthetic gas exchange properties and carbon isotope ratios of some mangroves in North Queensland. In: Teas HJ (ed) Physiology and management of mangroves. W Junk, The Hague, pp 15–23

Ansari TA (1987) Pakistan. In: Mangroves of Asia and the Pacific – status and management. UNDP/UNESCO Project Research and Training Pilot Program on Mangrove Ecosystems. RAS/79/002, UNDP/UNESCO, Manila, p 151–174

Antczak M, Antczak A (1987) Algunos problemas de la intervención humana en el ecosistema del Parque Nacional Archipiélago de Los Roques en la epoca pre-hispánica. Unpublished manuscript, Caracas, sp

Antunes P, Santos R (1999) Integrated environmental management of the oceans. Ecol Econ 31:215–226
Araújo DSD, Maciel NC (1979) Os manguezais do recôncavo da baia de guanabara. Fundação Estadual de Engenharia do Meio Ambiente, Rio de Janeiro
Arruda Leme MH, Negreiros-Fransozo ML (1998) Reproductive patterns of *Aratus pisonii* (Decapoda: Grapsidae) from an estuarine area of São Paulo northern coast, Brazil. Rev Biol Trop 46:673–678
Aschbacher J, Ofren R, Delsol JP, Suselo TB, Vibulsresth S, Charrupat T (1995) An integrated comparative approach to mangrove vegetation mapping using advanced remote sensing and GIS technologies: preliminary results. Hydrobiologia 295:285–294
Attiwill PM, Clough BF (1978) Productivity and nutrient cycling in the mangroves and seagrass communities of Westernport Bay. School of Botany, University of Melbourne Report to Ministry of Conservation, Victoria
Augustinus PGEF (1978) The changing shoreline of Suriname (South America). Uitgaven Natuurwetenschappelijke Studierkring vorr Suriname en de Nederlandse Antillen 95:1–232
Austin HM (1971) A survey of the ichthyofauna of the mangroves of western Puerto Rico during December 1967–August 1968. Caribb J Sci 11:427–455
Aveline LC (1980) Fauna dos manguezais Brasileiros. Rev Bras Geogr 42:786–821
Azfal-Rafii Z, Dodd RS, Fauvel MT (1999) A case of natural selection in Atlantic-East-Pacific *Rhizophora*. Hydrobiologia 413:1–9
Bacon PR (1970) The ecology of Caroni Swamp, Trinidad. Spec Publ Central Statistics Office, Port of Spain
Bacon PR (1987) Wetlands for tourism in the insular Caribbean. Ann Tourism Res 14:104–117
Bacon PR (1993) Conservation and utilization of mangroves in the Lesser Antilles, Jamaica and Trinidad and Tobago. In: Lacerda LD (ed) Conservation and sustainable utilization of mangrove forests in Latin America and Africa regions. Inter Soc Mangrove Ecosystems, Okinawa, pp 155–210
Bacon PR (1994) Template for evaluation of impacts of sea level rise on Caribbean coastal wetlands. Ecol Engineering 3:171–186
Bacon R (1997) The role of the Ramsar Convention in mangrove management. Intercoast Network: Internat Newslet Coast Manag Spec Ed 1:25–26
Baker JM (1982) Mangrove swamps and the oil industry. Oil and Petrochemical Polutt 1:5–22
Baksha MW (1996) Beehole borer infestation in coastal plantations in Bangladesh and possible management options. Wallaceana 77:17–20
Baksha MW, Islam MR, Chowdhurdy SH (1990) Notes on keora beehole. A pest on keora in Bangladesh. Proc 7th Nat Zool Conf, Bangladesh, p 113–114
Balachandran L (1988) A comprehensive account of the mangrove vegetation of Adaman and Nicobar Islands. Indian Forest 114:741–751
Ball MC, Farquahar GD (1984a) Photosynthetic and stomatal responses of two mangrove species, *Aegiceras corniculatum* and *Avicennia marina*, to long-term salinity and humidity conditions. Plant Physiol 74:1–6
Ball MC, Farquahar GD (1984b) Photosynthetic and stomatal responses of the grey mangrove, *Avicennia marina*, to transient salinity conditions. Plant Physiol 74:7–11
Ball MC, Munns R (1992) Plant responses to salinity under elevated atmospheric concentrations of CO_2. Austr J Bot 40:515–525
Ballment ER, Smith-III TJ, Stoddart JA (1989) Sibling species in the mangrove genus, *Ceriops* Arn. (Rhizophoraceae), detected using biochemical genetics. Aust Syst Bot 1:391–397
Bandyopadhyay AK (1986a) Soil and water characteristics of the mangrove forest of the Sundarbans (India). Indian Forest 112:58–65
Bandyopadhyay AK (1986b) The Sundarbans mangrove forest experience on reclamation. Proc workshop conversion Mangrove Areas to Paddy Cultivation. UNDP/UNESCO RAS/79/002, New Dehli, p 1–3
Banyikwa FF, Semesi AK (1986) Endangered mangrove ecosystems. The case of the Kunduchi and Mbweni mangrove forests. In: Mainoya JR, Siegel PJ (eds) Proc Workshop Save the Mangrove Ecosystems in Tanzania. Dar es Salaam, pp 103–132
Baran E, Hambrey J (1998) Mangrove conservation and coastal management in southeast Asia: What impact on fishery resources? Mar Pollut Bull 37:431–440
Barbier EB (1987) The concept of sustainable economic development. Environm Conserv 14:101–110
Barbier EB (1994) Valuing environmental functions: tropical wetlands. Land Economics 70:155–173
Barbier EB, Strand I (1997) Valuing mangrove-fishery linkages: a case study of Campeche, Mexico. IDRC Environmental and Resource Economics Course, EEPSEA Program, Singapore
Barbier EB, Acreman M, Knowler D (1997) Economic valuation of wetlands: a guide for policy makers and planner. Ramsar Convention Bureau, Gland
Barbier EB, Burgess JC, Folke C (1999) Paradise lost? The ecological economics of biodiversity. Earthscan, London

Barreto MB (1988) Estudio preliminar de las comunidades de hongos de los manglares en el Parque Nacional Laguna Tacarigua. Acta Cient Venez 39:1–59

Barth H (1982) The biogeography of mangroves. In: Sen DN, Rajpurohit KS (eds) Ecology of halophytes. Dr. W. Junk Publishers, The Hague, pp 35–60

Bastardo H (1993) Decomposition process in *Avicennia germinans*, *Rhizophora mangle* and *Laguncularia racemosa* under oil spill. Acta Biol Venez 14:53–60

Batibasaqa K, Overton J, Horsley P (1994) Vanua: land, people and culture in Fiji. In: South GR, Goulet D, Tuqiri C, Church M (eds) Traditional marine tenure and sustainable management of marine resources in Asia and the Pacific. International Ocean Institute (South Pacific), Suva, Fiji

Batista AC, Vital AF, Maia HS, Lima IH (1955) Coletánea de novas especies de fungos. An Soc Biol Pernambuco 13:187–224

Beever JW, Simberloff D, King LL (1979) Herbivory and predation by the mangrove tree crab *Aratus pisonii*. Oecologia 43:317–28

Beltrão MC (1976) Documentos sobre a pre-historia dos estados do Rio de Janeiro e da Guanabara (1500–1963). Col Museu Paulista Ser Arqueologia 2:1–143

Benton MJ (1995) Diversification and extinction in the history of life. Science 268:52–58

Bingham BL (1992) Life histories in an epifaunal community: coupling of adult and larval processes. Ecology 73:2244–2259

Bisbal FJ (1989) Distribution and habitat association of the carnivores in Venezuela. Adv Neotropical Mammal 2:339–362

Bishop RC (1978) Endangered species and uncertainty: the economics of a safe minimum standard. Amer J Econom 57:10–18

Blasco F (1983) Mangroves du Senegal et de Gambie. Status écologique – evolution. Final Draft CNRS ER n 73, Université de Toulouse III, Toulouse

Blasco F (1984) Climatic factors and the biology of mangrove plants. In: The mangrove ecosystem: research methods. UNESCO, Paris, pp 18–35

Blasco F, Aizpuru M (1999a) Recent data on the mangroves of the Bay of Bengal. Indian J Mar Sci (submitted)

Blasco F, Aizpuru M (1999b) Evolutionary trends of the mangroves of Asia. Wetland Ecol Mangt (submitted)

Bodero A (1993) Mangrove ecosystems of Ecuador. In: Lacerda LD (ed) Conservation and sustainable utilization of mangrove forests in Latin America and Africa regions. Part I. Latin America. ISME, Okinawa, pp 55–74

Braga MR, Fujji M, Yokoya NS, Eston VR, Plastino EM, Cordeiro-Marino M (1990) Macroalgal reproductive patterns in mangroves of Ilha do Cardoso, SP, Brazil. An II Symp Ecosys Coast Sul-sudeste Brasil. Academia de Ciências do Estado de São Paulo. São Paulo, pp 209–217

Breteler FJ (1969) The Atlantic species of *Rhizophora*. Acta Bot Neerl 18:434–441

Brinkman R, Pons LJ (1968) A pedo-geomorphological classification and map of the Holocene sediments in the coastal plain of the three Guianas. Netherlands Soil Survey Institute, Wageningen

Brinson MM, Brinson LG, Lugo AE (1974) The gradient of salinity, its seasonal movement and ecological implications for the Lake Izabal-Rio Dulce ecosystem. Bull Mar Sci 24:335–344

Bromley DW (1989) Economic interests and institutions: the conceptual foundations of public policy. Blackwell, Oxford

Bunt JS (1987) Country profile – Philippines. In: Mangroves of Asia and the Pacific. Status and Management Technical Report of the UNDP/UNESCO Research and Training Pilot Program on Mangrove Ecosystems in Asia and the Pacific (RAS/79/002), Manila, p 263–280

Bunt JS (1995) Continental scale patterns in mangrove above-ground biomass litter fall. Hydrobiologia 295:135–140

Burns KA, Garrity SD, Levings S (1993) How many years until mangrove ecosystems recover from catastrophic oil spills. Mar Pollut Bull 26:239–248

Calderón-Saenz E (1983) Hallazgo de *Pelliciera rhizophorae* Tr and Pl. en la costa del Atlántico con observaciones taxonómicas y biogeográficas preliminares. Bol. Museo del Mar Bogotá 11:100–111

Calderon-Saenz E (1984) Occurrence of the mangrove, *Pelliciera rhizophorae*, on the Caribbean coast of Colombia with biogeographical notes. Bull Mar Sci 35:105–110

Camilleri J (1989) Leaf choice by crustaceans in a mangrove forest in Queensland. Mar Biol 102: 453–459

Canniccia S, Ruwaa RK, Ritossaa S, Vannini M (1996) Branch-fidelity in the tree crab *Sesarma leptosoma* (Decapoda, Grapsidae). Oceanographic Literature Review 43:10–36

Canonizado JA, Hssain MA (1998) Integrated forest management plans for the Sundarbans Reserved Forest. Mandala Agricultural Development Corporation and The Forest Department, Dhaka

Carrera CJ (1975) Efectos históricos de la interacción entre los sistemas socio-económicos y los manglares: El caso de Puerto Rico. Mem II Simp Latinoamer Oceanogr Biol, Caracas, pp 153–161

Carter MR, Burns LA, Cavinder TR, Dugger KR, Fares PL, Hicks DB, Revells HT, Schimidt TW (1973) Ecosystem analysis of the Big Cypress Swamp and estuaries. USEPA 90419-74-002, Atlanta
Cassells D, Hamilton LS, Saplaco SR (1991) Understanding the role of forests in watershed protection. In: Carpenter RA (ed) Natural systems for development: what planners need to know. Macmillan Publ, New York, p 52–98
Castaño-Uribe C (1989) Introdución. In: Villegas B (ed) Manglares de Colombia. Villegas Ed and Banco de Occidente, Bogotá, pp 1–13
Cervigón F, Gómez A (1986) Las Lagunas Litorales de la Isla de Margarita. Centro de Investigaciones Científicas, Universidad de Oriente, Nueva Esparta
Chaffey DR, Miller FR, Sandom JH (1985) A forest inventory of the Sundarbans, Bangladesh. Overseas Development Organization, London
Chambers R (1997) Whose reality counts? Putting the first last. Intermediate Technology Publications, London
Chan EI (1977) Oil pollution and tropical littoral communities: biological effects of the 1975 Florida Keys oil spill. Proc. 1977 Oil Spill Conf, Am Petrol Inst, Washington, pp 539–542
Chan HT (1987) Country profile – Malaysia. In: UNDP/UNESCO (ed) Mangroves of Asia and the Pacific Status and Management Technical Report of the UNDP/UNESCO Research and Training Pilot Program on Mangrove Ecosystems in Asia and the Pacific (RAS/79/002), Manila, p 178–189
Chan HT (1996) Mangrove reforestation in peninsular Malaysia: a case study of matang. In: Field CD (ed) Restoration of mangrove ecosystem. International Society for Mangrove Ecosystem. Okinawa, Japan, p 123–134
Chan HT (1999) Some environmental consequences of mangrove conversion in Malaysia. ISME Mangrove Ecosystems Proc 4:19–21
Chapman VJ (1974) Mangrove biogeography. In: Walsh G, Snedaker S, Teas H (eds) Proc Inter Symp Biol Mangt Mangroves. Honolulu, pp 3–22
Chapman VJ (1976) Mangrove vegetation. Cramer Vaduz, 447 p
Chapman VJ (1977) Ecosystems of the World. 1. Wet coastal ecosystems. Elsevier. Amsterdam
Cheeseman JM, Clough BF, Carter DR, Lovelock CE, Ong JE, Sim RG (1991) The analysis of photosynthetic performance of leaves under field conditions: a case study using *Bruguiera* mangroves. Photosynthetic Res 29:11–22
Chinnaraj S (1993) Manglicolous fungi from Maldive Atolls, Indian Ocean. Indian J Mar Sci 22:141–142
Chong A (1979) Mangrove management planning and implementation, Sierra Leone. MANR/FAO/FA: DP/88/008, Field Document 6
Chong VC (1995) The prawn-mangrove connection-fact or fallacy? Proceedings of Seminar on Sustainable Utilization of Coastal Ecosystem for Agriculture, Forestry and Fisheries in Developing Regions. Malaysian Fisheries Department, Kuala Lampur, p 3–20
Chopra ARN, Nayar SL, Chopra IC (1956) Glossary of Indian medicinal plants. CSRI, New Dehli
Choudhury AM (1968) Working plan of the Sundarbans Forest Division for the period 1960–61 to 1979–80. East Pakistan Government Press, Dhaka
Choudhury AM (1971) Coastal afforestation and its techniques in East Pakistan. Forest Dale News 4:1–21
Choudhury AM, Ahmed I (1994) History of forest management. In: Hussain Z, Acharya G (eds) Mangroves of the Sundarbans. IUCN, Bangkok, p. 155–180
Choudhury AM, Baksha MW (1983) Studies on the role of *Chrysocroa* sp. in the initiation of top dying of sundri. Bano Biggyan Patrika 12:30–34
Choudhury AM, Quadir DA, Islam J (1990) Study of the Chokoria Sundarbans using remote sensing techniques. Space Research and Remote Sensing Organization, Dhaka
Chow BA, Bacon PR (1992) Mangrove forests and faecal rain. Caribb Mar Stud. 3:57–58
Christensen B (1982) Management and utilisation of mangroves in Asia and the Pacific. FAO Technical Report, Rome
Cintrón-Molero G, Schaeffer-Novelli Y (1983) Introducción a la Ecología del Manglar. UNESCO, Montevideo
Cintrón-Molero G, Schaeffer-Novelli Y (1992) Ecology and management of New World mangroves. In: Seeliger U (ed) Coastal plant communities of Latin America. Academic Press, San Diego, pp 233–258
Cintrón-Molero G, Lugo AE, Pool DJ, Morris G (1978) Mangroves of arid environments in Puerto Rico and adjacent islands. Biotropica 10:110–121
Cintrón-Molero G, Lugo AE, Martínez RF (1985) Características estructurales y funcionales de los bosques de manglar. Bol Soc Cienc Nat 40:399–442
Clarck MW, McConchie D, Saenger P, Pillsworth M (1997) Hydrological controls on copper, cadmium, lead and zinc concentrations in an anthropogenically polluted mangrove ecosystem. J Cost Res 13:1150–1158
Clarck MW, McConchie D, Lewis DW, Saenger P (1998) Redox stratification and heavy metal partitioning in *Avicennia* dominated mangrove sediments: a geochemical model. Chem Geol 149:147–171

Clarke PJ (1995) The population dynamics of the mangrove *Avicennia marina*; demographic synthesis and predictive modeling. Hydrobiologia 295:83–88

Clarke PJ, Myerscough PJ (1991) Floral biology and reproductive phenology of *Avicennia marina* in south-eastern Australia. Austr J Bot 39:283–293

CLIRSEN (1984) Aplicación de sensores remotos en el estudio del ecosistema del manglar. Centro de Levantamientos Integrados de Recursos Naturales por Sensores Remotos, Guayaquil

Clough BF, Andrews TJ, Cowan IR (1982) Physiological processes in mangroves. In: Clough BF (ed) Mangrove ecosystems in Australia: structure, function and management. Australian National University Press, Canberra, pp 193–210

Colonnello B G (1996) Aquatic vegetation of the Orinoco River Delta (Venezuela) – An overview. Hydrobiologia 340:109–113

Colonnello G, Medina E (1998) Vegetation changes induced by dam construction in a tropical estuary: the case of the Minamo River, Orinoco Delta (Venezuela). Plant Ecol 139:145–154

Conde JE (1989) Ecología poblacional del cangrejo de mangle *Aratus pisonii* (H. Milne-Edwards) (Brachyura: Grapsidae) en hábitats extremos. (PhD Dissertation. Instituto Venezolano de Investigaciones Científicas, Caracas

Conde JE (1990) Ecología poblacional del cangrejo de mangle *Aratus pisonii* (H. Milne-Edwards) (Brachyura: Grapsidae) en habitats extremos. Dissertation. Instituto Venezolano de Investigaciones Científicas, Caracas

Conde JE (1995) Quando a ficção vira realidade. Ciência Hoje 18:70–73

Conde JE, Alarcón C (1993) A review of current knowledge and status of Venezuelan mangroves. In: Lacerda LD (ed) Conservation and sustainable utilization of mangrove forests in Latin America and Africa regions. Part I. Latin America, ISME, Okinawa, pp 211–244

Conde JE, Díaz H (1985) Diseño de muestreo aleatorio estratificado aplicado al estudio de poblaciones del género *Uca* (Brachyura, Ocypodidae). Investigación Pesquera 49:567–579

Conde JE, Díaz H (1989a) The mangrove tree crab *Aratus pisonii* in a tropical estuarine coastal lagoon. Estuar Coast Shelf Sci 28:639–650

Conde JE, Díaz H (1989b) Productividad del habitat e historias de vida del cangrejo de mangle *Aratus pisonii* (H. Milne-Edwards) (Brachyura: Grapsidae). Bol Inst Oceanogr Venez 28:113–120

Conde JE, Díaz H (1992a) Variations in intraspecific relative size at onset of maturity (RSOM) in *Aratus pisonii* (H. Milne Edwards 1837) (Decapoda, Brachyura, Grapsidae). Crustaceana 62:214–216

Conde JE, Díaz H (1992b) Extension of the stunting range in ovigerous females of the mangrove crab *Aratus pisonii* (H. Milne Edwards 1837) (Decapoda, Brachyura, Grapsidae). Crustaceana 62:319–322

Conde JE, Díaz H, Rodríguez G (1989) Crecimiento reducido en el cangrejo de mangle *Aratus pisonii* (H. Milne-Edwards) (Brachyura: Grapsidae). Acta Cient Venez 40:159–160

Conde JE, Díaz H, Lacerda LD (1993) Caranguejos anões. Ciência Hoje 15:9–11

Conde JE, Alarcón C, Flores S, Díaz H (1995) Nitrogen and tannins in mangrove leaves might explain interpopulation variations in the crab *Aratus pisonii*. Acta Cient Venezol 46:303–304

Conde JE, Tognella MMP, Paes ET, Soares MLG, Louro IA, Schaeffer-Novelli Y (2000) Population and life history features of the crab *Aratus pisonii* (Decapoda: Grapsidae) in a subtropical estuary. Interciencia 25:151–158

Constanza R (1999) The ecological, economic, and social importance of oceans. Ecol Economics 31: 199–213

Constanza R, Wiagner L, Folke C, Maler KG (1993) Modelling complex ecological economic systems: towards an evolutionary dynamic understanding of people and nature. Bioscience 43:545–555

Constanza R, d'Arge R, Groot R de, Farber S, Grasso M, Hannon B, Limburg K, Naeem S, O'Neil RV, Paruelo J, Raskins RG, Sutton P, Belt M van den (1997) The value of the world's ecosystem services and natural capital. Nature 387:253–260

Constanza R, Andrade F, Antunes P, Belt M van den, Boesch D, Boersma D, Catarino F, Hanna S, Limburg KM, Low B, Molitor M, Pereira JG, Rayner S, Santos R, Wilson J, Young MD (1999) Ecological economics and sustainable governance of the oceans. Ecol Economics 31:171–187

Cook RG, Ranere AJ (1992) The relation of fish resources to the location, diet breadth, and procurement technology of a preceramic and a ceramic site in an estuarine embayment on the Pacific coast of Panama. In: Blake M (ed) The evolution of archaic and formative societies on the Pacific coast of Latin America. Washington University Press, Washington, pp 234–248

Cordeiro-Marino M, Braga MRA, Eston VR, Fujii MT, Yokoya NS (1992) Mangrove macroalgal communities of Latin America: The state of the art and perspectives. In: Seeliger U (ed), Coastal plant communities of Latin America. Academic Press, San Diego, pp 51–64

Cormier-Salem MC (1999) The mangrove: an area to be cleared for social scientists. Hydrobiologia 413:47–52

Corredor J, Morell J, Klekowski Jr Ej, Lowenfeld R (1995) Mangrove genetics. III. Pigment fingerprints of chlorophyll-deficient mutants. Int J Plant Sci 156:55–60

Costa CSB, Davy AJ (1992) Coastal salt marsh communities of Latin America. In: Seeliger U (ed), Coastal plant communities of Latin America. Academic Press, San Diego, pp 179–199
Cowan JM (1921) The flora of the Chakaria Sundarbans. Records of the Botanical Survey of India, 197:1–225
Craeger DB (1962) A new Cercospora on *Rhizophora mangle*. Mycologia 14:536–539
Cruxent JM, Rouse I (1958) An archeological chronology of Venezuela. Organization of American States, Washington
CSERGE (1995) Coastal zone resource assessment guidelines. Centre for Social and Economic Research on the Global Environment, University of East Anglia and University of London, London
D'Croz L (1993) Mangrove uses and conservation in Panama. In: Lacerda LD (ed) Conservation and sustainable utilization of mangrove forests in Latin America and Africa regions. Part I. Latin America, ISME, Okinawa, pp 115–128
D'Croz L, Kwiecinski B (1980) Contribucion de los Manglares a las Pesquerias de la Bahia de Panama. Rev Biol Trop 28:13–29
Dagar JC (1986) Ethnobotany of the canoes of the Nicobarese Tribals. Indian Forest 112:174–179
Dagar JC, Mongia AD, Bandyopadhyay AK (1991) Mangroves of the Adaman and Nicobar Islands, IBH Publ Co, Calcutta
Dalmacio MV, Rahman Z, Ahmad IU (1991) Coast afforestation management manual. Forest Departament, Dhaka
Das S, Siddiqi NA (1985) The mangroves and mangrove forests of Bangladesh. UNDP/FAO Project BGD/79/017, Rome
Davidson J (1984) Assistance to the forest sector of Bangladesh. Field document N° 4, FAO/UNDP Project BGD/79/017, Rome
Davie JDS (1984) Structural variation, litter production and nutrient status of mangrove vegetation in Moreton Bay. In: Coleman RJ, Covacevich J, Davie P (eds.), Focus on Stradbroke: new information on North Stradbroke Island and surrounding areas, 1974–1984. Stradbroke Island Management Organisation and Boolarong Publications, Brisbane. p. 208–223
Davie PJF (1994) Variations in diversity of mangrove crabs in tropical Australia. Memoirs Queensland Mus 36:55–58
Davis T (1996) Consultation, public participation, and integration of multiple interests into policy making. Report Prepared for the OECD, Paris (unpublished)
Day JW, Conner WH, Day RH, Ley-Lou F, Navarra AM (1988) Productivity and composition of mangrove forests at Boca Chica and Estero Pargo. In: Yanez-Arancibia A, Day JW (eds) Ecology of coastal ecosystems in the Southern Gulf of Mexico: The Terminos Lagoon region. Ed Universitaria, Mexico, pp 237–258
DeFontaubert AC, Downes DR, Agardy TS (1996) Biodiversity in the seas: implementing the convention on biological diversity in marine and coastal habitats. IUCN Gland and Cambridge
Degens ET (1990) Perspectives on biogeochemistry. Springer-Verlag, Berlin
DeGeorges PA, Samba I (1982) Compte rendu de mission sur les mangrove de Gambie. Rapport de mission OMVG 17
Diallo A (1991) Structure anatomique de la feuille des plantes de la mangrove de Guinee. Dissertation, Universite de Saint-Petersbourg
Diallo A (1992) Mangrove of Guinea. In: Conservation and sustainable utilization of mangrove forests in Latin America and Africa regions. Mangrove Ecosystems Technical Reports 3:27–46
Díaz H, Conde JE (1988) On the food sources for the mangrove tree crab *Aratus pisonii* (Brachyura: Grapsidae). Biotropica 20:348–350
Díaz H, Conde JE (1989) Population dynamics and life history of the mangrove crab *Aratus pisonii* (Brachyura: Grapsidae) in a marine environment. Bull Mar Sci 45:148–163
Díaz H, Conde JE, Bevilacqua M (1983) A volumetric method for estimating fecundity in Decapoda. Mar Ecol Progr Ser 10:203–206
Díaz H, Bevilacqua M, Bone D (1985) Esponjas en manglares del Parque Nacional Morrocoy. Fondo Editorial Acta Científica, Caracas, Venezuela
Díaz H, Conde JE, Orihuela B (1992) Estimating the species number and cover of a mangrove-root community: a comparison of methods. Austr J Mar Freshwat Res 43:707–714
Díaz-Piferrer M (1962) Effects of an oil spill on the shore of Guanica, Puerto Rico. Deep-Sea Res 2: 855–856
Diop ES (1990) La côte Ouest-Africaine: du Saloum (Sénégal) à la Mellacorée (Rep de Guinée). In: Coll Etudes et Thèses, Editions de l'ORSTOM. Paris
Diop ES (1992) Mangroves of Africa, status and management in mangroves. ISME Newsletter 5:6–7
Diop ES (1993) Status of mangrove ecosystems in Africa: In: Lacerda LD, Field CD (eds) Conservation and sustainable utilization of mangrove forests in Latin America and Africa regions. ISME, Okinawa, pp 21–22

Diop ES, Barusseau J, Ausseil-Badie P (1991) Holocene deltaic sequence in the Saloum Estuary (Senegal). Quaternary Res 36:178–194

Diop ES, Soumaré A, Diallo N, Guissé A (1995) Observations à propos des facteurs limitants le développement de la mangrove dans le Saloum (Ile du Gandoul). Rapport final de l'EPEEC-10, Dakar

Diop ES, Soumaré A, Diallo N, Guissé A (1996a) Dynamique de la mangrove des îles du Gandoul occidental (du nord de l'île de Guissanor au sud du bôlon de Niodior, Saloum/Sénégal). Rapport final de l'EPEEC-8, Dakar

Diop ES, Soumaré A, Diallo N, Guissé A (1996b) Mangrove restoration through replantation in Senegal, West Africa. ISME Newsletter 117:4–5

Diop ES, Soumaré A, Diallo N, Guissé A (1996c) Suivi de la station expérimentale de reboisement par la mangrove dans la lagune de la Somone. Rapport final de l'EPEEC-4, Dakar

Diop ES, Soumaré A, Diallo N, Guissé A (1997) Recent changes of the mangroves of Saloum River estuary (Senegal, West Africa). Mangroves and Salt Marshes 1:163–172

Dixon J, Lal P (1994) The management of coastal wetlands: economics analysis of combined ecological-economic systems. In: Dasgupta P, Maler KG (eds) The environment and emerging development issues. Oxford University Press, Clarendon

Doddema-de-Guia MB (1997) A preliminary exploratory survey of the mangrove vegetation in the northern part of Sofala, Mozambique: methods and information requirement for planning and management purposes. SAREC-SIDA Regional Workshop on Mangrove Ecology, Physiology and Management, Zanzibar

Dolianiti E (1955) Frutos de Nypa no Paleoceno de Pernambuco, Brasil. Bol Div Geol DNPM Brasil 158:1–36

Dover S (1999) Public policy and institutional R&D for natural resource management: issues and directions for LWRRDC. In: Mobbs C, Dover S (eds) Social, economic, legal, policy and institutional R&D for natural resource management: issues and directions for LWRRDC. Land and Water Research and Development Corporation, Canberra, p. 78–105

Doyen A, Agboba C (1985) La mangrove a usages multiples de l'estuaire du Saloum (Senegal). UNESCO-MAB-COMAR, Paris

Drake BG (1992) A field study of the effects of elevated CO_2 on ecosystem processes in a Chesapeake Bay wetland. Austr J Bot 40:579–595

Drigo D, Latif MA, Choudhury AM, Shaheduzzaman M (1987) The maturing mangrove plantations of coastal afforestation project. Field document N° 2, UNDP Project BGD/85/085, Dhaka

Dueñas H (1983) Fluctuaciones del nivel del mar durante el depósito de los sedimentos basales de la formación Ciénaga de Oro. Rev Acad Colomb de Ciencias 15:67–76

Dugan PJ (1990) Wetland conservation – a review of current issues and required action. IUCN, Gland

Duke NC (1988) Phenologies and litter fall of two mangrove trees, *Sonneratia alba* Sm. and *S. caseolaris* (L.) Engl., and their putative hybrid, *S.* × *gulngai* N. C. Duke. Austr J Bot 36:473–482

Duke NC (1990) Phenological trends with latitude in the mangrove tree *Avicennia marina*. J Ecol 78:113–133

Duke NC (1991) *Nypa* in the mangroves of Central América: introduced or relict? Principes 35:127–132

Duke NC (1992) Mangrove floristics and biogeography. In: Robertson AI, Alongi DM (eds) Tropical mangrove ecosystems. American Geophysical Union, Washington, pp 43–100

Duke NC (1994) A mangrove hybrid, *Sonneratia xurama* (Sonneratiaceae) from northern Australia and southern New Guinea. Aust Syst Bot 7:521–526

Duke NC (1995) Genetic diversity, distributional barriers and rafting continents – more thoughts on the evolution of mangroves. Hydrobiologia 295:167–181

Duke NC, Bunt JS (1979) The genus *Rhizophora* (Rhizophoraceae) in north-eastern Australia. Aus J Bot 27:657–678

Duke NC, Jackes BR (1987) A systematic revision of the mangrove genus *Sonneratia* (Sonneratiaceae) in Australasia. Blumea 32:277–302 Project BGD/79/017, Rome

Duke NC, Bunt JS, Williams WT (1984) Observations on the floral and vegetative phenologies of north-eastern Australian mangroves. Aus J Bot 32:87–99

Duke NC, Ball MC, Ellison JC (1998) Factors influencing biodiversity and distributional gradients in mangroves. Global Ecology and Biogeography Letters, Mangrove Special Issue, 7:27–47

Dunson WA (1970) Some aspects of electrolyte and water balance in three estuarine reptiles, the diamond back terrapin, American and "salt water" crocodiles. Compar Biochem Physiol 47A:1245–1255

Earthwatch Summit (1997) Shrimp sentinel online: national report – Malaysia, http://www.earthsummitwatch.org/shrimp/national_reports/thai96.html (21 December 1999)

Echevarria J, Sarabia J (1993) Mangroves of Peru. In: Lacerda LD (ed) Conservation and sustainable utilization of mangrove forests in Latin America and Africa regions. Part I. Latin America, ISME, Okinawa, pp 43–54

Edwards RRC (1978) The fishery and fisheries biology of penaeid shrimp on the Pacific coast of Mexico. Oceanogr Mar Biol A Rev 1:145–180
Ellison AM, Farnsworth EJ (1990) The ecology of Belizean mangrove-root fouling communities: I. Epibenthic fauna are barriers to isopod attach of red mangrove roots. J Exp Mar Biol Ecol 142:91–104
Ellison AM, Farnsworth EJ (1992) The ecology of Belizean mangrove-root fouling communities: II. Patterns of epibiont distribution and abundance, and effects of root growth. Hydrobiology 247: 87–98
Ellison AM, Farnsworth EJ (1996) Anthropogenic disturbance of Caribbean mangrove ecosystems: past impacts, present trends, and future predictions. Biotropica 28:549–565
Ellison AM, Farnsworth EJ, Twilley RR (1996) Facultative mutualism between red mangroves and root-fouling sponges in Belizean mangal. Ecology 77:2431–2444
Ellison JC (1993a) Mangrove retreat with rising sea-level, Bermuda. Estuar Coast Shelf Sci 37:75–87
Ellison JC (1993b) Mangroves of Bermuda and the Cayman Islands. In: Lacerda LD, Field CD (eds) Conservation and sustainable utilization of mangrove forests in Latin America and Africa regions. ISME, Okinawa, pp 2
Ellison JC (1998) Impacts of sediment burial on mangroves. Mar Pollut Bull 37:8–12
Ellison JC, Stoddart DR (1991) Mangrove ecosystem collapse during predicted sea-level rise: Holocene analogues and implications. J Coast Res 7:151–165
Elouard P, Rosso JC (1977) Biogéographie et habitat des mollusques actuels laguno-marins du delta du Saloum. Geobios 10:275–299
Emmerson WD, McGwynne LE (1992) Feeding and assimilation of mangrove leaves by the crab *Sesarma meinerti* de Man in relation to leaf-litter production in Mgazana, a warm-temperature southern African mangrove swamp. J Exper Mar Biol Ecol 157:41–53
Fairbridge RW (1976) Shellfish-eating preceramic Indians in coastal Brazil. Science 191:353–359
FAO (1979) Land resources survey-Sierra Leone. Freetown DP/SIL/73/002, Field Document, Rome
FAO (1997) State of the World's forests. Rome
Farnsworth EJ, Ellison AM (1991) Patterns of herbivory in Belizean mangrove swaps. Biotropica 23: 555–567
Farnsworth EJ, Ellison AM (1997) The global conservation status of mangroves. Ambio 26:328–334
Febvay G, Kermarrec A (1978) Quelques parametres physicochimiques de la foret littorale (mangrove et foret palustre) et leur evolution. Bull Liaison Groupe Travail Mangrove et Zone Cotiere 4:70–73
Fell JW, Master IM (1973) Fungi associated with the degradation of mangrove (*R. mangle*) leaves in south Florida. In: Stevenson HL, Colwell RR (eds), Estuarine microbial ecology. Univ. South Carolina Press. Columbia, pp 455–466
Feller IC (1995) Effects of nutrient enrichment on growth and herbivory of dwarf red mangrove (*Rhizophora mangle*). Ecol Monogr 65:477–505
Field CD (1995a) Impact of expected climate change on mangroves. Hydrobiologia 295:75–81
Field CD (1995b) Journey amongst mangroves. ISME, Okinawa
Field CD (1996) Restoration of mangrove ecosystem. International Society for Mangrove Ecosystem. Okinawa, Japan
Field CD (1998) Rehabilitation of mangrove ecosystems. Mar Pollut Bull 37:383–392
Field CD (1999) Mangrove rehabilitation: choice and necessity. Hydrobiologia 413:47–52
Figueroa D, Seijas AE (1986) Venezuela. In: Scott DA, Carbonell M (eds) Inventario de Humedales de la Region Neotropical. IWRB, Slimbridge and UICN, Cambridge, pp 292–317
Finlayson CM, Spiers AG (1999) Global review of wetland resources and priorities for wetland inventory. http//www.ramsar.org.html (9 November 1999)
Flores C (1977) Recursos Acuaticos. Direcciones de Publicaciones, Instituto Oceanografico, Universidad de Oriente, Cumana
Flores-Verdugo F (1986) Ecologia de los manglares y perfil de comunidades en os sistemas lagunares de Agua Brava y Marismas Nacionales, Nayarit Tech Rep Cons Nac Cienc Tecnol Clave PCECBNA-022068, Mexico
Flores-Verdugo F, Day JW, Briceno-Duenas R (1987) Structure, litter fall, decomposition and detritus dynamics in a Mexican coastal lagoon with an ephemeral inlet. Mar Ecol Progr Ser 35:83–90
Flores-Verdugo F, Gonzalez-Farias F, Zamorano DS, Ramirez Garcia P (1992). Mangrove ecosystems of the Pacific coast of Mexico: distribution, structure, litterfall, and detritus dynamics. In: Seeliger U (ed) Coastal plant communities of Latin America. Academic Press, San Diego, pp 269–288
Fondo EN, Martens EE (1998) Effects of mangrove deforestation on macrofaunal densities, Gazi Bay, Kenya. Mangroves and Salt Marshes 2:75–83
Fouda MM, Al-Muharrami M (1996) Significance of mangroves in the arid environment of the Sultanate of Oman. Sultan Qaboos Univ J Sci Res Agr Sci 1:41–49
Freeman III AM (1991) Valuing environmental resources under alternative management regimes. Ecol Economics 3:247–256

Frusher SD, Giddins RL, Smith TJ (1994) Distribution and abundance of grapsid crabs (Grapsidae) in a mangrove estuary: effects of sediment characteristics, salinity tolerances, and osmoregulatory ability. Estuaries 17:647–654

Fuchs HP (1970) Ecological and palynological notes on *Pelliciera rhizophora*. Acta Bot Neerl 19:884–894

Galloway RW (1982) Distribution and physiographic patterns of Australian mangroves. In: Clough BF (ed) Mangrove ecosystems in Australia: structure, function and management. Australian National University Press. Canberra, pp 31–54

Gaudian G, Koyo A, Wells S (1995) Marine region 12. East Africa. A global representative of system of marine protected areas. Vol. III. Central Indian Ocean, Arabian Seas, East Africa and East Asian Seas. Great Barrier Reef Marine Park Authority, the World Bank and IUCN, pp 71–101

Gentry AH (1981) Phytogeographical patterns as evidence for a Chocó refugee. In: Prance GT (ed) The biological model of diversification in the tropics. Columbia University Press, New York, pp 112–135

Geohidra (1996) Estudio de impacto ambiental para el Bloque guarapiche, Delta del Orinoco. British Petroleum. Caracas

Geohidra (1997) Estudio de impacto ambiental, bloque Este, Punta Pescador. AMOCO Venezuela Petroleum Company. Caracas

Germeraad JH, Hopping CA, Muller J (1968) Palynology of Tertiary sediments from tropical areas. Rev Palaeobot Palynol 6:189–348

Getter CD, Ballou TG (1985). Field experiments on the effects of oil and dispersant on mangroves. Proc 1985 Oil Spill Conf, Amer Petrol Inst, Washington DC, pp 533–538

Getter CD, Michel JM, Brown MS (1981) The effects of oil spills on mangrove forests: a comparison of five oil spill sites in the Gulf of Mexico and Caribbean Sea. Proc 1981 Oil Spill Conf Amer Petrol Inst, Washington DC, pp 535–540

Getter CD, Ballou TG, Koons CB (1985) Effects of dispersed oil in mangroves: synthesis of a seven year study. Mar Pollut Bull 16:318–324

Giddins RL, Lucas JS, Neilson MJ, Richards GN (1986) Feeding ecology of the mangrove crab *Neosarmatium smithi* (Crustacea: Decapoda: Sesarmidae). Mar Ecol Progr Ser 33:147–155

Gilbert AJ, Janssen R (1998) Use of environmental functions to communicate the values of a mangrove ecosystem under different management regimes. Ecol Economics 25:323–346

Gilfillan ES, Page DS, Gerber RP, Hansen S, Cooley J, Hotham J (1981) Fate of the "Zoe Colocotroni" oil spill and its effects on infaunal communities associated with mangroves. Proc 1981 Oil Spill Conf Amer Petrol Inst, Washington DC, pp 353–360

Gill AM, Tomlison PB (1977) Studies on the growth of red mangrove (*Rhizophora mangle* L.) 4. The adult root system. Biotropica 9:145–155

Gines H, Angell CL, Méndez-Arocha M, Rodríguez G, Febres GA, Gómez R, Rubio J, Pastor G, Otaola JR (1972) Carta pesquera de Venezuela. I. Areas del Nororiente y Guayana. Fundación La Salle de Ciencias Naturales. Monografia nº 16. Caracas

Gningue I (1991) Variations spatio-temporelles des facteurs physico-chimiques dans l'estuaire du Saloum. Rapport Final EPEEC, UNESCO, Division des Sciences de la mer, ROSTA, Dakar, pp 52–65

Golley FB, Odum HT, Wilson R (1962) A synoptic study of the structure and metabolism of a red mangrove forest in southern Puerto Rico in May. Ecology 43:9–18

Golley FB, McGinnis JT, Clements RG, Child GI, Duever MI (1975) Mineral cycling in a tropical moist forest ecosystem. University Georgia Press, Athens, Georgia

Gong WK, Ong JE (1995) The use of demographic studies in mangrove silviculture. Hydrobiologia 295:255–261

Goulter PFE, Allaway WG (1979) Litterfall and decomposition in a mangrove stand, *Avicennia marina* (Forsk.) Vierh., in Middle Harbour, Sydney. Austr J Mar Freshwat Res 30:541–546

Graham A (1977) New records of *Pelliciera* (Theaceae/Pelliceriaceae) in the Tertiary of the Caribbean. Biotropica 9:48–52

Graham A (1989) Paleofloristic and paleoclimatic changes in the tertiary of northern Latin America. Rev Palaeobot Palynol 60:283–293

Graham A (1992) Utilization of the isthmian land bridge during the Cenozoic: Paleobotanical evidence for timing, and the selective influence of altitude and climate. Rev Palaeobot Palynol 72:119:128

Graham A (1995) Diversification of Gulf/Caribbean mangrove communities through Cenozoic time. Biotropica 27:20–27

Graham J (1929) Mangrove of East Africa. J East Africa and Uganda Nat Hist 29:157

Green EP, Mumby PJ, Edwards AJ, Clark CD, Ellis AC (1998) The assessment of mangrove areas using high resolution multispectral airborne imagery. J Coast Res 14:433–443

Gremone C, Cervigón F, Gorzula S, Medina G, Novoa D (1980) Fauna de Venezuela. Vertebrados. Editorial Biósfera, Caracas

Guba EF (1961) Monograph of Monocaetia and Pestalotia. Harvard University Press, Cambridge

Gundlach ER (1987) Oil-holding capacities and removal coefficients for different shoreline types to computer simulate spills in coastal waters. Proc 1987 Oil Spill Conf, Amer Petrol Inst, Washington, pp 451–457
Hamilton LS, Snedaker SC (1984) Handbook for mangrove area management. Environment and Policy Institute, East West Centre, Honolulu
Hammen T van der, Wijmstra T (1994) Palynological study on the Tertiary and upper Cretaceous of British Guiana. Leid Geol Med 30:183–241
Hanna S (1999) Strengthening governance of ocean fishery resources. Ecol Economics 31:275–286
Haq BV (1981) Paleogene paleoceanography: Early Cenozoic oceans revisited. Oceanologica Acta No sp 71–82
Haq BV (1984) Paleoceanography: a synoptic overview of 200 million years of ocean history. In: Haq BV, Milliman JD (eds) Marine geography and oceanography of the Arabian Sea and coastal Pakistan. Van Nostrand Reinhold Co., New York, pp 201–231
Hardin G (1968) The tragedy of the commons. Science 162:1243–1248
Hasselman K (1979) Liberia:geographical mosaic of the land and people. MICAT, Monrovia
Hatton J, Massinga A (1994) The natural resources of Mecufi District. Mecufi Coastal Zone Management project. Ministry of Co-operation of Environmental Actions, Maputo
Hendrichs H (1975) The status of the tiger *Panthera tigris* (Linne, 1758) in the Sundarbans mangrove forest (Bay of Bengal). Saugetierk Mitteil 23:161–199
Herz R (1991) Manguezais do Brasil. Instituto Ocenografico da Universidade de São Paulo, São Paulo
Hickey F (1994) Traditional marine tenure in Vanuatu. In: South R, Goulet G, Tuqiri S, Church M (eds) Proc intern workshop traditional marine tenure and sustainable management of marine resources in Asia, Suva, p. 157–158
Hogarth PJ (1999) The biology of Mangroves. Oxford University Press, Oxford
Hong PN, San HT (1993) Mangroves of Vietnam, IUCN, Bangkok
Hoorn C (1993) Marine incursions and the influence of Andean tectonics on the Miocene depositional history of northwestern Amazonia: results of a palynostratigraphic study. Palaeogeogr Palaeoclimatol Palaeoecol 105:267–309
Horna R, Medina F, Macías R (1980) Breve estudio sobre el ecosistema del manglar en la costa ecuatoriana. ESPOL, Guayaquil
Hou D (1960) A review of the genus *Rhizophora*, with special reference to the Pacific species. Blumea 10:625–634
Huber O, Alarcón C (1988) Mapa de la vegetación de Venezuela. Escala 1:2000000. Ministerio de los Recursos Naturales Renovables/The Nature Conservancy. Oscar Toddman Ed., Caracas
Hudson BJ (1983) Wetland reclamation in Jamaica. Caribb Geogr 1:75–88
Hueck K (1972) As florestas da America do Sul. Ed Polígono SA, São Paulo
Hussain Z, Ahmed I (1994) Management of forest resources. In: Hussain Z, Achorya G (eds) Mangroves of the Sundarbans. IUCN, Bangkok. p. 78–84
Hutchings P, Saenger P (1987) Ecology of mangroves. University of Queensland Press, Queensland
Hyde KD, Lee SY (1995) Ecology of mangrove fungi and their role in nutrient cycling: what gaps occur in our knowledge. Hydrobiologia 295:107–118
Hyviding E (1996) Guardians of Marovo Lagoon; practice, place, and politics. Maritime Melanesia, Pacific Islands Monograph Series vol. 14, University of Hawaii Press, Honolulu
Inoue Y, Hadiyati O, Affendi HMA, Sudarma KR, Budiana IN (1999) Sustainable management models for mangrove forests. Indografika Utama, Bali
Islam MR, Khan MAS, Siddiqi NA, Saenger P (1990) Optimal planting season for Keroa (*Sonneratia apetala*). Bangladesh J Forest Sci 19:1–9
Islam MR, Khan MAS, Siddiqi NA, Saenger P (1991) Effect of thinning on Keroa (*Sonneratia apetala*). Survival and growth. Bangladesh J Forest Sci 20:8–24
Islam SS, Wazihullah, AKM, Islam MR, Rahman F, Das S (1988) Infestation of stem borer in keora plantations of Bangladesh. Bangladesh J Forest Sci 17:1–8
ITTO/ISME (1993) Conservation and sustainable utilization of mangrove forests in Latin America and Africa regions. II Africa. Inter Tropical Timber Org/Inter soc Mangrove Ecosystems, Okinawa
IUCN/ UNEP/WWF (1991) Caring for the Earth. IUCN, Gland
Jabbar MA, Rahman A, Kalam A (1992) A study on coastal morphology and coastal afforestation in Bangladesh using remote sensing techniques. Proc Workshop Coastal Zone Management in Bangladesh. UNESCO/BNCU/MAB, Dhaka
Jackson JB, Cubit JD, Keller BD, Burns K, Caffey HM, Caldwell RL, Garrity SD, Getter CD, Gonzalez C, Guzman HM, Kaufmann KW, Knap AH, Levings SC, Marshall MJ, Steger R, Thompson RC, Weil E (1989) Ecological effects of a major oil spill on Panamanian coastal marine communities. Science 243:37–44
Jacobi CM, Schaeffer-Novelli Y (1990) Oil spills in mangroves: a conceptual model based on long-term field observations. Ecol Modelling 32:53–59

Jagtap TG, Untawale AG (1999) Atoll mangroves and associated flora from the Republic of Maldives, Indian Ocean. ISME Mangrove Ecosystems Occ Pap 5:17–25

Janssen R, Padilla JE (1997) Mangroves or fishponds? Valuation and evaluation of alternative uses of a mangrove forest in the Philippines. Institute for Environmental Studies, Amsterdam

Jaramillo C (1998) Evolución del ecosistema manglar durante el Cuaternario en la Ensenada de Utría y el Golfo de Tribugá. Fundación Natura, Santa Fe de Bogotá DC, Inf Técnico 1:1–16

Jernelov A, Linden O, Rosenblum J (1976) The "St Peter" oil spill: an ecological and socio-economic study of effects, Colombia-Ecuador. Inst Vatten-Och Luftuards, Publ. N^{o} B334, Stockholm

Jiménez JA (1984) A hypothesis to explain the reduced distribution of the mangrove *Pelliciera rhizophorae* Tr and Pl. Biotropica 16:304–308

Jiménez JA (1992) Mangrove forests of the Pacific coast of Central America. In: Seeliger U (ed), Coastal plant communities of Latin America. Academic Press, San Diego, pp 259–267

Johannes R (1994) Design of tropical near-shore fisheries extension beyond the 1990s. In: South R, Goulet G, Tuqiri S, Church M (eds) Proc inter workshop traditional marine tenure and sustainable management of marine resources in Asia, Suva. p. 162–174

John DM, Lawson GW (1990) Review of mangrove and coastal ecosystem in West Africa and their possible relationships. Estuar Coast Shelf Sci 31:505–518

Johnston IM (1949) The botany of San Jose Island. Sargentia 8:1–298

Johnstone IM (1981) Consumption of leaves by herbivores in mixed mangrove stands. Biotropica 13:252–259

Jones DA (1984) Crabs in the mangal ecosystem. In: Por FD, Dor I (eds), Hydrobiology of the Mangal. Dr. W. Junk Publ., Den Hague, p 89–109

Julius A (1997) A study of the microbial population and the decomposition of leaf litter in the mangrove ecosystem along the Dar es salaam coast. Dissertation, University of Dar es Salaam

Kairo JG (1997) Mangrove forests of Kenya, status and management. SAREC-SIDA Regional workshop on Mangrove ecology, Physiology and Management. Zanzibar

Kairo JG, Gwada PO (1997) Structure and regeneration of mangrove forests at Mida creek, Kenya. SAREC-SIDA Regional Workshop on Mangrove Ecology, Physiology and Management. Zanzibar

Karanja FK (1997) Environmental impacts of saltwork and aquaculture development on mangrove ecosystem at Ngomeni, Kenya. SAREC-SIDA Regional workshop on Mangrove ecology, Physiology and Management. Zanzibar

Karim A (1994) Vegetation. In: Hussain Z, Achorya G (eds) Mangroves of the Sundarbans. IUCN, Bangkok. p. 43–74

Kawaguchi S, Meepol W, Baba S (2000) Nutrient dynamics of mudlobster mounds in *Rhizophora apiculata* dominated mangrove forest. Mangroves & Salt Marshes (in press)

Khan MS, Shafi M (1992) Use of korea trees with branches, tops, bark and keora thinning for hardboard manufacture. Bangladesh J Forest Sci 21:32–35

Kimball BA (1983) Carbon dioxide and agricultural yield: an assemblage of 770 prior observations. United States Department of Agriculture, Agricultural Research Service, Water Conservation Laboratory, Report 14. Phoenix

Kitheka JU (1999) Coastal tidally-driven circulation and the role of water exchange in the linkage between tropical coastal ecosystems. Estuar Coast Shelf Sci 45:177–187

Kjerfve B (1981) Tides of the Caribbean Sea. J Geophys Res 86:4243–4247

Kjerfve B (1990) Manual for investigation of hydrological processes in mangrove ecosystems. UNESCO/UNDP. New Delhi

Kjerfve B, Lacerda LD (1993) Management and status of the mangroves of Brazil. In: Lacerda LD (ed) Conservation and sustainable utilization of mangrove forests in Latin America and Africa regions. Inter Soc Mangrove Ecosystems, Okinawa, pp 245–272

Kjerfve B, Lacerda LD, Rezende CE, Ovalle ARC (1999) Hydrological and hydrogeochemical variations in mangrove ecosystems. In: Yáñez-Arancibia A, Lara-Dominquez AL (eds) Mangrove ecosystems in tropical America: structure, function and management. INECOL (Mexico), IUCN/ORMA (Costa Rica), NOAA/NMFS (Beaufort, NC, USA), pp 71–81

Kjerfve B, Perillo GME, Gardner LR, Rine JM, Dias GTM, Mochel FR (2001) Morphodynamics of muddy environments along the Atlantic coasts of North and South America. In: Wang Y, Healy T (eds) Muddy coasts and relative sea level change. CRC Press, Boca Raton, in press

Klekowski EJ, Corredor JE Morellr J, Del Castillo J (1994a) Petroleumpollution and mutation in mangroves. Mar Pollut Bull 28:166–169

Klekowski EJ, Corredor JE, Lowenfeld R, Klekowski EH, Morell J (1994b) Using mangroves to screen for mutagens in tropical marine environments. Mar Pollut Bull 28:346–350

Kneip LM, Palestrini L (1984) Restingas do Estado do Rio de Janeiro (Niteroi a Cabo Frio): 8 mil anos de ocupacao humana. In: Lacerda LD, Araujo DSD, Cerqueira R, Turcq B (eds) Restingas Brasileiras: Origem, Estrutura, Processos. EDUFF, Niteroi, pp 147–153

Kneip LM, Monteiro AMF, Palestrini L, Crancio F (1984) Informacao arqueo-geologicas sobre sambaquis da plan'icie de mare de Guaratiba, Rio de Janeiro. In: Lacerda LD, Araujo DSD, Cerqueira R, Turcq B (eds) Restingas Brasileiras: Origem, Estrutura, Processos. EDUFF, Niteroi, pp 139–146
Kobuski CE (1951) Studies in the *Theaceae*, XXII. The genus *Pelliciera*. J Arnold Arb. 32:256–262
Kogo M, Tsuruda K (1996) Species selection for mangrove planting: a case study of Ras-al-Khafji, Saudi Arabia. In: Field CD (ed) Restoration of mangrove ecosystems. Inter Soc mangrove Ecosystems, Okinawa, p. 197–206
Kohlmeyer J (1968) Marine fungi from the tropics. Mycologia 60:252–270
Kohlmeyer J (1969) Ecological notes on fungi in mangrove forests. Trans Br Mycol Soc 53:237–250
Kohlmeyer J, Kohlmeyer E (1979) Marine mycology. The higher fungi. Academic Press, New York
Komiyama A, Ogino K, Aksornkoae S, Sabhasri S (1987) Root biomass of a mangrove forest in Southern Thailand. 1. Estimation of the trench method and the zonal structure of root biomass. J Trop Ecol 3:87–1
Kwok PW, Lee SY (1995) The growth performances of two mangrove crabs, *Chiromanthes bidens* and *Parasesarma plicata* under different leaf litter diets. Hydrobiologia 295:141–148
Lacerda LD (1981) Mangrove wood pulp, an alternative food source for the tree-crab *Aratus pisonii*. Biotropica 13:317
Lacerda LD (1984) Manguezais: florestas de beira-mar. Ciência Hoje 13:62–70
Lacerda LD (1993) Conservation and sustainable utilization of mangrove forests in Latin America and Africa regions. Inter Soc Mangrove Ecosystems, Okinawa
Lacerda LD (1998) Biogeochemistry of trace metals and diffuse pollution in mangrove ecosystem. International Society for Mangrove Ecosystems, Okinawa
Lacerda LD, Abrão JJ (1984) Heavy metals concentrations in mangrove and salt marsh sediments. Rev Bras Bot 7:49–52
Lacerda LD, Hay JD (1982) Evolution of a new community type after the destruction of a mangrove ecosystem. Biotropica 14:238–239
Lacerda LD, Rezende CE (1993) Environmental status of mangrove forests in the State of Rio de Janeiro, Brazil. In: Lacerda LD, Field CD (eds) Conservation and sustainable utilization of mangrove forests in Latin America and Africa regions. ISME, Okinawa, pp 15–18
Lacerda LD, Schaeffer-Novelli (1992) Mangroves of Latin America: the need for conservation and sustainable utilization. ISME Mangroves Newslett 5:4–6
Lacerda LD, Rezende CE, Martinelli LA, Ovalle ARC, Mozetto AA, Nogueira FB, Victoria RL, Aragon GT, Cunha CT, Silva CAR (1986a) Composição isotópica de carbono em componentes de um manguezal na Baía de Sepetiba, RJ. Ciênc Cult 38:1714–1717
Lacerda LD, Rezende CE, José DMV, Francisco MCF, Wasserman JC, Martins JC (1986b) Leaf chemical characteristics affecting herbivory in a New World mangrove forest. Biotropica 18:350–355
Lacerda LD, José DMV, Francisco MCF (1988a) Nutritional status and chemical composition of mangrove seedlings during development. Rev Brasil Biol 48:401–405
Lacerda LD, Martinelli LA, Rezende CA, Mozetto AA, Ovalle ARC, Victoria RL, Silva CAR, Nogueira FB (1988b) The fate of trace metals in suspended matter in a mangrove creek during a tidal cycle. Sci Tot Environm 75:249–259
Lacerda LD, Silva CAR, Rezende CE, Martinelli LA (1991) Food sources for the mangrove tree crab *Aratus pisonii*: a carbon isotopic study. Rev Brasil Biol 51:685–687
Lacerda LD, Conde JE, Alarcón C, Alvarez-León R, Bacon PR, D'Croz L, Kjerfve B, Polanía J, Vannucci M (1993) Mangrove ecosystems of Latin America and the Caribbean: a summary. In: Lacerda LD (ed) Conservation and sustainable utilization of mangrove forests in Latin America and Africa regions. I Latin America. Mangrove Ecosystems Technical Reports. Inter Soc Mangrove Ecosys. Okinawa, Japan, p. 1–42
Lacerda LD, Tanizaki KF, Carvalho CEV, Ovalle ARC, Rezende CE (1993) The biogeochemistry and trace metal distribution of mangrove rhizosphere. Biotropica 25:252–257
Lacerda LD, Ittekkot V, Patchineelam SR (1995) Biogeochemistry of mangrove soil organic matter. A comparison between *Rhizophora* and *Aviccenia* soils in southeastern Brazil. Estar Coast Shelf Sci 40:713–720
Laegdsgaard P, Johnson CR (1995) Mangrove habitats as nurseries: unique assemblages of juvenile fish in subtropical mangroves in eastern Australia. Mar Ecol Progr Ser 126:67–81
Lal PN (1983) Institutional aspects of mangrove management in Fiji. Proc Interdepartmental Workshop on Mangrove Resource Management, Suva. p. 125–167
Lal PN (1989) Utilisation and management of coastal wetland resources in Kosrae. University of Hawaii Sea Grant Program Working Paper, Honolulu
Lal PN (1990a) Conservation or conversion of mangroves in Fiji – an ecological economic analysis. Occ Pap 11, Environment and Policy Institute, East West Centre, Honolulu
Lal PN (1990b) Mangrove management issues; strategies adopted in the Pacific Islands. National Centre of Development Studies, ANU Working Paper 91/3, Canberra

Lal PN (1991) Macroenvironmental standards approach to resource allocation and management: the case of Kosrae's coastal wetlands. National Centre of Development Studies, Australian National University Working Paper 91/1, Canberra

Lal PN, Minerbi L (1985) Environmental management for local and regional development in Fiji, Paper presented at the International Workshop on Environmental Management for Local and Regional Development, Nagoya

Lamberti A (1969) Contribuição ao conhecimento da ecologia das plantas do manguezal de Itanhaem. Bol Fac Filos Ciênc Univ São Paulo Bot 23:1–217

Lamparelli CC, Rodrigues FO, Moura DO (1997) A long-term assessment of an oil spill in a mangrove forest in São Paulo, Brazil. In: Kjerfve B, Lacerda LD, Diop ES (eds) Mangrove ecosystem studies in latin america and Africa. UNESCO/International Society for Mangrove Ecosystems. Paris. pp 191–203

Laroche J, Baran E, Rasoanandrasana NB (1997) Temporal patterns in a fish assemblage of a semiarid mangrove zone in Madagascar. J Fish Biol 51:3–20

Larsen S (1994) Mangrove harvesting and transportation in Sundarbans. BDG/84/056 Field document no. 2. Integrated Resource Development of the Sundarbans Reserved Forest, Bangladesh. Khulna

Latif MA, Del Castillo RA (1990) Growth and yield of keora (*Sonneratia apetala*) in the coastal plantations of Bangladesh. Bangladesh J Forest Sci 19:11–18

Latif MA, Del Castillo RA (1991) Rotation ages of keora (*Sonneratia apetala*) in the coastal plantations of Bangladesh. Bangladesh J Forest Sci 20:1–7

Latif MA, Rahman MF, Das S, Siddiqi NA (1992) Diameter increments of six mangrove tree species in the Sundarbans forest of Bangladesh. Bangladesh J Forest Sci 21:7–12

Lawson GW (1986) Coastal vegetation, plant ecology in West Africa: system and process. John Wiley & Sons, Chischester

Lebigre JM (1999) Natural spatial dynamics of mangals through their margins: diagnostic elements. Hydrobiologia 413:103–113

Lee SY (1991) Herbivory as an ecological process in a *Kandalia candel* (Rhizophoraceae) mangal in Hong Kong. J Trop Ecol 7:337–348

Lee SY (1995) Mangrove outwelling: a review. Hydrobiologia 295:203–212

Lee SY (1998) Ecological role of grapsid crabs in mangrove ecosystems: a review. Mar Fresh Res 49: 335–343

Lefebvre G, Poulin B (1997) Bird communities in Panamanian black mangroves: potential effects of physical and biotic factors. J Trop Ecol 13:97–113

Lemaitre-Velez R, Alvarez-Leon R (1992) Crustáceos decápodos del Pacífico colombiano: lista de especies y consideraciones zoogeográficas. An Inst Invest Mar Punta de Betín 21:33–76

Lescure J-P (1977) La mangrove guyanaise: architeture des jeunes stades et vie anciene. Cah ORSTOM Ser Biol 12:361–376

Linden O, Lundin CG (1996) Integrated coastal zone management in Tanzania. SIDA, Stockholm

Little MC, Reay PJ Grove SJ (1988) The fish community of an East African mangrove creek. J Fish Biol 32:729–747

Londoño MS, Martín Y (1988) Estudio preliminal sobre la microfauna asociada a bromelíaceas en Vigía (Costa Pacífica Colombiana). In: Steer-Ruiz R (ed) Mem VI Sem Natl Cienc Tecnol Mar, CCO/UBJTL, Bogotá DF, pp 102–105

Louis M, Bouchon C, Bouchon-Navaro Y (1995) Spatial and temporal variations of mangrove fish assemblages in Martinique (French West Indies). Hydrobiologia 295:275–284

Lovejoy NR, Bermingham E, Martin AP (1998) Marine incursions into South America. Nature 396: 421–422

Lugo AE (1980) Mangroves ecosystems: successional or steady state? Biotropica (Supplement) 12:65–72

Lugo AE (1981) The inland mangroves of Inagua. J Nat History 15:845–852

Lugo AE (1998) Mangrove forests: a tough system to invade but an easy one to rehabilitate. Mar Pollut Bull 37:427–430

Lugo AE, Snedaker SC (1974) The ecology of mangroves. Ann Rev Ecol Syst 5:39–63

Lugo AE, Evink G, Brinson MM, Snedaker SC (1975) Diurnal rates of photosynthesis, respiration and transpiration in mangrove forests of South Florida. In: Golley FB, Medina E (eds) Tropical ecological systems. Springer-Verlag, New York, pp 335–350

Lugo AE, Brown S, Brinson MM (1990) Concepts in wetland ecology. In: Lugo AE, Brinson MM, Brown S (eds) Forest wetlands. Ecosystems of the World, N° 15. Elsevier Sci Publ Comp, Amsterdam, pp 53–85

Machiwa JF (1999) Lateral fluxes of organic carbon in a mangrove forest partly contaminated with sewage wastes. Mangroves and Salt Marshes 3:95–104

Macnae W (1968) A general account on the fauna and flora of mangrove swamps and forests in the Indo-West Pacific region. Adv Mar Biol 6:73–270

Macnae W (1974) Mangrove forest and fisheries. FAO/UNDP Indian Ocean Fishery Programme. Indian Ocean Fishery Commission. Publication IOFCDev/74/34, Rome

MAG (1991) Inventario de los manglares del Ecuador continental. Ministerio de Agricultura y Ganaderia. Quito

Malley DF (1978) Degradation of mangrove leaf litter by the tropical sesarmid crab *Chiromanthes onychophorum*. Mar Biol 49:377–386

MARNR (1986) Conservación y manejo de los manglares costeros en Venezuela y Trinidad and Tobago (Sintesis). Serie Informes Técnicos DG 11 A/ti/259, Ministerio de Agricultura y Recursos Naturales Renovables, Caracas

MARNR (1991) Conservación y manejo de los manglares de Venezuela y Trinidad and Tobago. PF. 11-05-81-01 (2038). Sector Delta del Orinoco, Territorio Delta Amacuro (PT). Serie Informes Técnicos, DGS11, ANTI. 256, Caracas

Marten GG, Polovina JJ (1982) A comparative study of fish yields from various tropical ecosystems. In: Murphy GI, Pauly D (eds) Theory and management of tropical fisheries. ICLARM/ CSIRO, p 255–285

Martosubroto P, Naamin N (1977) Relationship between tidal forests (mangrove) and commercial shrimp production in Indonesia. Mar Res Indonesia 18:81–86

Mathews M (1992) Reproductive and ecological studies on *Avicennia* species. Ph.D. thesis, National University of Singapore, Department of Botany

Matta SB (1997) Species and structural composition of natural mangrove forests and the implications on management practices: a case study of the Rufiji Delta, Tanzania. Dissertation, Sokoine University of Agriculture

Matthes H, Kapetsky JM (1988) Worldwide compendium of mangrove-Associated aquatic species of economic importance. FAO Fishery Circular 814:1–238

Matthijs S, Tack J, Speybroeck D Van, Koedam N (1999) Mangrove species zonation and soil redox state, sulphide concentration and salinity in Gazi Bay (Kenya), a preliminary study. Mangroves and Salt Marshes 3:243–249

May JD (1999) Spatial variation in litter production by the mangrove *Avicennia marina* var. *australasica* in Rangaunu Harbour, Northland, New Zealand. New Zealand J Mar Freshwat Res 33:163–172

Mayo EM (1965) Algunas características ecológicas de los bosques inundables de Darién, Panamá, con vistas a su posible utilización. Turrialba 15:336–347

Mazda Y, Nanazawa K, Wolanski E (1995) Tidal asymmetry in mangrove creeks. Hydrobiologia 295:51–58

Mazzotti FJ, Dunson WA (1984) Adaptations of *Crocodylus acutus* and *Alligator* for life in saline water. Compar Biochem Physiol 79A: 641–646

McConchie D (1990a) Draft report on land stability problems affecting coastal plantations. UNEP/ FAO Project BGD/85/085, Dhaka

McConchie D (1990b) Draft report on progress toward minimizing damage caused to coastal plantations by changes in land stability. UNEP/FAO Project BGD/85/085, Dhaka

McCoy ED, Heck KL (1976) Biogeography of corals, seagrasses and mangroves: an alternative to the center of origin concept. Syst Zool 25:201–210

McCusker A (1977) Seedling establishment in mangrove species. Geol Ecol Trop 1:23–33

McGuinness KA (1997) Seed predation in a tropical mangrove forest: a test of the dominance-predation model in northern Australia. J Trop Ecol 13:293–302

McIvor CC, Smith TJ (1995) Differences in the crab fauna of mangrove areas at a Southwest Florida and a Northeast Australia location: implications for leaf litter processing. Estuaries 18:591–597

McKee KL (1993) Soil physicochemical patterns and mangrove species distribution – reciprocal effects? J Ecol 81:477–487

McMillan C (1975) Adaptive differentiation to chilling in mangrove populations. In: Walsh GE, Snedaker SC, Teas HJ (eds) Proc Inter Symp Biol Managt Mangroves. University of Florida, Gainville pp 275–281

Medina E, Cuevas E, Popp M, Lugo AE (1990) Soil salinity, sun exposure, and growth of *Achrosticum aureum*. Bot Gaz 151:41–49

MENR (1994) Kenya forestry master plan 1995–2020 executive summary. Ministry of Environment and Natural Resources, Nairobi

Mepham RH (1983) Mangrove floras of the southern continents. Part 1. The geographical origin of the Indo Pacific mangrove genera and the development and present status of the Australian mangroves. S Afr J Bot 2:1–8

Mepham RH, Mepham JS (1985) The flora of tidal forests – a rationalization of the use of the term “mangroves”. S Afr J Bot 51:77–99

Mgaya YD, Muruke MHS, Chande AI, Semesi AK (1999) Evaluation of crustacean resources in Bagamoyo District. In: Howell KM, Semesi AK (eds) Coastal resources of Bagamoyo District Tanzania. Faculty of Science, University of Dar es Salaam, pp 55–64

Micheli F (1993a) Effect of mangrove litter species and availability on survival, moulting, and reproduction of the mangrove crab *Sesarma messa*. J Exper Mar Biol Ecol 171:149–163

Micheli F (1993b) Feeding ecology of mangrove crabs in North Eastern Australia: mangrove litter consumption by *Sesarma messa* and *Sesarma smithii*. J Exper Mar Biol Ecol 171:165–186
Micheli F, Gherardi F, Vannini M (1991) Feeding and burrowing ecology of two east African mangrove crabs. Mar Biol 111:247–254
Middelburg JJ, Nieuwenhuize J, Slim FJ, Ohowa B (1996) Sediment biogeochemistry in an East African mangrove forest (Gazi Bay, Kenya). Biogeochemistry 34:133–155
Mochel FR (1993) Mangroves of Maranhão State, North Brazil. In: Lacerda LD, Field CD (eds) Conservation and sustainable utilization of mangrove forests in Latin America and Africa. ISME, Okinawa
Mochel FR (1999) Caracterização espectral e mapeamento dos manguezais por sensoramento remoto na Baía de Turuaçu, Maranhão, Brasil. Dissertation, Universidade Federal Fluminense, Niteroi
Morton RM (1990) Community structure, density and standing crop of fishes in a subtropical Australian mangrove area. Mar Biol 105:385–394
Muller J (1961) A palynological contribution of the history of mangrove vegetation. Pac Sci Congr Abstr 1961:146–147
Muller J (1980) Palynological evidence for Paleogene climatic changes. Mem Mus Nat Hist Nat Paris, ns 27:211–218
Muller J, Caratini C (1977) Pollen of *Rhizophora* (Rhizophoraceae) as a guide fossil. Pollen et Spores 19:361–369
Munro-Faure PW (1991) The marine resource and its management: attitudes and approaches to the leasing and licensing of tides and submerged lands. University of Reading, The College of Estate Management, Reading
Myers N (1997) Consumption in reaction to population, environment and development. Environmentalist 17:33–34
Nadeau RJ, Bergqvist ET (1977) Effects of the 18 March 1973 oil spill near Cabo Rojo, Puerto Rico on tropical marine communities. In: Proc 1977 Oil Spill Conf, Amer Petrol Inst, Washington DC, pp 535–538
Nadkarni KM (1976) India Materia Medica. Popular Prakashan, Bombay
Naidoo G, Rogalla H, Willert DJ von (1998) Field measurements of gas exchange in *Avicennia marina* and *Bruguiera gymnorrhiza*. Mangroves and Salt Marshes 2:99–107
Navid D (1993) Introduction. In: Davis TJ (ed) Towards the wise use of wetlands. Ramsar Convention Bureau, Gland, p. 1
Naylor R, Drew M (1998) Valuing mangrove resources in Kosrae, Micronesia. Environ Developt Economics 3:471–490
Neilson MJ, Richards GN (1989) Chemical composition of degrading mangrove leaf litter and changes produced after consumption by the mangrove crab *Neosarmatium smithi* (Crustacea: Decapoda: Sesarmidae). J Chem Ecol 15:1 267–1 283
Ngunyali HR (1997) Mangroves must live in Tanzania. SAREC-SIDA Regional Workshop on Mangrove ecology, Physiology and Management. Zanzibar
Nickerson DJ (1999) Trade-offs of mangrove area development in the Philippines. Ecol Economics 28:279–298
Nickerson NH, Thibodeau FR (1985) Association between pore water sulfide concentrations and the distribution of mangroves. Biogeochemistry 1:183–192
Nilsson A (1992) Greenhouse Earth. Wiley, Chichester
Nishat A, Haque M (1984) Sedimentation in coastal areas of Bangladesh. UNDP/UNESCO, Dhaka
Noakes DSP (1952) A working plan for the Matang mangrove forest. Perak State Forestry Department, Kuala Lumpur
North DC (1990) Institutions, institutional change and economic performance. Cambridge University Press, Cambridge
Norton B, Constanza R, Bishop R (1998) The evolution of preferences – why 'sovereign' preferences may not lead to sustainable policies and what to do about it. Ecol Economics 24:193–211
Nuruzzaman M (1962) Creation of a protected belt of forest along the coast of Bay of Bengal by raising keora and baen plantations and the technique of their natural regeneration. Proc Bangladesh 2nd Nat Conf Forestry, Dhaka. P. 159–169
Odum WE, Heald EJ (1972) Trophic analysis of an estuarine mangrove community. Bull Mar Sci 22: 671–738
Odum WE, Heald EJ (1975) Mangrove forests and aquatic productivity. In: Hasler AD (ed) Coupling of land and water systems. Springer-Verlag, New York, pp 129–136
Odum WE, McIvor CC, Smith III TJ (1982) The ecology of mangroves of South Florida: a community profile. US Fish and Wildlife Ser Off Biol Ser, FWS/OBS-81/24, Washington
Ohn U (1999) Significance of mangrove ecosystems for the coastal people in Ayeyarwaddy Delta, Myanmar. ISME Mangrove Ecosystems Proc 4:63–74

Olafsson E (1995) Meiobenthos in mangrove areas in eastern Africa with emphasis on assemblage structure of free-living marine nematodes. Hydrobiologia 312:47–57
Oliveira Filho EC (1984) Mangrove algae from Cananeia. In: Por FD, Dor I (eds) Hydrobiology of the Mangal, Dr. W. Junk Publ., Den Hague, pp 55–66
Ong JE (1995) The ecology of mangrove conservation and management. Hydrobiologia 295:343–351
Onuf CP, Teal JM, Valiela I (1977) Interactions of nutrients, plant growth and herbivory in a mangroves ecosystem. Ecology 58:514–526
Orihuela B, Díaz H, Conde JE (1991) Mass mortality in a mangrove roots fouling community in a hypersaline tropical lagoon. Biotropica 23:592–601
Osore MKW, Tackx MLM, Daro MH (1997) The effect of rainfall and tidal rhythm on the community structure and abundance of the zooplankton of Gazi Bay, Kenya. Hydrobiologia 356:117–126
Ovalle ARC, Rezende CE, Lacerda LD, Silva CAR (1990) Factors affecting the hydrochemistry of a mangrove tidal creek in Sepetiba Bay, Rio de Janeiro, Brazil. Estuar Coast Shelf Sci 31:639–650
Oxloj RM (1987) Los manglares de el Salvador. University of El Salvador, manuscr unpubl
Padrón CM, Llorente S, Menendez L (1993) Status Report on the mangroves of Cuba. In: Lacerda LD (ed) Conservation and sustainable utilization of mangrove forests in Latin America and Africa regions. Inter Soc Mangrove Ecosystems, Okinawa, pp 147–154
Pagavino M (1983) Condición Actual de las Pesquerías de las Lagunas Costeras de Tacarigua, Unare y Páritu, Venezuela. Dissertation. Universidad de Oriente, Cumaní
Page DS, Mayo DW, Cooley JF, Sorenson E, Gilfillan ES, Hanson SA (1979) Hydrocarbon distribution and weathering characteristics at a tropical oil spill site. Proc 1979 Oil Spill Conf Amer Petrol Inst, Washington, pp 700–712
Palis HG (1998) The Bangladesh Forest Research Institute: Mangrove research. Mandala Agricultural Development Corporation, Dhaka
Pannier F, Pannier RF (1985) La vegetacion de manglares de la cuenca del Lago Maracaibo. Memorias de la Sociedad Venezolana de Ciencias Naturales XL 143:327–346
Pannier F, Pannier RF (1989) Manglares de Venezuela. Cuadernos LAGOVEN-PDVSA, Caracas
Pantoja CN (1993) Mangroves of Pará State, North Brazil. In: Lacerda LD, Field CD (eds) Conservation and sustainable utilization of mangrove forests in Latin America and Africa. ISME, Okinawa, pp 13–14
Parkinson RW, DeLaune RD, White JR (1994) Holocene sea-level rise and the fate of mangrove forests within the wider Caribbean region. J Coast Res 10:1077–1086
Pasqualini V, Iltis J, Dessay N, Lointier M, Guelorguet O, Polidori L (1999) Mangrove mapping in North-Western Madagascar using SPOT-XS and SIR-C radar data. Hydrobiologia 413:103–113
Patterson CS, Mendelssohn IA, Swenson EM (1993) Growth and survival of *Avicennia germinans* seedlings in a mangal community in Louisiana, USA. J Coast Res 9:801–810
Pauly D, Ingles J (1986) The relationship between shrimp yield and intertidal (mangrove) areas: a reassessment. Proc IOC/FAO Workshop on Recruitment in Tropical Coastal Demersal Communities, UNESCO, Paris, p. 227–284
Pedrini A (1980) Algas Marinhas Bentonicas da Baia de Sepetiba e seus arredores (Rio de Janeiro). Dissertation, Universidade Federal Rio de Janeiro, Rio de Janeiro
Perdomo-Rojas L (1978) Manual de arqueologia Colombiana. Carlos Valencia Editores, Bogotá
Pereira DN, Vannucci M (1991) The hydromedusae and water masses of the Indian Ocean. Bolm Inst Oceanogr S Paulo 39:25–60
Peria LCS, Fernandes PPCP, Menezes GV, Grasso M, Tognella MMP (1990) Estudos estruturais comparativos entre bosques de mangue impactados (Canal de Bertioga) e não impactados (Ilha do Cardoso), Estado de São Paulo. In: An II Symp Ecossis Costa Sulsudeste Bras, pp 154–178
Perry DM (1988) Effects of associated fauna on growth and productivity in the red mangrove. Ecology 69:1064–1075
Philippines National Mangrove Management Committee (1987) Country Profile – Philippines. In: Mangroves of Asia and the Pacific. Status and Management Technical Report of the UNDP/UNESCO Research and training Pilot Program on Mangrove Ecosystems in Asia and the Pacific (RAS/79/002), Manila, p. 234–243
Phillips PC (1981) Diversity and community structure in a Central American mangrove embayment. Estuaries 29:227–236
Phillips RC (1992) The seagrass ecosystems and resources in Latin America. In: Seeliger U (ed), Coastal plant communities of Latin America. Academic Press, San Diego, pp: 259–267
Pinto-Nolla F (1996) El mangle nato, *Mora oleifera* (Triana) Ducke (Caesalpinaceae): Síntesis bioecológica. Proy. PD 171/91 Rev. 2 (F) Fase I Conservación y Manejo para el Uso Múltiple y el Desarrollo de los Manglares en Colombia. Santa Fe de Bogotá D. C. Inf. Técnico, 6:1–19
Ponte AC, Fonseca IAZ, Marques M, Freitas ML, Claro SMCA (1984) Produção e decomposição do material foliar em ecossistema de mangue. An IV Congr Soc Bot São Paulo, São Paulo, 1:103–107

Prahl H, Cantera JR, Contreras R (1990) Manglares y hombres del Pacífico Colombiano. Fondo FEN Colombia, Bogotá

Prance GT, Silva MF, Albuquerque BW, Araujo IJS, Correira LMM, Braga MMV, Macedo M, Conceição PN, Lisboa PLB, Braga PI, Lisboa LCL, Vilhena RCQ (1975) Revisão taxonômica das espécies amazônicas de Rhizophoraceae. Acta Amazônica 5:5–22

Pretty J (1995) Regenerating agriculture: politics and practice for sustainability and self-reliance. Earthscan, London

Primavera JH (1993) A critical review of shrimp pond culture in Philippines. Rev Fish Sci 1:151–201

Primavera JH (1997) Socio-economic impacts of shrimp culture. Aquaculture Res 28:815–827

Primbert M (1999a) Multifunctional character of agriculture and land. Paper presented at the FAO/ Netherlands Conference on "Multifunctional Character of Agriculture and Land, 12–17 September, Maastricht

Primbert M (1999b) Sustaining the multiple functions of agricultural biodiversity. IIED Gatekeeper Series vol 88, London

Qureshi MT (1990) Experimental plantation for rehabilitation of mangrove forests in Pakistan. UNDP/ UNESCO Regional Mangrove Project for Asia and the Pacific, Occ Pap N° 4, New Dehli

Qureshi MT (1996) Restoration of mangroves in Pakistan. In Field CD (ed) Restoration of mangrove ecosystems. Inter Soc Mangrove Ecosystems, Okinawa, p 126–142

Qureshi MT (1999) Significance of mangrove ecosystems for people in Pakistan. ISME Mangrove Ecosystems Proc 4:85–90

Rahman MA, Khisa SK, Bask AC (1983) Top dying of sundri in the Sundarbans. Bano Biggyan Patrika 16:67–71

Ramírez BN, Núñez BMG (1988) Producción de detritus de *Rhizophora mangle* en la Barra de Santiago de El Salvador. Dissertation, Universidad de El Salvador

Ramsar Bureau (1971) Convention on wetlands of intentional importance especially as waterfowl habitat (Ramsar, Iran). Gland

Randall A (1987) Resources economics: an economic approach to natural resource and environmental policy. Wiley, New York

Rao AN (1987) Mangrove ecosystems of Asia and the Pacific. Mangrove Ecosystems of Asia and the Pacific: Status and Management. Tech Rep, UNDP/UNESCO Research and Training Pilot Program on Mangrove Ecosystems in Asia and the Pacific. UNESCO-COMAR-UNDP, Philippines, pp 1–33

Rao RG, Woitchik AF, Goeyens L, Riet A van, Kazungu J, Dehairs F (1994) Carbon, nitrogen contents and stable isotope abundance in mangrove leaves from an east African Coastal Lagoon (Kenya). Aquat Bot 47:175–183

Rashid SMA, Khan A, Akonda AW (1994) Fauna. In: Hussain Z, Achorya G (eds) Mangroves of the Sundarbans. IUCN, Bangkok. p. 115–132

Rehm A (1976) The effects of the wood-boring isopod *Sphaeroma terebrans* on the mangrove communities of Florida. Environ Conserv 3:47–57

Rehm A, Humm HJ (1973) *Sphaeroma terebrans*: a threat to the mangroves of southwestern Florida. Science 182:173–174

Reichel-Dolmatoff G (1965) Colombia: ancient people and places. Glyn Daniel Ed. Thames and Hudson, London

Reichel-Dolmatoff G (1985) Monsú. Un sitio arqueológico. Publ Banco Popular, Bogotá

Reitz PR (1961) Vegetação da zona maritima de Santa Catarina. Sellowia 13:17–115

Reyes R, Campos NH (1992) Macroinvertebrados colonizadores de raíces de *Rhizophora mangle* en la bahía de Chengue, Caribe colombiano. An Inst Invest Mar Punta Betín 21:101–116

Rezende CE, Lacerda LD, Ovalle ARC, Silva CAR, Martinelli LA (1990) Nature of POC transport in a mangrove ecosystem: a carbon isotopic study. Estuar Coast Shelf Sci 30:641–645

Ribi G (1981) Does the wood boring isopod *Sphaeroma terebrans* benefit red mangroves crab (*Rhizophora mangle*)? Bull Mar Sci 31:925–928

Richards BN, Hassan MM (1988) A co ordinated forest soil research programme of Bangladesh. Working paper N° 4, Second Agriculture Research Project, FAO/BFRI, Dhaka

Rico-Gray V, Lot A (1983) Producción de hojarasca del manglar de la Laguna de la Machaca, Veracruz, Mexico. Biotica 8:295–301

Rincón RRG, Mendoza REB (1984) Estudio taxonómico y morfológico de los manglares de Panamá. Natura 4:1–5

Robertson AI (1986) Leaf-burying crabs: their influence on energy flow and export from mixed mangrove forests (*Rhizophora* spp.) in northeastern Australia. J Exper Mar biol Ecol 102: 237–248

Robertson AI (1991) Plant animal interactions and the structure and function of mangrove forests ecosystems. Austr J Ecol 16:433–443

Robertson AI, Blaber SJM (1992) Plankton, epibenthos and fish communities. In: Robertson AI, Alongi DI (eds) Tropical mangrove ecosystems. Coastal and Estuarine Studies American Geophysical Union, Washington, DC, p. 172–224
Robertson AI, Daniel PA (1989). The influence of crabs on litter processing in high intertidal mangrove in tropical Australia. Oecologia 78:191–198
Robertson AI, Duke NC (1989) Insect herbivory on mangrove leaves in North Queensland. Austr J Ecol 12:1–7
Robertson AI, Duke NC (1990) Mangrove fish-communities in tropical Queensland, Australia: spatial and temporal patterns in densities, biomass and community structure. Mar Biol 104:369–379
Robertson AI, Phillips MJ (1995) Mangroves as filters of shrimp pond effluent: predictions and biogeochemical research needs. Hydrobiologia 295:311–321
Robertson AI, Alongi DI, Boto, KG (1992) Food chains and carbon fluxes. In: Robertson AI, Alongi DI (eds) Tropical mangrove ecosystems. Coastal and Estuarine Studies American Geophysical Union, Washington, DC, p. 293–326
Rodríguez C, Stoner AW (1990) The epiphyte community of mangrove roots in a tropical estuary: distribution and biomass. Aquat Bot 36:117–126
Rodríguez G (1987) Structure and production in neotropical mangroves. Trends Ecol Evol 2:264–267
Rodríguez R (1984) Los Manglares de Venezuela. Instituto Forestal Latino Americano, Mérida
Ronback J (1999) The ecological basis for economic value of seafood production supported by mangrove ecosystem. Ecol Economics 29:235–252
Rosolofo MV (1997a). Mangroves and coastal aquaculture development in Madagascar. A paper presented in SAREC-SIDA Regional Workshop on Mangrove ecology, Physiology and Management. Zanzibar
Rosolofo MV (1997b) Use of mangroves by traditional fishermen in Madagascar. Unpubl
Roth LC (1992) Hurricanes and mangrove regeneration: effects of hurricane Joan, October 1988, on the vegetation of Isla del Venado, Bluefields, Nicaragua. Biotropica 24:373–384
Roth LC, Grijalva A (1991) New records of the mangrove *Pelliciera rhizophorae* (Theacea) on the Caribbean coast of Nicaragua. Rhodora 21:249–251
Rubec C (1999) Guidelines for developing and implementing national wetland policies http//www. http://ramsar.org/key_guide_nwp_ehtm (16 December 1999)
Ruddle K, Johannes RE (1993) The traditional knowledge and management of coastal systems in Asia and the Pacific. Paper presented at UNESCO-ROSTSEA Regional Seminar, 5–9 December, UNESCO, Jakarta
Ruitenbeck HJ (1992) Mangrove management: an economic analysis of mangrove options with a focus on Bintuni Bay, Irian Jaya. Environment Management and Development in Indonesia Project (EMDI), Jakarta and Halifax
Rull V (1998) Evolucion de los manglares neotropicales: La crisis del Eoceno. Interciencia 23:355–362
Rull V, Vegas-Villarœbia T, Espinoza de Perna N (1999) Palynological record of an Early-Mid Holocene mangrove in estaern Venezuela. Implications for sea-level rise and disturbance history, pp 496–504
Ruwa, RK (1995) Mangroves of Kenya. In: Diop ES (ed) Conservation and sustainable utilization of mangrove forests in Latin America and Africa. Mangroves Ecosystems Technical Reports 3:227–243
Saenger P (1995) The mangrove vegetation of the Atlantic Coast of Africa – a review. ICIV, University of Toulouse, Toulouse
Saenger P, Siddiqi NA (1993) Land from the sea: the mangrove afforestation program of Bangladesh. Ocean Coast Managt 20:23–39
Saenger P, Snedaker SC (1993) Pantropical trends in mangrove above-ground biomass and annual litterfall. Oecologia 96:293–299
Saenger P, Specht MM, Specht RL, Chapman VJ (1977) Mangal and coastal saltmarsh communities in Australia. In: Chapman VJ (ed) Wet coastal ecosystems. Elsevier, Amsterdam, p 293–345
Saenger P, Hegerl EJ, Davie JDS (1983) Global status of mangrove ecosystems. Environmentalist 3: 1–88
Said MI, Othman MS, Khan N (1992) The socioeconomic value of wetland plant species and recommendations to improve wetland conservation in Peninsular Malaysia. Asian Wetland Bureau, WWF and the Institute of Advanced Studies, University of Malaysia, Kuala Lumpur
Saintilan N, Hashimoto TR (1999) Mangrove-saltmarsh dynamics on a bay-head delta in the Hawkesbury River estuary, New South Wales, Australia. Hydrobiologia 413:95–102
Saket M, Mattusse RV (1994) A study for the determination of the rate of deforestation of the mangrove vegetation in Mozambique. FAO/PNUD/MOZ/92/103. Forest Dept, Forests Inventory Unit, Maputo
Salvatierra M (1983) La fauna. Ambiente MARNR, Caracas 5:23–25
Samb B, Dioh BC (1996) Maîtrise concertée de la connaissance, de la valorisation et de la gestion des ressources halieutiques. Projet PNUD/CSRP, Dakar

Sánchez-Páez H, Alvarez-Leon R, Pinto-Nolla F, Sanchez-Alferez AS, Pino-Rengifo JC, Acosta-Peñaloza MT, Garcia-Hansen I (1997a) Diagnóstico y zonificación preliminar de los manglares del Caribe de Colombia. In: Sánchez-Páez H, Alvaraz-León R (eds) Proy PD 171/91 Rev 2 (F) Fase I Conservación y Manejo para el Uso Múltiple y el Desarrollo de los Manglares en Colombia. Santa Fe de Bogotá DC, pp 1–511

Sánchez-Páez H, Alvarez-Leon R, Guevara-Mancera OA, Zamora-Guzman A, Rodriguez-Cruz H, Bravo-Pazmiño HE (1997b) Diagnóstico y zonificación preliminar de los manglares del Pacífico de Colombia. In: Sánchez-Páez H, Alvarez-León R (eds) Proy PD 71/91 Rev. 2 (F) Fase I Conservación y Manejo para el Uso Múltiple y el Desarrollo de los Manglares en Colombia. Santa Fe de Bogotá DC pp 1–343

Sanoja M (1992) Wetland ecosystems and the management of cultural heritage. In: Lugo AE, Bayle B (eds) Wetlands management in the Caribbean and the role of forestry and wetlands in the economy. Institute of Tropical Forestry, Southern Forest Experiment Station, and the Caribbean National Forest, Region 8. Rio Piedras, Puerto Rico, pp 66–73

Santos MCFV (1986) Considerações sobre a ocorrência de *Rhizophora harrisonii* Leechman e *Rhizophora racemosa* G.F.W. Meyer, no litoral do Estado do Maranhão, Brasil. Bol Lab Hidrob São Luis 7:71–91

Santos MCFV (1989) Structural patterns of hyperhalinity-stressed mangrove forests in the State of Maranhão, Northern Brazil. Dissertation. Colorado School of Mining, Bolder

Santos MCFV, Zieman JC, Cohen RRH (1997) Interpreting the upper mid-littoral zonation patterns of mangroves in Maranhão (Brazil) in response to microtopography and hydrology. In: Kjerfve B, Lacerda LD, Diop ES (eds) Mangrove Ecosystem Studies in Latin America and Africa. UNESCO/ International Society for Mangrove Ecosystems, Paris pp 127–144

Sathirathai S (1998) Economic valuation of mangroves and the roles of local communities in the conservation of natural resources: case study of Surat Thani, South of Thailand. Economy and Environment Program for Southeast Asia, Report 51, Singapore

Saur E, Imbert D, Etienne J, Mian, D (1999) Insect herbivory on mangrove leaves in Guadeloupe: effects on biomass and mineral content. Hydrobiologia 413:89–93

Savage T (1972) Florida mangroves as shoreline stabilizer. Florida Dept Nat Resour Prof Pap 19:1–46

Schaeffer-Novelli Y, Cintrón-Molero G (1990) Status of mangrove research in Latin America and the Caribbean. Bol Inst Oceanogr 38:93–97

Schaeffer-Novelli Y, Cintrón-Molero G (1999) Brazilian mangroves: a historical ecology. Ciênc Cult 51:274–283

Schrijvers J, Van Gansbeke D, Vincx M (1995) Macrobenthic infauna of mangroves and surrounding beaches at Gazi Bay, Kenya. Hydrobiologia 306:53–66

Schrijvers J, Schallier R, Silince J, Okondo JP, Vincx M (1997) Interactions between epibenthos and meibenthos in a high intertidal *Avicennia marina* mangrove forest. Mangroves and Salt Marshes 1:137–154

Schrijvers J, Camargo MG, Pratiwi R, Vincx M (1998) The infaunal macrobenthos under East African *Ceriops tagal* mangroves impacted by epibenthos. J Exper Mar Biol Ecol 222:175–193

Schubbart O (1955) Historia da Química no Brazil. Universidade de São Paulo, São Paulo

Seim HE, Kjerfve B, Sneed JE (1987) Tides of Mississippi Sound and the adjacent continental shelf. Estuar Coast Shelf Sci 25:143–156

Selvam V, Somasundaram R, Ansari Y, Ragu J (1999) Internal report. M. S. S. Swaminatham Foundation, Madras

Semeniuk V (1994) Predicting the effect of sea-level rise on mangroves in northwestern Australia forests. J Coast Res 10:1050–1076

Semesi AK (1986) Zonation and vegetation structure of mangrove communities in Tanzania. In: Mainoya JR, Siegel PR (eds) Status and utilisation of mangroves. Proceedings of a Workshop on "Save the Mangrove Ecosystems in Tanzania", Faculty of Science, University of Dar es Salaam, pp 15–36

Semesi AK (1988) Status and utilization of mangroves along the coast of Tanga, Tanzania. In: Mainoya JR, Siegel PR (eds) Status and utilisation of mangroves. Proceedings of a Workshop on "Save the Mangrove Ecosystems in Tanzania", Faculty of Science, University of Dar es Salaam, pp 174–181

Semesi AK (1991a) Management plan for the mangrove ecosystem of mainland Tanzania. Ministry of Tourism, Natural Resources and Environment, Forest and Beekeeping Division, Dar es Salaam

Semesi AK (1991b) The mangrove resources of Rufiji Delta, Tanzania. Proc Wetlands Conserv Conf Southern Africa, Gaborone, Botswana, pp 157–172

Semesi AK (1992) Developing management plans for the mangrove forest reserves of mainland Tanzania. Hydrobiologia 247:1–10

Semesi AK (1998) Mangrove management and utilization in Eastern Africa. Ambio 27:620–626

Semesi AK, Howell K (1992) The mangroves of the Eastern African region. UNEP, Nairobi

Semesi AK, Muruke MHS, Mgaya YD (1999) Mangroves of Ruvu River and Kaole, Bagamoyo District. In: Howell KM, Semesi AK (eds) Coastal resources of Bagamoyo District Tanzania. Faculty of Science, University of Dar es Salaam pp 17–26

Sen A (1967) Isolation, assurance, and the social discount rate of discount. Quart J Econ 81:112–24

Sen A (1982) Approaches to the choice of discount rates for social benefit cost analysis. In: Lind RC (ed) Discounting for time and risk in energy policy. Resources for the Future, Washington, DC

Seymour AB (1929) Host index of the fungi of North America. Harvard University Press, Cambridge

Shafi M (1992a) Adverse effects of Farakka on the forest of southwest region of Bangladesh (Sundarbans). Proc 2nd Nat Conf Forestry, Dhaka, p 30–57

Shafi M (1992b) Artificial regeneration of keroa Bangladesh. Proc 2nd Nat Conf Forestry, Dhaka, p 136–139

Shah NJ (1995) Coastal zone management in Seychelles. In: Lundin CG, Linden O (eds) Integrated coastal zone management in the Seychelles. SIDA, Stockholm, pp 14–125

Shah NJ, Linden O, Lundin CG, Johnstone R (1997) Coastal management in Eastern Africa: status and future. Ambio 26:227–234

Shahidullah M, Siddiqi NA (1994) Notes on the feasibility of reforestation in the Chakaria Sundarbans mangroves. Bangladesh J Forest Sci 23:65–66

Sheridan PF (1992) Comparative habitat utilization by estuarine macrofauna within the mangrove ecosystem of Rookery Bay, Florida. Bull Mar Sci 50:21–39

Sherrod CL, McMillan C (1981) Black mangrove *Avicennia germinans* in Texas: past and present distribution. Contr Mar Sci 44:115–131

Siddiqi AN (1986) Preliminary trial of mangrove and mainland species in the Sundarbans highlands. Bano Biggyan Patrika 15:25–30

Siddiqi AN (1987) Observation on initial spacing in a Keora (*Sonneratia apelata*) plantation along the coastal belt of Bangladesh. Malaysian forester 50:204–216

Siddiqi AN (1988) Growth, natural thinning and wood production in a Keora (*Sonneratia apelata*) stand. Bano Biggyan Patrika 17:91–93

Siddiqi AN (1993) Need for popularizing Korea (*Sonneratia apelata*) as fuelwood. Aranya 2:15–17

Siddiqi AN (1994) Natural regeneration. In: Hussain Z, Achary G (eds) Mangroves of the Sundarbans. Vol 2, Bangladesh. IUCN, Bangkok, p. 75–99

Siddiqi AN (1995) Site suitability for raising *Nypa fruticans* in the Sundarbans mangroves. J Trop Forest Sci 7:405–411

Siddiqi AN (1998) Enrichment planting in the mangrove of Sundarbans. Bangladesh J Forest Sci 27: 34–47

Siddiqi AN, Das S (1988) Effect of time and method of storage on survival of uprooted Keora (*Sonneratia apelata*) seedlings. Indian J Forest 11:326–330

Siddiqi AN, Islam MR (1988) Studies on fruit size, seed production and viability of seeds of Keora (*Sonneratia apelata*). Bano Biggyan Patrika 17:15–19

Siddiqi AN, Khan MAS (1990) Growth performance of mangrove trees along the coastal belt of Bangladesh. UNDP/UNESCO/RAS Mangrove Ecosystems Occ Pap 8:5–14

Siddiqi AN, Khan MAS (1996) Planting techniques for mangroves on new accretions in the coastal areas of Bangladesh. In: Field CD (ed) Restoration of mangrove ecosystems. Inter Soc Mangrove Ecosystems, Okinawa, p. 143–159

Siddiqi AN, Shahjalal MAH (1997) Feasibility of underplanting in the mangrove plantations along the eastern Bangladesh. Bangladesh J Forest Sci 26:76–78

Siddiqi NA, Islam MR, Khan MAS (1989) Effect of salinity on germination success in keora (*Sonneratia apelata*). Bano Biggyan Patrika 18:57–62

Siddiqi NA, Shahidullah M, Shahjalal MAH (1991) Studies on seeds, germination success and raised seedlings of *Nypa fruticans*. Indian Forest 111:553–559

Siddiqi AN, Khan MAS, Islam MR, Hoque AKF (1992) Underplanting – a means to ensure sustainable mangrove plantations in Bangladesh. Bangladesh J Forest Sci 21:1–6

Siddiqi NA, Hoque AKF, Alam MS (1993a) The performance of some non-mangrove species in the coastal areas of Bangladesh. Bangladesh J Forest Sci 22:71–72

Siddiqi NA, Islam MR, Khan MAS, Shahidullah M (1993b) Mangrove nurseries in Bangladesh. ISME Mangrove Ecosystems Occas Pap 1:1–14

Siddiqi NA, Shahidullah M, Hoque AKF (1994) Present status of chakaria Sundarbans – the oldest mangrove forest of the sub-continent. Bangladesh J Forest Sci 23:26–34

Silva CAR (1988) Distribuição e Ciclagem Interna de Metais Pesados em um Ecossistema de Manguezal Dominado por *Rhizophora mangle*, na Baia de Sepetia, Rio de Janeiro. Dissertation, Universidade Federal Fluminense, Niteroi

Silva CAR, Lacerda LD, Silva LFF, Rezende CE (1991) Forest structure and biomass distribution in a red mangrove stand in Sepetiba Bay, Rio de Janeiro. Revta Brasil Bot 14:21–25

Simao D (1993) Mangroves of Guinea Bissau. In: Diop ES (ed) Conservation and sustainable utilization of mangrove forests in Latin America and Africa regions. Part II Africa. ISME Mangrove Ecosystems Tech Rep, Okinawa, pp 27–46
Simberloff D, Brown BJ, Lowrie (1978) Isopod and insect root borers may benefit Florida mangroves. Science 201:630–632
Singh N, Steinke TD (1992) Colonization of decomposing leaves of *Bruguiera gymnorrhiza* by fungi and *in vitro* cellulolytic activity of the isolates. S. Afr. J. Bot 58:525–529
Slima FJ, Hemmingaa MA, Ochiengb C, Janninka NT, Cocheret de la Morinièrea E, Veldec G van der (1997) Leaf litter removal by the snail *Terebralia palustris* (Linnaeus) and sesarmid crabs in an East African mangrove forest (Gazi Bay, Kenya). J Exper Mar Biol Ecol 215:35–48
Smith DA (1998) The Honduran shrimp farming industry: social and environmental impacts http://earthsummitwatch.org/shimp/national_reports/crhond 1.htmp, 21 December 1999
Smith TJ (1987a) Effects of seed predators and light level on the distribution of *Avicennia marina* (Forks.) Vierh. in tropical, tidal forests. Estuar Coast Shelf Sci 25:43–51
Smith TJ (1987b) Seed predation in relation to tree dominance and distribution in mangrove forests. Ecology 68:266–273
Smith TJ, Chan HT, McIvor CC, Robblee MB (1989) Comparisons of seed predation in tropical, tidal forests from three continents. Ecology 70:146–151
Snedaker SC (1984) Mangroves: a summary of knowledge with emphasis on Pakistan. In: Haq BU, Milliman JD (eds) Marine geology and oceanography of Arabian Sea and coastal Pakistan. Van Nostrand Reinhold Co, New York pp 255–262
Snedaker SC (1995) Mangroves and climate change in the Florida and Caribbean region: scenarios and hypothesis. Hydrobiologia 295:231–241
Snedaker SC (1999) The future of mangrove research. Hydrobiologia 413:ix–x
Soepadamo E, Wong KM (1995) The flora of Sabah and Saravak. ITTO-ODA, Manila
Soumare A (1996) Etude comparative de l'évolution géomorphologique des bas-estuaires du Sénégal et du Saloum. Approche par les données de terrain et la télédétection. Dissertation, Université Cheikh Anta Diop, Dakar
South RG, Goulet D, Tuqiri S, Church M (1994) Traditional marine tenure and sustainable management of marine resources in Asia – Proc Internat Workshop, University of the South Pacific Suva
Spalding M, Blasco F, Field CD (1997) World mangrove atlas. Inter Soc Mangrove Ecosystems, Okinawa
Specht RL (1981) Biogeography of halophytic angiosperms (salt-marsh, mangrove and sea-grass). In: Keast A (ed) Ecological biogeography of Australia. W Junk, The Hague
SPREP (1994) An action plan for wetland conservation in the South Pacific. Oceanic Program of the Asian Wetland Bureau and SPREP, Apia
Steenis CGGJ van (1962) The distribution of mangrove plant general and its significance for paleogeography. Proc Kon Nederl Akad Wetensch C 65:64–169
Steinke TD (1999) Mangroves of South African estuaries. In: Allanson BR, Baird D (eds) Estuaries of South Africa. Cambridge University Press, pp 119–140
Steinke TD, Jones EBG (1993) Marine and mangrove fungi from the Indian Ocean Coast of South Africa. S Af Tydsk Plantk 59:386–389
Steinke TD, Rajh A, Holland AJ (1993) The feeding behaviour of the red mangrove crab *Sesarma meinerti* De Man 1887 (Crustacea: Decapoda: Grapsidae) and its effect on the degradation of mangrove leaf litter. South African J Mar Sci 13:151–160
Steinke TD, Ward CJ, Rajh A (1995) Forest structure and biomass of mangroves in the Mgeni estuary, South Africa. Hydrobiologia 295:159–166
Stevens FL (1920) New or noteworthy Puerto Rican fungi. Bot Gaz 70:399–402
Stewart RW, Kjerfve B, Milliman J, Dwivedi SN (1990) Relative sea level change: a critical evaluation. UNESCO Rep Mar Sci 54:1–22
Stieglitz T, Ridd P, Müller P (2000) Passive irrigation and functional morphology of crustacean burrows in a tropical mangrove swamp. Hydrobiologia 421:69–76
Stoddart DR, Bryan GW, Gibbs PE (1973) Inland mangroves and water chemistry, Barbuda, West Indies. J Nat History 7:33–46
Stoffers AL (1956) Studies on the flora of Curacao, and other Caribbean islands. Vol. 1. The vegetation of the Netherlands Antilles. Natuurwetenschappelijke Studiekring voor Suriname en de Nederlandse Antillen, Utrecht
Stoner AW (1985) Macrobenthic consumers in Joyuda Lagoon: residence and transients. In: Stoner AW, López J M (eds) Cycling and transport processes in a tropical lagoon. Proc Workshop on the Joyuda Lagoon Ecosystem. Report of the Center for Energy and Environment Research, University of Puerto Rico, Mayaguez. pp 1–86
Stoner AW (1986) Community structure of the demersal fish species of Laguna Joyuda, Puerto Rico. Estuaries 9:142–152

Stoner AW (1988) A nursery ground for four tropical Penaeus species: Laguna Joyuda, Puerto Rico. Mar Ecol Progr Ser 42:133–141

Stromberg H, Pettersson C, Johnstone R (1998) Spatial variations in benthic macrofauna and nutrient dynamics in a mangrove forest subject to intense deforestation: Zanzibar, Tanzania. Ambio 8:743–739

Sutherland SC (1980) Dynamics of the epibenthic community of roots of the mangrove *Rhizophora mangle*, at Bahía de Buche, Venezuela. Mar Biol 58:75–84

Tacconi L (1998) Participation: a review of the concept and its adoption in research, development and conservation activities, ACIAR Project N° 9 643 – Participatory Land Use Planning for Forest and Agricultural Resources in Papua New Guinea, Australian Centre for International Agricultural Research, Canberra

Tam NFY, Wong YS (1995) Mangrove soils as sinks for wastewater-borne pollutants. Hydrobiologia 295:231–241

Tan CGS, Ng PKL (1994) An annotated checklist of mangrove brachyuran crabs from Malaysia and Singapore. Hydrobiologia 285:75–84

Teas HT, McVan RI (1982) An epidemic dieback gall disease of *Rhizophora* mangroves in Gambia, West Africa. Plant Disease 66:2–7

Tejera E (1977) Indigenismos. Tomo II. Editora de Santo Domingo SA, Santo Domingo

Thanikaimoni G (1987) Mangrove palynology. UNDP/UNESCO and French Institute of Pondichery

Thayer GW, Colby DR, Hettler WF (1987) Utilization of the red mangrove prop root habitat by fishes in south Florida. Mar Ecol Progr Ser 35:25–38

Thibodeau FR, Nickerson NH (1986) Differential oxidation of mangrove substrate by *Avicennia germinans* and *Rhizophora mangle*. Amer J Bot 73:512–516

Thibodeau PM, Gardner LR, Reeves HW (1998) The role of groundwater flow in controlling the spatial distribution of soil salinity and rooted macrophytes in a southeastern salt marsh, USA. Mangroves and Salt Marshes 2:1–13

Thom BG (1967) Mangrove ecology and deltaic geomorphology, Tabasco, Mexico. J Ecol 55:301–343

Thrupp LA, Cabarle B, Zazueta A (1994) Participatory methods in planning and political processes: linking the grassroots and policies for sustainable development. Agriculture and Human Values 11:77–84

Tomlinson PB (1986) The botany of mangroves. Cambridge Tropical Biology Series. Cambridge University Press, Cambridge

Tomlinson PB (1994) The botany of mangroves. Updated Paperback Edition. Cambridge University Press, Cambridge

Troup RS (1921) The silviculture of Indian trees. Clarendon Press, Oxford

Turner RE (1977) Intertidal vegetation and commercial yields of penaeid shrimp. Trans Ameri Fish Soc 106:411–416

Turner RE, Jones T (1990) Wetlands: markets and intervention failures. Earthscan Publication, London

Twilley RR (1985) The exchange of organic carbon in basin mangrove forests in a southwest Florida Estuary. Est Coast Shelf Sci 20:543–557

Twilley RR (1998) Mangrove wetlands. In: Messina MG Conner WH (eds) Southern forested wetlands – ecology and management. Lewis Publishers, Boca Raton, pp 445–473

Twilley RR, Zimmerman R, Solorzano L, Rivera-Monroy V, Bodero A, Zambrano R, Pozo M, Garcia V, Loor K, Garcia R, Cardenas W, Gaibor N, Espinoza J, Lynch J (1990) The importance of mangroves in sustaining fisheries and controlling water quality in coastal ecosystems. Interim Report, USAID, Program on Science and Technology Cooperation, Washington

Twilley RR, Chen RH, Hargis T (1992) Carbon sink in mangroves and their implications to carbon budget of tropical coastal ecosystems. Water Air Soil Pollut 64:265–288

Twilley RR, Bodero A, Robadue D (1993) Mangrove ecosystem biodiversity and conservation in Ecuador. In: Potter CS, Cohen JI, Janezewski D (eds) Perspectives on Biodiversity: Case studies of genetic resource conservation and development. American Association for the Advancement of Science, Washington, DC. p. 105–127

Twilley RR, Snedaker SC, Yáñez-Arancibia A, Medina E (1996) Biodiversity and ecosystem processes in tropical estuaries: Perspectives of mangrove ecosystems. In: Mooney HA, Cushman JH, Medina E, Sala OE, Schulze ED (eds) Functional roles of biodiversity: a global perspective. John Wiley & Sons, Chichester, p. 327–370

Twilley RR, Rivera-Monroy VH, Chen R, Botero L (1998) Adapting an ecological mangrove model to simulate trajectories in restoration ecology. Marine Pollution Bulletin 37:404–419

UNDP/UNESCO (1991) The integrated multidisciplinary survey and research programme of the Ranong mangrove ecosystem. RAS/86/120, National Research Council of Thailand, Bangkok

UNEP (1982) Environmental problems of the East African region. UNEP Regional Seas Rep St N° 12, Nirobi

Untawale AG (1986) Mangroves of India. Mangroves of Asia and the Pacific – Status and Management, UNDP/UNESCO Project Research and Training Pilot Program on Mangrove Ecosystems. RAS/79/002, UNDP/UNESCO, Manila, p. 471–512

Untawale AG (1987) Some aspects of mangrove afforestation along the central west coast of India. In: Rana RS (ed) Afforestation of salt affected soils. CSSRI, Karnal, p 133–150

Upoff N (1991) Fitting projects to people. In: Cernea MM (ed) Putting people first: sociological variables in rural development. The World Bank, Washington, DC

Vannucci M (1987) Conversion of mangroves to other uses: the Cochin backwaters. UNDP/UNESCO Project Research and Training Pilot Program on Mangrove Ecosystems. RAS/79/002, UNDP/UNESCO, Manila, p. 331–336

Vannucci M (1999) Os manguezais e nós. EDUSP, São Paulo

Vannucci M (2000) What is so special about mangroves? In: Barros H (ed) Conservation and managment of tropical estuaries. Univ Rural Pernambuco, Recife. CD-ROM 4p

Vietnam Government (1999) Vietnam national report. 7th Meeting of the Conference of the Contracting Parties to the Convention on Wetlands http.//www.iucn.org/themes/ramsar/cop7_nr_vietnam. (16 December 1999)

Vizioli J (1923) Some Pyrenomycetes of Bermuda. Mycologia 15:107–119

Von Mitzlaff U (1989) Coastal communities in Tanzania and their mangrove environment. The Catchment Forestry Project, Dar es Salaam

Walker LR, Brokaw NVL, Lodge DJ, Waide RB (1991) Ecosystem, plant, and animal responses to hurricanes in the Caribbean. Biotropica. 23:1–521

Warren JH (1990) Role of burrows as refuges from subtidal predators of temperate mangrove crabs. Mar Ecol Progr Ser 67:295–299

Walsh GE (1974) Mangroves: a review. In: Reinbold R, Queen W (eds) Ecology of halophytes. Academic Press, New York, pp 51–174

Walter H, Steiner M (1936) The ecology of East African mangroves. Z Bot 30:65–191

Wanless HR, Parkinson RW, Tedesco LP (1994) Sea level control on stability of Everglades wetlands. In: Everglades: The ecosystem and its restoration. St Lucie Press, Delray Beach, Florida, USA. pp 199–223

Waqaraitu S (1994) The delimitation of the traditional fishing grounds: the Fiji experience. In: South GR, Goulet D, Tuqiri C, Church M (eds) Traditional marine tenure and sustainable management of marine resources in Asia and the Pacific. International Ocean Institute, South Pacific, Suva

Warner GF (1967) The life history of the mangrove crab, *Aratus pisoni*. J Zool 153:321–335

Warner GF (1969) The occurrence and distribution of crabs in a Jamaican mangrove swamp. J Anim Ecol 38:379–389

Warner M (1997) Consensus' participation: an example of protected area planning. Public Administration and Development 17:413–432

Warren JH, Underwood AJ (1986) Effects of burrowing crabs on the topography of mangrove swamps in New South Wales. J Exper Mar Biol Ecol 102:223–235

Watling D (1985) A mangrove management plan for Fiji – Ba, Labasa and Rewa Deltas. Mangrove Management Committee and the South Pacific Commission, Suva

Watling D (1986) A mangrove management plan for Fiji-Nadi Bay and Suva-Navua Locales. Mangrove Management Committee and the South Pacific Commission, Suva

West RC (1956) Mangrove swamps of the Pacific coast of Colombia. Ann Assoc Amer Geogr 46:1

West RC (1977) Tidal salt-marsh and mangal formations of Middle and South America. In: Chapman VJ (ed) Wet coastal ecosystems. Ecosystems of the World Nº 1. Elsevier Sci Publ Comp, Amsterdam, pp 193–213

Whitten AJ, Damanik SJ (1986) Mass defoliation of mangroves in Sumatra, Indonesia. Biotropica 18:176

Whyte J, Siwatibau S, Tapisuwe A (1998) Participatory resource management in Vanuatu: An Australian Centre for International Agricultural Research (ACIAR) research report. Foundation of the People of the South Pacific International, Port Vila, Vanuatu

Widmer RJ (1988) The evolution of the Calusa. The University of Alabama Press, Tuscaloosa and London

Wigley TML, Raper SCB (1992) Implications for climate and sea level of revised IPCC emissions scenarios. Nature 357:293–300

Wijmstra TA (1968) The identity of *Psilatricolporites* and *Pelliciera*. Acta Bot Neerl 17:114–116

Wijmstra TA (1969) Palynology of the Alliance Well. Geol Mijnbouw 48:125–133

Wilson KA (1989) Ecology of mangrove crabs: predation, physical factors and refuges. Bull Mar Sci 44:263–273

Winograd M (1983) Observaciones sobre el hallazgo de *Pelliciera rhizophoerae* (Theaceae) en el Caribe Colombiano. Biotropica 15:297–298

Woitchik AF, Ohowa B, Kazungu JM, Rao RG, Goeyens L, Dehairs F (1997) Nitrogen enrichment during decomposition of mangrove leaf litter in an east African coastal lagoon (Kenya): Relative importance of biological nitrogen fixation. Biogeochemistry 39:15–35

Wolanski E (1995) Transport of sediment in mangrove swamps. Hydrobiologia 295:31–42
Wolanski E, Sarsenski J (1997) Larvae dispersion in coral reefs and mangroves. Ambio 85:236–243
Wolanski E, Mzada Y, Ridd P (1992) Mangrove hydrodynamic. In: Robertson AI, Alongi MD (eds) Tropical Mangrove Ecosystem. Amer Geographys Union, Washington DC, pp 43–62
Woodroffe CD (1987) Pacific Island mangroves: distribution and environmental settings. Pacific Sci 41:166–185
Woodroffe CD (1999) Response of mangrove shorelines to sea-level change. Tropics 8:159–177
Woodroffe CD, Grindrod J (1991) Mangrove biogeography: the role of quaternary environmental and sea-level change. J Biogeogr 18:479–492
Wright SJ, Carrasco C, Calderón O, Paton S (1999) The El Niño Southern Oscillation, variable fruit production and famine in a tropical forest. Ecology, 5:768–797
Wunderle JM, Lodge DJ, Waide RB (1992) Short-term effects of hurricane Gilbert on terrestrial bird populations on Jamaica. Auk 109:148–166
Yáñez-Arancibia A, Amezcua-Linares F, Day JW (1980) Fish community structure and function in Terminos Lagoon, a tropical estuary in the southern Gulf of Mexico. In: Kennedy VS (ed) Estuarine perspectives. Academic Press, New York. pp: 465–482
Yáñez-Arancibia A, Lara-Domingues AL, Zapata GJV, Arriaga ER (1993) Mangroves of Mexico. In: Lacerda LD, Field CD (eds) Proc. workshop on conservation and sustainable utilization of mangroves in Latin America and Africa regions. Okinawa, pp 3–4
Young MD (1992) Sustainable investment and resource use – equity, environmental integrity and economic efficiency. UNESCO Man and the Biosphere Vol. 9, UNESCO and the Parthenon Publishing Group, Paris
Zack A, Roman-Mas A (1988) Hydrology of the Caribbean Island wetlands. Acta Científica 2:65–73
Zamorano D (1983) Productividad del manglar y su importancia para el pelagial de la Ciénaga Grande de Santa Marta. Dissertation, Universidad Nacional de Colombia, Bogota

Subject Index

A

B

C

G

H

I

J

K

L

M

N

O

P

Taxonomic Index

A

B

Printing (Computer to Film): Saladruck, Berlin
Binding: Stürtz AG, Würzburg